Pseudocereals

As the global demand for sustainable and nutrient-rich crops grows, pseudocereals are emerging as a valuable alternative to traditional cereals. This book brings together recent scientific advances and practical insights on the chemistry, nutritional quality, processing technologies, and industrial applications of pseudocereals such as quinoa, amaranthus, buckwheat, and chia.

It explores their unique nutritional and bioactive components, innovative milling and extraction methods, and their potential to support food security, health, and sustainable development. The chapters provide a detailed examination of their techno-functional properties, utilization in food and non-food industries, and growing relevance in functional food markets.

This book also addresses consumer acceptance, market strategies, and safety considerations, offering a balanced perspective that connects traditional knowledge with modern food science. It serves as a comprehensive reference for researchers, food technologists, nutritionists, and industry professionals looking to harness the potential of pseudocereals in product development and industrial applications.

Features

- Examines the chemistry and nutritional composition of key pseudocereals
- Explores techno-functional properties of dietary fibers and their role in food innovation
- Covers emerging milling techniques and valorization of pseudocereal by-products
- Discusses bioactive compound extraction, health benefits, and functional food applications
- Reviews starch and protein extraction methods, functional roles, and industrial uses
- Addresses anti-nutritional factors and safety strategies for improved utilization

Ideal for researchers, food technologists, nutritionists, and industry professionals, this book offers a clear and comprehensive resource on the science and applications of pseudocereals. It is equally valuable for students and early-career scientists seeking to deepen their understanding of novel crops, innovative processing methods, and their role in developing sustainable, functional food systems.

Pseudocereals

Nutritional, Technological, and Industrial Perspectives

Edited by
Amir Gull, Gulzar Ahmad Nayik,
Jahangir Ahmad Rather, F.A. Masoodi,
and Charles Brennan

CRC Press
Taylor & Francis Group
Boca Raton London New York

CRC Press is an imprint of the
Taylor & Francis Group, an **informa** business

Designed cover image: Shutterstock

First edition published 2026
by CRC Press
2385 NW Executive Center Drive, Suite 320, Boca Raton FL 33431

and by CRC Press
4 Park Square, Milton Park, Abingdon, Oxon, OX14 4RN

CRC Press is an imprint of Taylor & Francis Group, LLC

ISBN: 9781032954233 (hbk)
ISBN: 9781032954288 (pbk)
ISBN: 9781003584858 (ebk)

DOI: 10.1201/9781003584858

Typeset in Times
by codeMantra

All the editors dedicate this book to their beloved families

Contents

Preface

In an era where global food security faces unprecedented challenges – climate change, population growth projected to exceed 9.7 billion by 2050, and persistent nutritional deficiencies – pseudocereals like quinoa, amaranth, buckwheat, and chia emerge as beacons of hope. *Pseudocereals: Nutritional, Technological, and Industrial Perspectives* is born from a collective vision to explore these underutilized crops as sustainable, nutrient-dense solutions to address hidden hunger and promote resilient agriculture. As researchers and educators deeply engaged in food science, we recognize the urgent need to diversify food systems beyond traditional cereals like wheat and rice, which face limitations in adaptability and nutritional scope.

Pseudocereals, though distinct from monocot cereals, mirror their versatility while offering superior nutritional profiles – rich in proteins (10%–20%), essential amino acids, minerals, and bioactive compounds like flavonoids and phytosterols. Their ability to thrive in marginal lands under abiotic stresses such as drought and salinity makes them uniquely suited for sustainable farming. This book delves into their biochemical composition, innovative processing techniques, and health benefits, including their gluten-free appeal for celiac patients and their role in combating chronic diseases like diabetes and cardiovascular issues. From milling methods to functional food applications, we aim to provide a holistic resource for harnessing pseudocereals in modern diets.

Our journey in crafting this book stems from a shared commitment to advancing food science and nutrition. We have meticulously compiled insights from global research, industry practices, and cultural traditions to highlight pseudocereals' potential in functional foods like noodles, energy bars, and beverages. This preface sets the stage for a comprehensive exploration of pseudocereal cultivation, processing, and product development, addressing challenges and future prospects. We intend this book to inspire graduate students, researchers, nutritionists, and food industry professionals to embrace pseudocereals as a cornerstone of sustainable nutrition.

This work is a collaborative effort, blending expertise from various institutions across the globe to bridge traditional knowledge with cutting-edge innovation. We invite readers to join us in envisioning a future where pseudocereals not only enrich diets but also foster social diversity and environmental sustainability. Through this book, we hope to ignite curiosity and action toward a healthier, more resilient world.

Dr. Amir Gull
Dr. Gulzar Ahmad Nayik
Dr. Jahangir Ahmad Rather
Prof. F.A. Masoodi
Prof. Charles Brennan

About the Editors

Amir Gull completed his Master's degree in Food Technology from the Islamic University of Science and Technology, Awantipora, J&K, India, and PhD from Sant Longowal Institute of Engineering and Technology, Longowal, Sangrur, Punjab, India. Gull has published 40+ peer-reviewed research papers and review papers in reputed journals. He has also edited five books and many book chapters and delivered a number of presentations in many national and international conferences. Amir Gull has worked as Senior Research Fellow in the Department of Food Technology, SKUAST-Kashmir, Srinagar, J&K, India and later also worked as Research Associate (ICMR-RA) in the Department of Food Technology at the University of Kashmir, Srinagar, J&K, India. Gull's main research activities include developing functional food products from millets, bioactive compounds, and edible coatings for fresh-cut fruits and vegetables. Gull is also serving as an editorial board member and reviewer of several journals. He is an active member of the Association of Food Scientist and Technologist, India. He is also a recipient of MANF from UGC, India.

Gulzar Ahmad Nayik completed his Master's degree in Food Technology from the Islamic University of Science and Technology, Awantipora, J&K, India, and earned his PhD from Sant Longowal Institute of Engineering and Technology, Longowal, Sangrur, Punjab, India. He has published more than 130 peer-reviewed research and review papers, contributed over 65 book chapters, and edited 20 books with leading publishers including Springer, Elsevier, the Royal Society of Chemistry, and Taylor & Francis. He has delivered presentations at numerous national and international conferences, seminars, workshops, and webinars. Nayik was shortlisted twice (2017 and 2018) for the Inspire-Faculty Award by the Indian National Science Academy, New Delhi, and was nominated for the INSA Medal for Young Scientists (2019–2020). In 2025, Nayik was featured among the top 2% cited scientists published on Elsevier Data Repository. He serves as editor, associate editor, assistant editor, and reviewer for several reputable journals in food science and technology. He has received multiple awards, appreciations, and recognitions, and is an active member of various international scientific societies and organizations. He is currently engaged in editing several book projects with Elsevier, Taylor & Francis, Springer Nature, and the Royal Society of Chemistry.

Jahangir Ahmad Rather is a researcher and academic in Food Science and Technology with expertise in sustainable food processing, functional ingredients, and innovative packaging technologies. He completed his MSc in Food Science and Technology at the University of Kashmir, J&K, India, and earned his PhD from the Islamic University of Science and Technology, Awantipora, J&K, India. His doctoral work focused on developing bionanocomposite films from poultry waste gelatin to extend the shelf life of foods. This research combined sustainability with material science, turning poultry by-products into biodegradable films that can replace synthetic plastics. Rather has worked on sustainable protein extraction, bioactive peptide development, nanoencapsulation, and nanotechnology-enabled active packaging. In 2021, he qualified the ASRB–ICAR National Eligibility Test and later worked as a Project Research Scientist-I at ICMR, linking food science with public health. He has published over 30 papers in reputed journals and contributed more than 10 book chapters. His work reflects resilience, innovation, and a strong commitment to sustainable food systems that support both nutrition and environmental responsibility.

F. A. Masoodi is an internationally recognized and distinguished academic in Food Science and Technology at the University of Kashmir, J&K, India, known for his expertise in the value addition and postharvest management of foods, particularly those native to the Himalayan region. Masoodi has over 31 years of experience in teaching, research, and extension. His dedication to academics was recognized with the Best Teacher Award of the University of Kashmir, J&K, India, in 2022. Masoodi has a robust record of mentorship, guiding numerous scholars in his areas of expertise. Consistently ranked among the top 2% scientists globally by Stanford University (2021–2025), his prolific research includes over 270 publications and an impressive h-index of 69. Masoodi's tenure includes significant administrative roles, such as Dean Academic Affairs and the founding Head of the Department of Food Science and Technology at the University of Kashmir, J&K, India. He has successfully secured over $800,000 in research funding and has also contributed to national scientific policy as a member of bodies like NAAC and the former FSSAI Expert Panel.

Charles Brennan is a Professor of Food Science at RMIT University, Australia, where he is the Dean of the School of Science. Brennan's research interests lie in the interface between food science and human nutrition with a particular interest in how bioactive ingredients and processing manipulate the nutritional quality of foods, in particular, the role of plant dietary fiber in manipulating the glycemic response. Brennan is a graduate of London University, Wye College (BSc in Applied Plant Science), and King's College London (PhD in Food Science and Nutrition). During his career, he has also worked at Durham University (UK), Plymouth University (UK), Massey University (NZ), and Manchester Metropolitan University (UK). He currently serves as Editor-in-Chief for the *International Journal of Molecular Sciences* and the *International Journal of Food Science & Technology* and in the Editorial Board of the *Journal of Bioactive Carbohydrates and Dietary Fibre.*

Contributors

Aminu Abdullahi
Department of Food Science and Technology
Bayero University, Kano
Kano, Nigeria

Muhammad Afzaal
Department of Food Science
Government College University Faisalabad
Faisalabad, Pakistan

Isam A. Mohamed Ahmed
Faculty of Agriculture, Department of Food Science and Technology
University of Khartoum
Shambat, Khartoum, Sudan

Muhammad Talha Akhtar
Department of Biotechnology
University of Sargodha
Sargodha, Pakistan
and
Department of Food Science and Nutrition
College of Food and Agricultural Sciences, King Saud University
Riyadh, Saudi Arabia

Najmeenah Akhter
Department of Food Technology
Islamic University of Science and Technology
Awantipora, J&K, India

Dilek Dülger Altıner
Tourism Faculty, Department of Gastronomy and Culinary Arts
Kocaeli University
Kocaeli, Türkiye

Kashif Ameer
Institute of Food Science and Nutrition
University of Sargodha
Sargodha, Pakistan

Maha Atiq
Department of Food Science
Government College University Faisalabad
Faisalabad, Pakistan

Emine Aydın
Faculty of Agriculture, Department of Agricultural Biotechnology
Düzce University
Düzce, Türkiye

Gitanjali Behera
Department of Life Science (Microbiology and Food Science & Technology)
GITAM (Deemed to be University)
Visakhapatnam, Andhra Pradesh, India

Deb Kumar Chakraborty
Department of Life Science (Microbiology and Food Science & Technology)
GITAM (Deemed to be University)
Visakhapatnam, Andhra Pradesh, India

Sobia Chashman
Department of Food Science
Government College University Faisalabad
Faisalabad, Pakistan

Basharat Nabi Dar
Department of Food Technology
Islamic University of Science and Technology
Awantipora, Kashmir, India

Umar Farooq
Department of Food Science and Technology
MNS-University of Agriculture
Multan, Pakistan

Nida Firdous
Department of Food Science and Technology
MNS-University of Agriculture
Multan, Pakistan

Umar Garba
Department of Food Science and Technology
Bayero University Kano
Kano, Nigeria
and
Ningbo Research Institute
Zhejiang University
Ningbo, China

Mariam Islam
Department of Food Science
Government College University Faisalabad
Faisalabad, Pakistan

Balarabe B. Ismail
Ningbo Research Institute
Zhejiang University
Ningbo, China

Hafizu Ibrahim Kademi
Department of Food Science and Technology
Aliko Dangote University of Science and Technology
Wudil, Kano, Nigeria

Mehvish Majeed
Department of Food Technology
Islamic University of Science and Technology
Awantipora, Kashmir, India

Darakshan Majid
Department of Food Technology
Islamic University of Science and Technology
Awantipora, Kashmir, India

Sajad Ahmad Mir
Department of Life Science (Microbiology and Food Science & Technology)
GITAM (Deemed to be University)
Visakhapatnam, Andhra Pradesh, India

Sabeera Muzzaffar
Department of Food Science and Technology
University of Kashmir
Hazratbal, Srinagar, J&K, India

Aliza Naeem
Department of Food Science
Government College University Faisalabad
Faisalabad, Pakistan

Zahrau Bamalli Nouruddeen
Department of Food Science and Technology
Aliko Dangote University of Science and Technology
Wudil, Kano, Nigeria

Zahida Naseem
Division of Food Science and Technology
SKUAST-K
Shalimar, Srinagar, J&K, India

Mariya Nayeem
Department of Food Technology
Islamic University of Science and Technology
Awantipora, Kashmir, India

Gulzar Ahmad Nayik
Department of Microbiology
Marwadi University Research Centre, Marwadi University
Rajkot, Gujarat, India

Ambreen Naz
Department of Home Sciences
MNS-University of Agriculture
Multan, Pakistan

Bushra Niaz
Department of Food Science
Government College University Faisalabad
Faisalabad, Pakistan

Dayanand Peter
Division of Food Processing Technology
Karunya Institute of Technology and Sciences
Coimbatore, Tamil Nadu, India

Huma Qayyum
Department of Human Nutrition and Dietetics
MNS-University of Agriculture
Multan, Pakistan

Amara Rasheed
Department of Food Science
Government College University Faisalabad
Faisalabad, Pakistan

Thottiam Vasudevan Ranganathan
Division of Food Processing Technology
Karunya Institute of Technology and Sciences
Coimbatore, Tamil Nadu, India

J. Sudhakara Rao
Department of Life Science (Microbiology and Food Science & Technology)
GITAM (Deemed to be University)
Visakhapatnam, Andhra Pradesh, India

Jahangir A. Rather
Department of Food Science and Technology
University of Kashmir
Hazratbal, Srinagar, J&K, India

Nighat Raza
Department of Food Science and Technology
MNS-University of Agriculture
Multan, Pakistan

Farhan Saeed
Department of Food Science
Government College University Faisalabad
Faisalabad, Pakistan

Bisma Shabbir
Department of Food Science and Technology
MNS-University of Agriculture
Multan, Pakistan

Muhammad Shahbaz
Department of Food Science and Technology
MNS-University of Agriculture
Multan, Pakistan

Aanchal Sharma
University Centre for Research and Development
Chandigarh University
Gharuan, Mohali, Punjab, India

Munazah Sidiq
Department of Food Science and Technology
University of Kashmir
Hazratbal, Srinagar, J&K, India

Gökçen Yıldız
Faculty of Engineering and Natural Sciences, Department of Food Engineering
Bursa Technical University
Bursa, Türkiye

Alkasim Kabir Yunusa
Department of Food Science and Technology
Aliko Dangote University of Science and Technology
Wudil, Kano, Nigeria

Hamna Zafar
Department of Food Science
Government College University Faisalabad
Faisalabad, Pakistan

1 Chemistry and Nutritional Constituents of Pseudocereals

Najmeenah Akhter, Jahangir A. Rather, Mehvish Majeed, Darakshan Majid, Mariya Nayeem, and Basharat Nabi Dar

1.1 INTRODUCTION

The most significant danger to the survival of humanity is the growing disparity between population expansion and food availability (Khatri et al., 2024). In terms of nutrition, the effectiveness of any grain as human food largely relies on both the quantity and quality of its protein content. Proteins are a crucial category of biomacromolecules that play vital roles in physiological functions (Day et al., 2022). Natural plant-based proteins are valuable due to their safety, high biocompatibility, nutritional benefits, and cost-effectiveness. Therefore, discovering new plant proteins that are abundant in essential amino acids (EAAs) is essential for both the food and pharmaceutical sectors (Rather et al., 2023). The functionality of grains largely relies on their genetic composition and the influence of environmental factors on key components such as carbohydrates, proteins, vitamins, minerals, and phenolic phytochemicals. As a result, many cereal crops may excel in one component while lacking in another (Akhter et al., 2024). Thus, to overcome this problem, a great focus has been placed on the exploration and use of unutilized food plants, such as pseudocereals. Nutritional benefits, rich phytochemical content, and important role in gluten-free products, make pseudocereals a great focus of interest (Singh & Sood, 2020). Pseudocereals are dicotyledonous grains that serve as a superior substitute to true cereals. The most widely used pseudocereals are buckwheat (*Fagopyrum esculentum*; family Polygonaceae), amaranth (including *Amaranthus hypochondriacus*, *Amaranthus caudatus*, and *Amaranthus cruentus*; family Amaranthaceae), and quinoa (*Chenopodium quinoa* subsp. quinoa; family Chenopodiaceae) (Morales et al., 2021). For the diversification and development of agricultural products and food, these new plant species play a vital function. To improve human health, different innovative products from these species can be developed providing an extensive range of health benefits. Scientific community, consumers, and food manufacturers are gradually appreciating such foods (Samtiya et al., 2021). Due to the presence of a wide range of bioactive compounds such as flavonoids, phenolic acids, trace elements, fatty acids, and vitamins, these pseudocereals such as buckwheat, amaranth, and quinoa are known to have good

DOI: 10.1201/9781003584858-1

effect on human health such as prevention and reduction of many degenerative diseases (Kaur et al., 2023; Manzoori, 2023). Different research studies have observed that the use of these whole grains reduces the chances of different diseases such as cardiovascular diseases (Zong et al., 2016; Tieri et al., 2020), diabetes (Makarem et al., 2016), cancer (Gaesser, 2020; Benisi-Kohansal et al., 2016), and hypertension (Chen et al., 2016). Although such foods provide excellent-quality minerals, protein, essential amino acids, and phytochemicals still their commercialization is quite limited due to a lack of research gap on their nutritional composition and processing technologies. To valorize the knowledge about the processing technologies to achieve attractive functional properties and nutritive value, different technology portfolios should be continued and expanded. In addition to their gluten-free nature and abundance of protein and essential nutrients, they also contain saponins that have various pharmaceutical, industry, and agriculture applications (Morya et al., 2024). Saponins have the ability to reduce the cholesterol levels in the blood, in addition to their significant hemolytic and antilipemic effects. This attribute of saponins is considered as most important characteristics, especially as there have been no negative impacts of saponins on protein digestibility (Shah et al., 2022). Pseudocereals are becoming a nutritious substitute to gluten-containing grains for making gluten-free food items. They also provide a significant amount of various essential nutrients beside being gluten-free (Thakur et al., 2021). Due to the rise in the individuals being diagnosed with celiac disease there is an increase in the demand for alternative gluten-free flours (Mazzola et al., 2024). High phenolic, nutritional, and phytochemical profiles make pseudocereals center of interest to scientists. In comparison to conventionally used cereals like wheat, rice, and maize, pseudocereals possess higher amount of EAAs, protein efficiency ratio, high biological value, and overall nutritional index; besides, they also offer substantial health advantages. The phenolic compounds found in pseudocereals have the potential to reduce oxidative stress and lower the risks of cancer, diabetes, hypertension, and cardiovascular diseases as per the current research (Thakur et al., 2021). This chapter will provide new innovation toward the utilization of different underutilized pseudocereals, their potential health benefits and nutritional properties for the development of novel and functional foods by keeping in view the importance of the abovementioned properties.

1.2 CHEMICAL AND NUTRITIONAL CONSTITUENTS OF PSEUDOCEREALS

The primary focus of this chapter is on the chemistry and nutritional value of various pseudocereals. Consumers with health-consciousness are showing a growing interest in value-added products, so to increase or entirely replace the conventional cereals like wheat, corn, and rice with more nutritious alternatives such as quinoa, amaranth, and buckwheat. Given the exceptional nutrient profiles of these pseudocereals, they can serve as ideal substitutes to fulfill the dietary needs of the selective populace. The composition of quinoa, buckwheat, and amaranth in terms of their chemical and nutritional properties is shown in Figure 1.1.

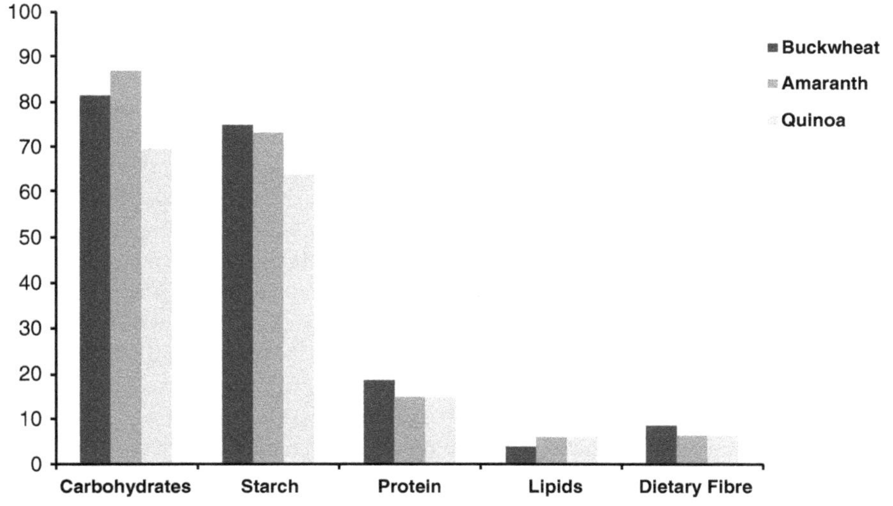

FIGURE 1.1 Chemical composition (%) of pseudo-cereals, (Jha et al., 2024; Kumar et al., 2024; Pathan et al., 2024; Sattar et al., 2024; Yang et al., 2024).

1.2.1 Proteins

The nutritious value of proteins is assessed by the inclusion of EAAs that the body cannot produce on its own (Rather et al., 2024a). If even one of these amino acids is lacking, the others will break down and be excreted, leading to inadequate growth in both livestock and humans, as well as a loss of nitrogen in the diet. From a nutritional standpoint, the most crucial aspect of protein is its essential amino acids, which must be obtained through dietary sources since they cannot be synthesized by humans (Yu & Fukagawa, 2020). EAAs play a crucial role in growth and fulfilling metabolic requirements compared to non-EAAs. Additionally, there is a third group known as "conditionally essential" amino acids. These are not typically needed but during certain physiological or pathological conditions, the body is unable to produce them in sufficient quantities, making their intake essential in those situations (Tessari, 2019). The quantity and quality of pseudocereal protein are superior to the conventional cereals. They are particularly abundant in lysine, the amino acid that is often lacking in cereals. Furthermore, the rich content of histidine and arginine, which is important for babies and toddlers, establishes quinoa and amaranth as crucial components of children's nutrition. Mir et al. (2018) in their study found that quinoa protein provides approximately 107.33% of isoleucine, 210% of the recommended histidine, 157.72% of methionine, 132.2% of lysine, plus tyrosine, 177.82% of threonine, plus cysteine, 247.63% of phenylalanine, 15.83% of tryptophan, and 107.43% of valine. In comparison to other cereal grains like sorghum, rice, and wheat, quinoa typically has greater levels of the sulfur-containing amino acids, methionine, and cysteine (Allai et al., 2022; Abugoch et al., 2009). Bhargava et al. (2005) reported that quinoa is regarded as top sources of leaf protein concentrate which could be used as an alternative of protein in diet, animal feed, and the medical sector. Valadez-Vega et al. (2022) investigated how the quality of amaranth protein is affected by fermentation and popping. They discovered that popping enhanced the overall lysine amount of the grains, leading to extents that were much greater than those typically seen in standard grains and similar to those found in legumes. Additionally, the concentrations of cysteine and methionine in amaranth were greater than what is typically present in proteins of legumes and cereals. Fermentation and popping processes help to reduce the levels of exogenous factors such as tannins, phytates, and trypsin inhibitors, which can negatively impact digestibility of protein. Taylor et al. (2016) indicated that rheological characteristics are influenced by prolamins, which are gluten subunits with high molecular weight. These prolamins transform from a β-turn structure to a β-sheet structure in response to mechanical stretching. Further beneficial connections between chains result from the reformation of disulfide bonds between proteins, which is facilitated by mechanical effort performed in the presence of air. The process enhances resistance to deformation during frequent stretching, ultimately resulting in the appropriate rheological characteristics needed to trap expanding gas bubbles. It has been established that other prolamins and globulins are ineffective at creating appropriate dough, even though a variety of substitutes are examined to mimic the qualities of wheat gluten. To tackle this problem, a number of researchers have used pseudocereals, with encouraging results.

1.2.1.1 Main Proteins of Pseudocereals

Before utilizing proteins in any activity, it is vital to take into account their properties. Generally, they are categorized into two types: prolamins, found in cereals, and a majority of globulins along with some albumins in pseudocereals (Mullaiselvan et al., 2024). The proteins that serve as storage belong to various groups and exhibit notably distinct structures and characteristics. The primary storage protein in pseudocereals is globulin, which can be classified into group's based on their sedimentation coefficients (which reflect their molecular masses). The molecular weight of typical 11S globulins varies from 250 to 400 kDa and is hexameric. With single disulfide link, the two chains that make up their subunits are basic and acidic. The molecular weight of 7S globulins is 150–190 kDa with no disulfide bonds and is trimeric. Albumins contain small and large subunits with molecular weight between 8 and 15 kDa, which are linked by two disulfide bonds. The primary storing proteins in pseudocereals resemble those in legumes, including the storage proteins 2S albumin and 11S globulin, as well as the 7S globulin present in buckwheat and amaranth, according to Taylor et al. (2016). Pseudocereals are associated in the formation of good-quality doughs as the globulin present in them has a predominantly β-sheet structure with β-barrel confirmation as noted by Mir et al. (2021). The 11S type globulins of rice, oats, and pseudocereals are polymerized by disulfide bonding. Although these storage proteins are similar to glutenin in a number of ways, they differ significantly in terms of their secondary, tertiary, and quaternary structures, amino acid makeup, and sequence.

1.2.2 Carbohydrates

One of the main biopolymeric substances in plants, including grains, seeds, and tubers, is starch which usually appears in a granular form with diverse shapes and sizes. It is typically classified as an accessible carbohydrate and is the primary energy source for human physiology. The seeds perisperm contains starchy materials, which might be spherical clusters or separate pieces. Jancurová et al. (2009) noted that the carbohydrate content of quinoa is between 67% and 74% of its dry weight, with nearly 11% of amylose content, that is very less than that found in cereals. Wang and Copeland (2013) examined quinoa's pasting and physicochemical properties and found that the size of its starch granules was lesser compared to wheat (2–40 μm) and maize (1–2 μm). Additionally, it has been noted that higher gelatinization temperatures are correlated with smaller starch granule sizes. This temperature ranges from 57°C to 64°C for quinoa. There are trace amounts of other carbohydrates, such as crude fiber (2.5–3.9 g/100 g), pentosans (2.9–3.6 g/100 g), monosaccharides (2 g/100 g), and disaccharides (2.3 g/100 g). Jan et al. (2017) discovered that the starch extracted from quinoa is suitable for better binding and lower breakability. These requirements are fulfilled as the size of quinoa's starch granule is small. They also pointed out that the extracted starch's swelling capacity was associated with noticeably less swelling, which is advantageous for food products like noodles and composite mixes. Buckwheat contains a higher percentage of starch than amaranth and quinoa. A hundred grams of buckwheat seeds, for instance, provide 343 calories of energy. In terms of calories, its seeds are comparable to those of main cereals like rice, wheat, and maize, as well as pulses like chickpeas, mung beans, and cowpeas (black-eyed peas).

1.2.3 Fiber

Although there is a growing attention in substitute grains, very less information is known about their dietary fiber content. Quinoa and amaranth are dicotyledonous grains that are now known to have positive nutritive properties and to be healthful. After analyzing the entire composition of amaranth grains, Malik et al. (2023) found that the seeds' dietary fiber content ranged from 8% to 16%, with 33%–44% being soluble fiber. Contrary to amaranth and quinoa, which have dietary fiber contents similar to those of conventional cereals, buckwheat seeds have a significantly greater dietary fiber concentration (Alvarez-Jubete et al., 2009). The composition of monosaccharides in the insoluble fiber of amaranth was determined by Scilingo and Sabbione (2024) to be as follows: fucose (1%), followed by mannose (2%), rhamnose (4%), galactose (6%), xylose (9%), arabinose (22%), and glucose (57%). Other studies have also shown that quinoa's overall dietary fiber level is similar to that of legumes and cereal grains. According to recent research, rhamnogalacturonan-I (RG-I), which has arabinan and galactan side chains, and linear arabinans comprised the pectic polysaccharides extracted from quinoa seeds. These pectic polysaccharides have a protective effect against ethanol-induced gastric lesions in rats (Chen et al., 2019). The investigation conducted by Lamothe et al. (2015) found that the insoluble fiber content of quinoa and amaranth is mainly composed of cellulose, xyloglucans that are mainly branched with di- and trisaccharide side chains, and homogalacturonans combined with RG-I segments that contain arabinan and galactan side chains. They found that there were minor differences in the soluble fractions between amaranth and quinoa, with amaranth having more soluble dietary fiber. Furthermore,

Lamothe et al. (2015), found that arabinans and homogalacturonans are main soluble fibers present in quinoa. However, the soluble fiber in amaranth was predominately composed of branched xyloglucans with a majority of disaccharide and trisaccharide side chains, as well as pectic polysaccharides. It also contains less amount of xylose than quinoa. As the composition of fibers found in quinoa and amaranth is different from that of cereals, further study is necessary to determine the potential physiological and functional characteristics of them as very less research is available on physiological and fermentative properties of xyloglucans.

1.2.4 Vitamins and Minerals

Vitamins are essential for most physiological processes of the body (Hanna et al., 2022). In terms of vitamin content, amaranth has a higher thiamine level compared to wheat. According to Olmos et al. (2022), both quinoa and amaranth contain good amounts of vitamin C, riboflavin, as well as notably high in vitamin E and folic acid. Quinoa and amaranth have folic acid contents of 78.1 μg/100 g and 102 μg/100 g, respectively, which is 2.5 times greater than wheat's 40 μg/100 g. Pirzadah et al. (2020) confirmed that buckwheat seeds contain vitamins B2 and B6 and investigated the levels of rutin and quercitrin in tartary buckwheat. With the largest concentrations of B vitamins found in the bran, they observed that tartary buckwheat has higher total vitamin B content than regular buckwheat. It was noticed that gluten-free items are deficient in magnesium, iron, and calcium, and that gluten-free diets also exhibit these deficiencies. According to Alvarez-Jubete et al. (2009), pseudocereals such as buckwheat, quinoa, and amaranth are excellent suppliers of iron, magnesium, calcium, and essential minerals. It is clear that amaranth is highly loaded with mineral content, persuade by buckwheat and quinoa. Notably, considering the recognized prevalence of osteopenia and osteoporosis in newly diagnosed celiac patients, the higher amount of calcium content in amaranth seeds is very vital for people dealing with celiac disease consequences (Alkalay, 2021).

1.2.5 Fat

Studies on fat content have revealed that pseudocereals have a greater fat content than most other types of cereals. In pseudocereals, the fat is distinguished by a significant presence of unsaturated fatty acids, particularly a high level of linolenic acid. There is a positive effect on cardiovascular health and enhancement in insulin sensitivity due to polyunsaturated fatty acids (Kharazmi-Khorassani et al., 2021; Rather et al., 2024b). The high amount of squalene in amaranth which is an open chain triterpene that is mostly present in the livers of deep-sea fish and other marine life. This squalene, which is extensively extracted from pseudocereals, is used in cosmetics and pharmaceutical industries. Amaranth and quinoa have lipid content twice or thrice than that of conventional grains like wheat (Alvarez-Jubete et al., 2009). The high level of unsaturation of these lipids found in the pseudocereals makes them desirable from nutritional perspective. Linoleic acid is the predominant polyunsaturated fatty acid, which is highly present in quinoa (Matías et al., 2022). A research conducted by Takic et al. (2022) found that a food with great content of α-linolenic acid is beneficial as it reduces biological signs linked to various deteriorating ailments, including osteoporosis, cancer, cardiovascular disease, and both autoimmune and inflammatory disorders. According to Wirkowska-Wojdyła et al. (2022), the lipids in quinoa and amaranth are usually resistant to oxidation, even though their fat content is high and they have a significant degree of unsaturation. In pseudocereals, the presence of tocopherols in large quantities is thought to have protective qualities, which accounts for this stability.

1.2.6 Bioactive Components

1.2.6.1 Phenolic Content of Pseudocereals

Benzoic and cinnamic acids are certainly the source of phenolic acids, which are found in practically all grains. On the basis of their molecular weight, they are easily divided into three categories: tannins, flavonoid-type chemicals, and phenolic acids. Phenolic acids, lignans, flavonoids, phenylpropanoids, coumarins, xanthones, quinones, phenols, and stilbenoids are among the various groups of phenolic chemicals that should be identified. The majority of them are flavonoids, which are additionally separated into flavonols (including proanthocyanidins), isoflavonoids, flavones, anthocyanins, dihydroflavonols, chalcones, dihydrochalcones, and flavanones. Furthermore, it is classified into hydroxyphenylacetic, hydroxybenzoic, hydroxyphenylpropanoic, and hydroxycinnamic acids. The hydroxybenzoic acids specifically include p-hydroxybenzoic, vanillic, protocatechuic acids, syringic, and gallic acid. The hydroxycinnamic acids, characterized by their C6–C3 structure, encompass caffeic, sinapic, ferulic, and coumaric acids. Phenolic acids found in grains are known to occur in free as well as in bound forms. The phenolic acid present in pseudocereals is discussed briefly as follows:

1.2.6.2 Overall Phenolic and Flavonoid Content of Pseudocereals

Piasecka-Jóźwiak et al. (2016) found that in buckwheat overall phenolic level varies between 29 and 1,371 mg/100 g, and this variation is influenced by the extraction method and the standard (either ferulic acid or gallic acid) used for calibration. According to the study investigated by Singh and Punia, (2020), measuring amaranth on dry weight basis, the plant phenolic level ranges from 2 to 24 mg/100 g, whereas its polyphenols content ranges from 2 to 300 mg/100 g. Furthermore, anthocyanin concentrations of 90 and 103 mg of cyanidin 3-glucoside equivalents per 100 g were investigated (Paśko et al., 2009).

1.2.6.3 Phenolic Acids and Flavonoids

Numerous beneficial flavonoids and phenolic acids are discovered in pseudocereals, showcasing their impressive nutritional profile. Cinnamic acid and benzoic acid derivatives make up the majority of phenolic acids, and they are generally found in bound forms, particularly the cinnamic acid versions. There are several different types of flavonoids, including anthocyanins, flavonols, flavanones, and flavones. Among these, the identification of 3-deoxyanthocyanins in pigmented sorghum is particularly noteworthy, as these compounds may offer significant health benefits, making pseudocereals an invaluable addition to a healthy diet. An overview of the different flavonoids and phenolic acids found in pseudocereals is provided in Table 1.1.

TABLE 1.1
Bioactive Components of Pseudocereals

Pseudocereal	Compound	Formula	Structure	Compound CID	Reference
Buckwheat	Vanillic acid	$C_8H_8O_4$	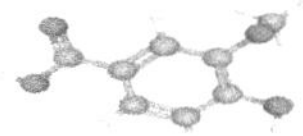	8468	Thakur et al. (2021) Kreft et al. (2006) Paśko et al. (2008)
	Coumaric acid	$C_9H_6O_2$	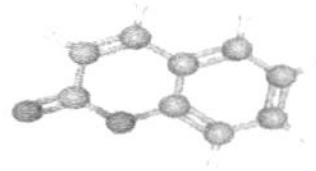	323	
	Gallic acid	$C_7H_6O_5$	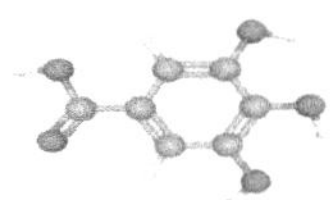	370	
	Protocatechuic acid	$C_7H_6O_4$	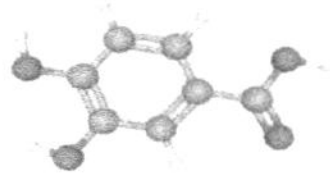	72	
	Rutin	$C_{27}H_{30}O_{16}$	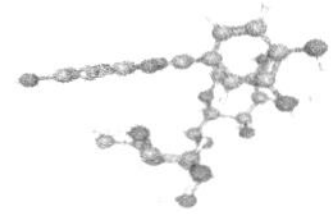	5280805	
Amaranth	Amaranthin betacyanin	$C_{30}H_{34}N_2O_{19}$	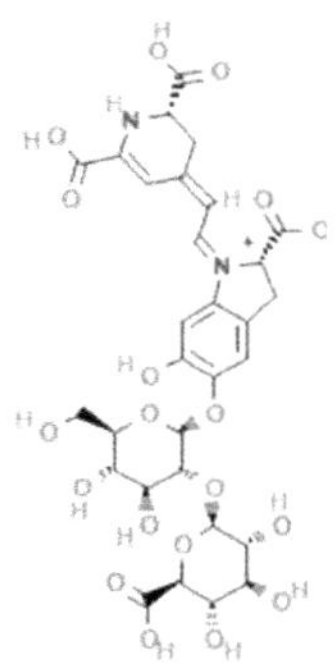	112500418	Tang et al. (2016) Tyszka-Czochara et al. (2016)
	Zeaxanthin	$C_{40}H_{56}O_2$		5280899	
	Lutein	$C_{40}H_{56}O_2$	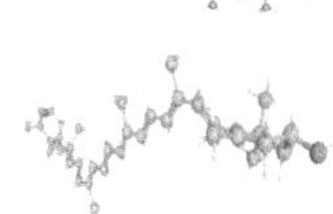	5281243	
	β-carotene	$C_{40}H_{56}$		5280489	
	Isoquercetin	$C_{21}H_{20}O_{12}$	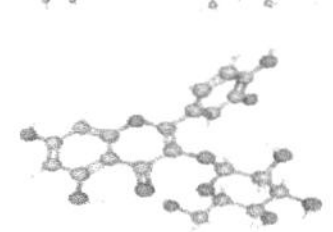	5280804	

(*Continued*)

TABLE 1.1 (*Continued*)
Bioactive Components of Pseudocereals

Pseudocereal	Compound	Formula	Structure	Compound CID	Reference
Quinoa	Nicotiflorine	$C_{27}H_{30}O_{15}$	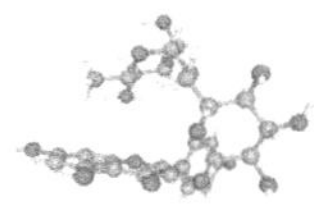	5318767	Repo-Carrasco-Valencia et al. (2010) sTang et al. (2015)
	Rutin	$C_{27}H_{30}O_{16}$	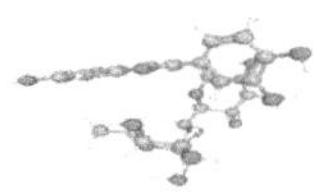	5280805	
	Ferulic acid-4–glucoside	$C_{16}H_{20}O_9$	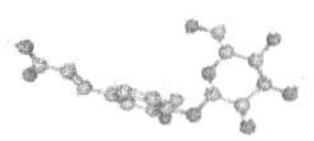	13916049	
	Betanin	$C_{24}H_{26}N_2O_{13}$	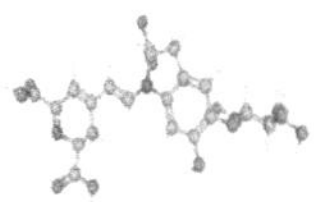	6540685	
	Myricetin	$C_{15}H_{10}O_8$	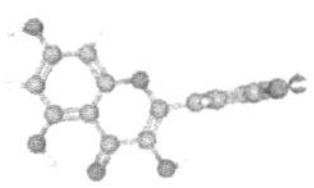	5281672	
	Quercetin	$C_{15}H_{10}O_7$	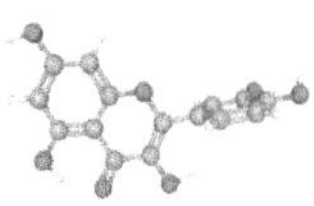	5280343	
	Isorhamnetin	$C_{16}H_{12}O_7$	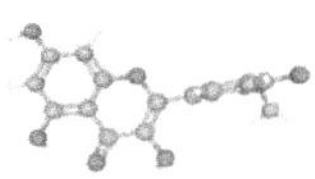	5281654	
	Kaempferol	$C_{15}H_{10}O_6$	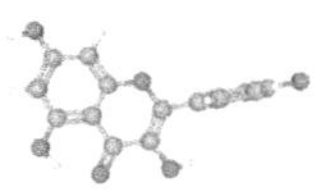	5280863	

1.3 ANTI-NUTRITIONAL FACTORS

Anti-nutritional factors are unsafe compounds that negatively impact the growth and health of living beings, commonly present in cereals, which significantly restrict their use in cereal-based food production. Among these, amaranth and quinoa are notable, as they also contain numerous anti-nutritional factors such as saponins, phytic acid, protease inhibitors, and tannins. Saponins are distinguished by their widespread distribution in the kingdom of plants and their glycosylated secondary metabolites (Rather et al., 2024a). The structural characteristics of this varied collection of compounds – which include either a steroidal or triterpenoid aglycone connected to more than one sugar chain – demonstrating the complexity and importance of understanding these factors in food development (Jan et al., 2023). Mohamed Ahmed et al. (2021) noted that saponins serve as anti-quality components in quinoa, yet moreover they exhibit certain intriguing biological characteristics. Saponins are extremely bitter surfactants that can cause a lot of foaming in solutions that contain water; they are capable of forming complexes with proteins and lipids, such as cholesterol, and they have a hemolytic effect. Research has shown that saponins are absorbed only in small quantities, and their primary impact remains localized to the intestinal tract; they can bind with zinc and iron, thereby decreasing their bioavailability (Zhang et al., 2024). The health benefits of saponins include antimicrobial, anti-inflammatory, cholesterol-lowering and anticancer effects. According to Valadez-Vega et al. (2022), there are just 0.09% of saponins in amaranth seeds, and because of their low toxicity and low concentration, there are no serious health risks for consumers. According to the literature, quinoa (whole seeds) has a saponin level of 0.03%–2.05%, which is still less than that of soybeans. Saponin concentration can also be significantly reduced by dehulling and washing. Additionally, saponins can be destroyed during processing; however, the amount is not as much as it is after washing or dehulling. By crossing a bitter (high-level saponin) quinoa with sweet (less content) quinoa, the saponin level of quinoa seeds can be reduced. One can classify a variety as sweet if its saponin content is less than 0.11%.

1.4 HEALTH BENEFITS OF PSEUDOCEREALS

Pseudocereals, rich in nutraceutical constituents, offer a range of health benefits, such as potential cancer-fighting properties, diabetes management, and cholesterol reduction. Bioactive constituents of pseudocereals and their nutritional properties along with health benefits are illustrated in Figure 1.2. The potential of pseudocereals to decrease cholesterol has been well investigated. Consuming pseudocereals such as buckwheat and buckwheat-based products can lower blood sugar levels by 14.5%, total cholesterol by 4%–13.56% low-density lipoprotein by 9.05% and triglycerides by 4.85% (Kaur et al., 2023). Buckwheat proteins' capacity to increase bile acid secretion is

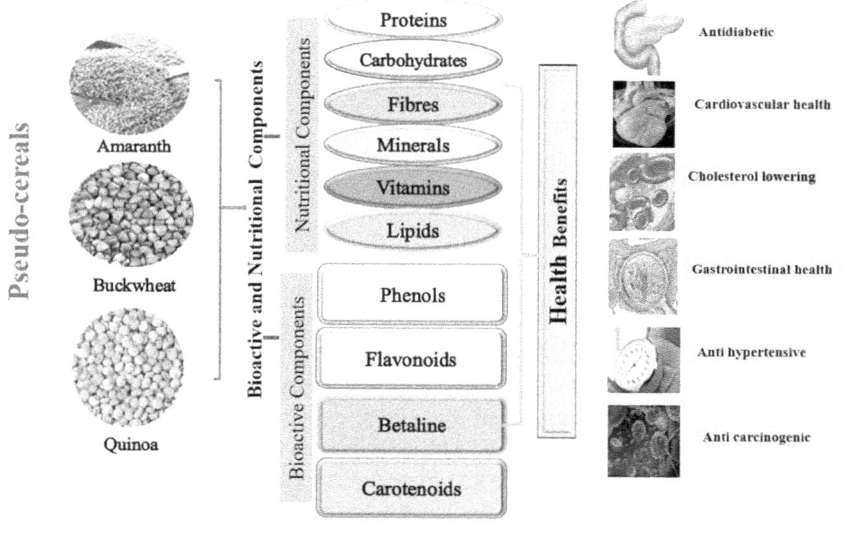

FIGURE 1.2 Health benefits of pseudo cereal bioactive components.

thought to be the cause of their lipid-lowering properties (Liu et al., 2021). The cholesterol levels in the gall bladder, serum, and liver of hamsters were decreased by the incorporation of buckwheat protein (Liu et al., 2021). Bektas and Kaptan (2023) reported the antiproliferative effects of lectin present in *Amaranthus caudatus* which can bind to a tumor factor. They deduced that diets rich in TF-binding (Thomsen–Friedenreich antigen-binding) lectins might serve as indicators of malignant gastrointestinal epithelial cell growth and could significantly contribute to the diagnosis of intestinal cancer. Although no diet can entirely stop or slow cancer, regular consumption of pseudocereals like buckwheat has been associated with a lower risk of developing certain types of cancer (Kreft et al., 2022). Eating pseudocereals that are high in antioxidants can help prevent cancer by safeguarding DNA from oxidative harm. Buckwheat is considered the best grains for reducing the levels of blood sugar thereby decreasing the glycemic index as demonstrated by numerous studies (Chiang et al., 2023).

1.5 CONCLUSION

In recent decades, underused grains like millets and pseudocereals have become more popular due to the increase in demand for wholesome gluten-free foods. Such cereals are currently becoming a strong alternative for gluten-rich grains in gluten-free food items. Rich in essential nutrients – proteins, vitamins lipids, phenolics, minerals, and bioactive compounds – pseudocereals offer significant health benefits. Pseudocereals are a rich source of bioactive substances such as carotenoids, phenolic compounds, betalains, and flavonoids, making them a powerhouse choice for people with conscious of health. However, it's also significant to consider that pseudocereals have certain antinutritional factors – such as phytic acid, saponins, and tannins – that can bind to micro- and macronutrients, limiting their availability to the body. Increasing the bioavailability of pseudocereals can be achieved through beneficial processing methods like soaking, fermentation, germination, and popping. These adaptable grains provide diverse health advantages, like lowering cholesterol, combating cancer, reducing inflammation, and managing diabetes, as evidenced by multiple studies. A large number of gluten-free products on the market depend on refined starches and artificial ingredients, causing a significant gap in key nutrients that are crucial for the growth and overall health of those with gluten sensitivities. By incorporating pseudocereals into gluten-free diets, they can address these nutrient gaps and support patients with wheat- or gluten-related disorders. Moreover, promoting the commercialization of these underutilized grains will not only enhance dietary options but also contribute to better overall health for consumers.

REFERENCES

Abugoch, L., Castro, E., Tapia, C., Añón, M. C., Gajardo, P., & Villarroel, A. (2009). Stability of quinoa flour proteins (Chenopodium quinoa Willd.) during storage. *International Journal of Food Science & Technology*, *44*(10), 2013–2020.

Akhter, N., Majid, D., Rather, J. A., Majeed, M., Mehvish, S. T., Dar, B. N., Makroo, H. A., Muzaffar, K., & Sofi, S. A. (2024). A comprehensive review on formulation of baby foods: nutraceutical potential of fruits, vegetables, cereals and probiotics in baby food preparation. *Current Food Science and Technology Reports*, 2, 255270.

Alkalay, M. J. (2021). Nutrition in patients with lactose malabsorption, celiac disease, and related disorders. *Nutrients*, *14*(1), 2.

Allai, F. M., Azad, Z., Gul, K., & Dar, B. N. (2022). Wholegrains: a review on the amino acid profile, mineral content, physicochemical, bioactive composition and health benefits. *International Journal of Food Science & Technology*, *57*(4), 1849–1865.

Alvarez-Jubete, L., Arendt, E. K., & Gallagher, E. (2009). Nutritive value and chemical composition of pseudocereals as gluten-free ingredients. *International Journal of Food Sciences and Nutrition*, *60*(sup4), 240–257.

Bektas, S., & Kaptan, E. (2023). Therapeutic potential of lectins in the treatment of breast cancer: a review. *International Journal of Biological Macromolecules*, *249*, 126073.

Benisi-Kohansal, S., Saneei, P., Salehi-Marzijarani, M., Larijani, B., & Esmaillzadeh, A. (2016). Whole-grain intake and mortality from all causes, cardiovascular disease, and cancer: a systematic review and dose-response meta-analysis of prospective cohort studies. *Advances in Nutrition*, *7*(6), 1052–1065.

Bhargava, A., Rana, T. S., Shukla, S., & Ohri, D. (2005). Seed protein electrophoresis of some cultivated and wild species of Chenopodium. *Biologia Plantarum*, *49*, 505–511.

Chen, G.-C., Tong, X., Xu, J.-Y., Han, S.-F., Wan, Z.-X., Qin, J.-B., & Qin, L.-Q. (2016). Whole-grain intake and total, cardiovascular, and cancer mortality: a systematic review and meta-analysis of prospective studies. *The American Journal of Clinical Nutrition*, *104*(1), 164–172.

Chen, H., Nie, Q., Xie, M., Yao, H., Zhang, K., Yin, J., & Nie, S. (2019). Protective effects of β-glucan isolated from highland barley on ethanol-induced gastric damage in rats and its benefits to mice gut conditions. *Food Research International*, *122*, 157–166.

Chiang, J. H., Hua, X. Y., Yu, A. H. M., Peh, E. W. Y., See, E., & Jeyakumar Henry, C. (2023). A review on buckwheat and its hypoglycemic bioactive components in food systems. *Food Reviews International*, *39*(9), 6362–6386.

Day, L., Cakebread, J. A., & Loveday, S. M. (2022). Food proteins from animals and plants: differences in the nutritional and functional properties. *Trends in Food Science & Technology*, *119*, 428–442.

Gaesser, G. A. (2020). Whole grains, refined grains, and cancer risk: a systematic review of meta-analyses of observational studies. *Nutrients*, *12*(12), 3756.

Hanna, M., Jaqua, E., Nguyen, V., & Clay, J. B. (2022). Vitamins: functions and uses in medicine. *The Permanente Journal*, *26*(2), 89–97.

Jan, K. N., Panesar, P. S., Rana, J. C., & Singh, S. (2017). Structural, thermal and rheological properties of starches isolated from Indian quinoa varieties. *International Journal of Biological Macromolecules*, *102*, 315–322.

Jan, N., Hussain, S. Z., Naseer, B., & Bhat, T. A. (2023). Amaranth and quinoa as potential nutraceuticals: a review of anti-nutritional factors, health benefits and their applications in food, medicinal and cosmetic sectors. *Food Chemistry: X*, *18*, 100687.

Jancurová, M., Minarovicová, L., & Dandar, A. (2009). Quinoa–a review. *Czech Journal of Food Sciences*, *27*(2), 71–79.

Kaur, H., Shams, R., Dash, K. K., & Dar, A. H. (2023). A comprehensive review of pseudo-cereals: nutritional profile, phytochemicals constituents and potential health promoting benefits. *Applied Food Research*, *3*(2), 100351.

Kharazmi-Khorassani, J., Zirak, R. G., Ghazizadeh, H., Zare-Feyzabadi, R., Kharazmi-Khorassani, S., Naji-Reihani-Garmroudi, S., Kazemi, E., Esmaily, H., Javan-Doust, A., & Banpour, H. (2021). The role of serum monounsaturated fatty acids (MUFAs) and polyunsaturated fatty acids (PUFAs) in cardiovascular disease risk. *Acta Bio Medica: Atenei Parmensis*, *92*(2), e2021049.

Khatri, P., Kumar, P., Shakya, K. S., Kirlas, M. C., & Tiwari, K. K. (2024). Understanding the intertwined nature of rising multiple risks in modern agriculture and food system. *Environment, Development and Sustainability*, *26*(9), 24107–24150.

Kreft, I., Fabjan, N., & Yasumoto, K. (2006). Rutin content in buckwheat (Fagopyrum esculentum Moench) food materials and products. *Food Chemistry*, *98*(3), 508–512.
Kreft, I., Vollmannová, A., Lidiková, J., Musilová, J., Germ, M., Golob, A., Vombergar, B., Kocjan Ačko, D., & Luthar, Z. (2022). Molecular shield for protection of buckwheat plants from UV-B radiation. *Molecules*, *27*(17), 5577.
Kumar, H., Guleria, S., Kimta, N., Dhalaria, R., Nepovimova, E., Dhanjal, D. S., Alomar, S. Y., & Kuca, K. (2024). Amaranth and buckwheat grains: nutritional profile, development of functional foods, their pre-clinical cum clinical aspects and enrichment in feed. *Current Research in Food Science*, *9*, 100836.
Lamothe, L. M., Srichuwong, S., Reuhs, B. L., & Hamaker, B. R. (2015). Quinoa (Chenopodium quinoa W.) and amaranth (Amaranthus caudatus L.) provide dietary fibres high in pectic substances and xyloglucans. *Food Chemistry*, *167*, 490–496.
Liu, J., Song, Y., Zhao, Q., Wang, Y., Li, C., Zou, L., & Hu, Y. (2021). Effects of tartary buckwheat protein on gut microbiome and plasma metabolite in rats with high-fat diet. *Foods*, *10*(10), 2457.
Makarem, N., Nicholson, J. M., Bandera, E. V, McKeown, N. M., & Parekh, N. (2016). Consumption of whole grains and cereal fiber in relation to cancer risk: a systematic review of longitudinal studies. *Nutrition Reviews*, *74*(6), 353–373.
Malik, M., Sindhu, R., Dhull, S. B., Bou-Mitri, C., Singh, Y., Panwar, S., & Khatkar, B. S. (2023). Nutritional composition, functionality, and processing technologies for amaranth. *Journal of Food Processing and Preservation*, *2023*(1), 1753029.
Manzoori, M. F. (2023). Pigmented pseudocereals: chemistry, functionality, and technological aspects in food systems. Edited by Taha Mehany, Ahmed Taha, Babatunde Olawoye and Sameh A. Korma. Royal Society of Chemistry. In *Pigmented Cereals and Millets: Bioactive Profile and Food* Applications (pp. 144–176).
Matías, J., Rodríguez, M. J., Granado-Rodríguez, S., Cruz, V., Calvo, P., & Reguera, M. (2022). Changes in quinoa seed fatty acid profile under heat stress field conditions. *Frontiers in Nutrition*, *9*, 820010.
Mazzola, A. M., Zammarchi, I., Valerii, M. C., Spisni, E., Saracino, I. M., Lanzarotto, F., & Ricci, C. (2024). Gluten-free diet and other celiac disease therapies: current understanding and emerging strategies. *Nutrients*, *16*(7), 1006.
Mir, N. A., Bharadwaj, M., Yousuf, B., Gul, K., Riar, C. S., & Singh, S. (2021). Food biopolymers: structural, functional, and nutraceutical properties: food proteins: an overview. Edited by Nisar Mir, Mamta Baharadwaj, Basharat Yousuf and Khalid Gul. Springer Nature. In *Food Biopolymers: Structural, Functional and Nutraceutical* Properties (pp. 211–229).
Mir, N. A., Riar, C. S., & Singh, S. (2018). Nutritional constituents of pseudo cereals and their potential use in food systems: a review. *Trends in Food Science & Technology*, *75*, 170–180.
Mohamed Ahmed, I. A., Al Juhaimi, F., & Özcan, M. M. (2021). Insights into the nutritional value and bioactive properties of quinoa (Chenopodium quinoa): past, present and future prospective. *International Journal of Food Science & Technology*, *56*(8), 3726–3741.
Morales, D., Miguel, M., & Garcés-Rimón, M. (2021). Pseudocereals: a novel source of biologically active peptides. *Critical Reviews in Food Science and Nutrition*, *61*(9), 1537–1544.
Morya, S., More, A., Neumann, A., & Chauhan, S. (2024). Pseudocereals nutraceuticals. *Herbal Nutraceuticals: Products and Processes*, 347–371.
Mullaiselvan, I., Nelluri, P., Nisha, B., & Saha, D. (2024). Properties and applications of proteins from pseudocereals. Edited by Indumathi Mullaiselvan, Puja Nelluri, Baghya Nisha and Debapam Saha. CRC Press. In *Pseudocereals* (pp. 132–153).
Olmos, E., Roman-Garcia, I., Reguera, M., Mestanza, C., & Fernandez-Garcia, N. (2022). An update on the nutritional profiles of quinoa (Chenopodium quinoa Willd.), amaranth (Amaranthus spp.), and chia (Salvia hispanica L.), three key species with the potential to contribute to food security worldwide. *JSFA Reports*, *2*(12), 591–602.

Paśko, P., Bartoń, H., Zagrodzki, P., Gorinstein, S., Fołta, M., & Zachwieja, Z. (2009). Anthocyanins, total polyphenols and antioxidant activity in amaranth and quinoa seeds and sprouts during their growth. *Food Chemistry*, *115*(3), 994–998.

Paśko, P., Sajewicz, M., Gorinstein, S., & Zachwieja, Z. (2008). Analysis of selected phenolic acids and flavonoids in Amaranthus cruentus and Chenopodium quinoa seeds and sprouts by HPLC. *Acta Chromatographica*, *20*(4), 661–672.

Pathan, S., Ndunguru, G., & Ayele, A. G. (2024). Comparison of the nutritional composition of quinoa (Chenopodium quinoa Willd.) inflorescences, green leaves, and grains. *Crops*, *4*(1), 72–81.

Piasecka-Jóźwiak, K., Słowik, E., Rozmierska, J., & Chabłowska, B. (2016). Development of organic buckwheat gluten-free bread, characterized by a high level of bioactive compounds. *Journal of Research and Applications in Agricultural Engineering*, *61*(4).

Pirzadah, T. B., Malik, B., Tahir, I., & Ul Rehman, R. (2020). Buckwheat journey to functional food sector. *Current Nutrition & Food Science*, *16*(2), 134–141.

Rather, J. A., Akhter, N., Akhter, S., Punoo, H. A., Manzoor, N., Muzaffar, H., & Dar, B. N. (2024). Processed meat products. In *Hand Book of Processed Functional Meat Products* (pp. 81–96). Springer.

Rather, J. A., Akhter, N., Ayaz, Q., Mir, S. A., Singh, A., Goksen, G., Majid, D., Makroo, H. A., & Dar, B. N. (2023). Fruit peel valorization, phytochemical profile, biological activity, and applications in food and packaging industries: comprehensive review. *Current Food Science and Technology Reports*, *1*(2), 63–79.

Rather, J. A., Akhter, N., Punoo, H. A., Haddad, M., Manzoor, N., Goksen, G., & Dar, B. N. (2024a). Sustainable algal proteins, novel extraction techniques and applications in the bakery, dairy and pharmaceutical industries: a comprehensive review. *Food Chemistry*, *465*, 141828.

Rather, J. A., Punoo, H. A., Akhter, N., Haddad, M., & Muzzaffar, S. (2024b). Optimization of soy-whey fortified pineapple juice beverage using response surface methodology. *LWT*, *207*, 116613.

Repo-Carrasco-Valencia, R., Hellström, J. K., Pihlava, J.-M., & Mattila, P. H. (2010). Flavonoids and other phenolic compounds in Andean indigenous grains: quinoa (Chenopodium quinoa), kañiwa (Chenopodium pallidicaule) and kiwicha (Amaranthus caudatus). *Food Chemistry*, *120*(1), 128–133.

Samtiya, M., Aluko, R. E., Dhewa, T., & Moreno-Rojas, J. M. (2021). Potential health benefits of plant food-derived bioactive components: an overview. *Foods*, *10*(4), 839.

Sattar, M., Saeed, F., Afzaal, M., Rasheed, A., Asif, A., Sharif, S., Hussain, M., Asad Ur Rehman, H., Raza, M. A., & Munir, H. (2024). An overview of the nutritional and therapeutic properties of amaranth. *International Journal of Food Properties*, *27*(1), 263–272.

Scilingo, A., & Sabbione, A. C. (2024). Pseudocereals dietary fiber. Amaranth, quinoa, and buckwheat fiber composition and potential prebiotic effect. *Revista de La Facultad de Ciencias Agrarias UNCuyo*, *56*(2), 163–180.

Shah, S., Khan, Y., & Yan, H. (2022). The phytochemical, pharmacological and medicina l evaluation of quinoa (Chenopodium quinoa Willd.). *Pakistan Journal of Weed Science Research*, *28*(2), 138.

Singh, A., & Punia, D. (2020). An inside review of amaranth seeds: a potential nutritive pseudo cereal of India, *6*(2): 378–383.

Singh, M., & Sood, S. (2020). *Millets and Pseudo Cereals: Genetic Resources and Breeding Advancements*. Woodhead Publishing.

Takic, M., Pokimica, B., Petrovic-Oggiano, G., & Popovic, T. (2022). Effects of dietary α-linolenic acid treatment and the efficiency of its conversion to eicosapentaenoic and docosahexaenoic acids in obesity and related diseases. *Molecules*, *27*(14), 4471.

Tang, Y., Li, X., Chen, P. X., Zhang, B., Hernandez, M., Zhang, H., Marcone, M. F., Liu, R., & Tsao, R. (2015). Characterisation of fatty acid, carotenoid, tocopherol/tocotrienol compositions and antioxidant activities in seeds of three Chenopodium quinoa Willd. genotypes. *Food Chemistry*, *174*, 502–508.

Taylor, J. R. N., Taylor, J., Campanella, O. H., & Hamaker, B. R. (2016). Functionality of the storage proteins in gluten-free cereals and pseudocereals in dough systems. *Journal of Cereal Science*, *67*, 22–34.

Tessari, P. (2019). Nonessential amino acid usage for protein replenishment in humans: a method of estimation. *The American Journal of Clinical Nutrition*, *110*(2), 255–264.

Thakur, P., Kumar, K., & Dhaliwal, H. S. (2021). Nutritional facts, bio-active components and processing aspects of pseudocereals: a comprehensive review. *Food Bioscience*, *42*, 101170.

Tieri, M., Ghelfi, F., Vitale, M., Vetrani, C., Marventano, S., Lafranconi, A., Godos, J., Titta, L., Gambera, A., & Alonzo, E. (2020). Whole grain consumption and human health: an umbrella review of observational studies. *International Journal of Food Sciences and Nutrition*, *71*(6), 668–677.

Tyszka-Czochara, M., Pasko, P., Zagrodzki, P., Gajdzik, E., Wietecha-Posluszny, R., & Gorinstein, S. (2016). Selenium supplementation of amaranth sprouts influences betacyanin content and improves anti-inflammatory properties via NFκB in murine RAW 264.7 macrophages. *Biological Trace Element Research*, *169*, 320–330.

Valadez-Vega, C., Lugo-Magaña, O., Figueroa-Hernández, C., Bautista, M., Betanzos-Cabrera, G., Bernardino-Nicanor, A., González-Amaro, R. M., Alonso-Villegas, R., Morales-González, J. A., & González-Cruz, L. (2022). Effects of germination and popping on the anti-nutritional compounds and the digestibility of Amaranthus hypochondriacus Seeds. *Foods*, *11*(14), 2075.

Wang, S., & Copeland, L. (2013). Molecular disassembly of starch granules during gelatinization and its effect on starch digestibility: a review. *Food & Function*, *4*(11), 1564–1580.

Wirkowska-Wojdyła, M., Ostrowska-Ligęza, E., Górska, A., & Bryś, J. (2022). Application of chromatographic and thermal methods to study fatty acids composition and positional distribution, oxidation kinetic parameters and melting profile as important factors characterizing amaranth and quinoa oils. *Applied Sciences*, *12*(4), 2166.

Yang, X., Fu, W., Xiao, L., Wei, Z., & Han, L. (2024). Nutrition, health benefits, and processing of sand rice (Agriophyllum squarrosum): comparisons with quinoa and buckwheat. *Food Science & Nutrition*, *12*(10), 7060–7074.

Yu, Y. M., & Fukagawa, N. K. (2020). Protein and amino acids. In *Present knowledge in nutrition* (pp. 1535). Academic Press.

Zhang, Y., Hao, R., Chen, J., Li, S., Huang, K., Cao, H., Farag, M. A., Battino, M., Daglia, M., & Capanoglu, E. (2024). Health benefits of saponins and its mechanisms: perspectives from absorption, metabolism, and interaction with gut. *Critical Reviews in Food Science and Nutrition*, *64*(25), 9311–9332.

Zong, G., Gao, A., Hu, F. B., & Sun, Q. (2016). Whole grain intake and mortality from all causes, cardiovascular disease, and cancer: a meta-analysis of prospective cohort studies. *Circulation*, *133*(24), 2370–2380.

2 Pseudocereal Dietary Fiber Techno-Functional Properties and Applications in Food Product Development

Kashif Ameer, Isam A. Mohamed Ahmed, Muhammad Talha Akhtar, Aanchal Sharma, and Gulzar Ahmad Nayik

2.1 INTRODUCTION

Pseudocereals are grains stemming from plants which generate seeds that are utilized in a similar method as true cereals such as wheat, rice, and corn even though these are not actual grains. Pseudocereals can be broadly defined to not be members of the grass species, but they can be prepared and consumed as grains and are nutritionally similar to grains as well. Some of the most common pseudocereals include quinoa, amaranth, buckwheat, and chia. These seeds are particularly known to contain relatively high amounts of proteins, essential amino acids, vitamins, minerals, as well as fiber (Martínez-Villaluenga et al., 2020; Ameer, Shahbaz, & Kwon, 2017).

Thus, the prospective of classification sets pseudocereals apart from the true cereals at botanical level. For crop quality: real cereals belong to the family Poaceae or grass family, while pseudocereals come from other families. For example, quinoa and amaranth are species of the Amaranthaceae family, buckwheat belongs to the Polygonaceae family, and chia belongs to the Lamiaceae family. However, pseudocereals are applied in the same way as true grains concerning cooking, baking or producing bread, flour, porridges, etc. (Langyan et al., 2024). Gluten, a protein present in many grains, is prohibited in diets assumed by people with celiac disease or gluten intolerance, and these seeds are widely used in gluten-free diets. Owing to their nutritional qualities and utility, they have gained importance as a functional and health ingredient in formulating new food products (Martínez-Villaluenga et al., 2020).

Novel pseudocereal dietary fiber has attracted significant attention in the formulation of functional foods because of its health-enhancing effects and functional

 DOI: 10.1201/9781003584858-2

properties. Pseudocereals such as quinoa, amaranth, buckwheat, and chia are other outstanding sources of fiber that is an important nutrient for digestion that if lacking or poorly consumed can lead to different diseases. In the current and future food markets, consumers continue to look for foods that could help them improve their health hence the use of pseudocereal fibers in foods (Mir et al., 2018).

The fiber present in pseudocereals includes both soluble and insoluble fiber. Soluble fiber is understood for controlling the blood sugar and cholesterol levels and on the other hand, insoluble fiber has importance to support the digestive system with usual bowel movement. Pseudocereal fibers also have prebiotic properties, that is, stimulate the growth of useful bacteria in the gut, which is essential for a healthy organism. This places pseudocereal fiber in a strategic position of being used in the formulation of foods for the management of digestion, diabetes, heart disease, as well as obesity (Langyan et al., 2024). When it comes to food product development, pseudocereal fibers are desirable in order to enhance texture and moisture maintaining capabilities in addition to the quality of a specific product. They can also thicken the texture of low fat or reduced calorie foods, and increase feelings of fullness, which is beneficial in the case of weight loss. These fibers also aid in bodying and gelling agent in products such as sauces, soups, and dairy product replacers for better stability of the product defects and shelf life (Mir et al., 2018).

Furthermore, pseudocereal fibers are particularly applicable in gluten-free food products. It enhances the doughs' gluten–like properties for better texture and elasticity in all gluten-free products. Since consumers' awareness of gluten-free products is rising with time, the addition of pseudocereal fibers contributes positively to the nutritional and functional properties of these products to meet the health-conscious population (Sarkar et al., 2024).

2.2 HEALTH BENEFITS OF PSEUDOCEREAL DIETARY FIBERS

Pseudocereal dietary fibers include quinoa, amaranth, buckwheat, and chia which have multiple health benefits of preserving good health. These fibers are full of soluble as well as insoluble fibers and are good for digestion, diseases, and weightless control. Another advantage of the pseudocereal fiber is that it has a health advantage of having a beneficial impact on the human digestion. The pectin present in insoluble fiber directly increases the mass of feces and eases bowel movement thereby likely to minimize chances of constipation (Ciudad-Mulero et al., 2019). This type of fiber may also help decrease the chances of getting gastrointestinal disease such as diverticulosis. In contrast, insoluble fiber has no such effect, while soluble fiber found in pseudocereals stimulates the growth of favorable bacteria in the digestive tract while serving as a prebiotic. The right and healthy combination of the bacteria in the human digestive system is crucial to digestion, immunity boosting, and healthy absorption of nutrients which, together with it, may reduce the chances of contracting diseases that are associated with inflammation (Sarkar et al., 2024).

Apart from the digestive health responsibility of pseudocereal fiber, it also contributes to a healthy heart. Thus, soluble fiber helps reduce LDL (bad) cholesterol in the body by grappling with the bile acids and expelling them from the body. It can help in preventing heart disease or atherosclerosis as part of the process. There is also evidence that indicates that pseudocereal fiber could also assist to control blood pressure level, which plays its part to cardiovascular health (Mir et al., 2018).

Thus, for those finding it necessary to control body weight, pseudocereal fiber may be useful. Fiber aids in reducing calorie intake since it encourages feelings of fullness during consumption of meals. Moreover, fiber makes carbs digest slowly, thereby causing a slow rise in blood sugar levels. This then makes pseudocereal fiber especially important for people who are overweight or have diabetes among other ailments. Except for the above advantages, pseudocereal fibers are endowed with antioxidants and anti-inflammatory substances that assist in shielding the body from oxidative tension and inflammation, which are extra health advantages (Ciudad-Mulero et al., 2019).

2.3 TECHNO-FUNCTIONAL PROPERTIES OF PSEUDOCEREAL DIETARY FIBER

Pseudocereal dietary fibers also have distinctive techno-functional commercial characteristics, including water absorption, fat binding, and gelation. These fibers make food products to have better texture, mouthfeel, and viscosity and increase product stability and shelf life. They also assist in emulsifying, maintaining moisture content, stabilizing, as well as providing form, thus are useful in food formulation.

2.3.1 Water Absorption and Retention

Various dietary fibers called as pseudocereals are used in functional foods originating from quinoa, amaranth, buckwheat, and chia which are involved in the enhancement of water-holding capacity and modification of texture in food systems. These fibers have very high water-absorption capacity as well as swelling capacity, which is of immense importance in food product reformulation. Their properties require for the conservation of the moisture content, formation of texture, and feel of food products that are regained by their special characteristics. A main function of the pseudocereal dietary fibers in food is their water absorption function. These fibers can hold somewhere between 4 and 20 times their own weight in water, turning into a type of gel (Torbica et al., 2022). This space hold capacity is very useful in products such as soups, sauces, and dressings where it enhances the thickness of the final product. In baked food products, the water holding capacity of the pseudocereal fiber prevents the food product from becoming dry or stale. This leads to products with better shelf stability as well as the texture of the final food product (Serna Saldívar & Hernández, 2020).

Besides, pseudocereal fibers can have a high-water absorption ability that supports the generation of soft, moist foods to the mouth. These fibers can, for instance, work miracles when applied in gluten-free preparations such as breads, cakes, and muffins, by replacing the structural role often played by gluten. This is especially desirable in gluten-free formulations because adding such ingredients provides dough with elasticity and cohesiveness, and therefore, it becomes of lower quality (Torbica et al., 2022; Serna Saldívar & Hernández, 2020).

In products such as dairy replacers, plant-based milks and smooths, pseudocereal fibers function as natural stabilizers. They thicken the texture of the liquid, making it creamier and much more enjoyable than liquid sweeteners which often require preservatives. Moreover, pseudocereal fibers have great gelling capacity useful in stabilizing products such as soy yogurt or soy ice cream by improving texture and moisture retention. In general, pseudocereal dietary fibers are useful in increasing water uptake and viscosity in food systems and therefore indispensable in controlling the stability and perceptual characteristics of multiple food systems and formulations (Torbica et al., 2022; Kaur et al., 2023).

Starch-like dietary fibers derived from quinoa, amaranth, and chia are particularly influential on viscosity and gelation in foods. Also, these fibers have demonstrated the very high potentials for holding water, and transforming the same to a gel form. This feature of gelling has the tendency to thicken the body of soups, sauces, and even dressings to provide a gelling effect. In baked products and dairy products' substitute, the fibers with pseudocereal properties increase viscosity, and therefore, stability. These gel-forming properties also play the role of contributing to formation of smooth, cream-like texture in products such as smoothies, plant-based yogurts, and ice creams while enhancing the moisture and product shelf life (Kaur et al., 2023; Ciudad-Mulero et al., 2019).

2.3.2 Fat Absorption and Emulsification

Pseudocereal dietary fibers which are derived from grains such as quinoa, amaranth, buckwheat, chia, and other have a lot of advantages when they are used in processed foods, mainly in fat binding and oil absorption. These fibers have special physicochemical characteristics as they can selectively gel and encapsulate fats, thereby minimizing oil uptake in food systems. This ability makes them especially valuable in the development for healthier, low-fat versions of traditionally "high fat" foods (Hitache et al., 2023). Due to its fat-binding capability, pseudocereal fibers are most beneficially employed in snacks, fried, and baked foods. These fibers act as a fat sink by binding to excess oils that are present in the food product, thus bringing improvement to the nutritional character of the product and therefore becoming attractive to the consumer looking to improve their level of health. For instance, in baked products, pseudocereals can replace or minimize on oil and fats, and this results to the production of food products, which are low in calorie intensity with appropriate texture. In fried foods, these fibers can absorb less oil during the cooking process turning foods that would ordinarily be soft and greasy into crispier snacks without any added fats (Sindhu & Khatkar, 2019).

Furthermore, pseudocereal fibers have the ability to serve the function of emulsifiers in processed foods. The stabilizers also known as emulsifiers refer to constituents that have the ability to help support combinations of oil and water from coalescing into two separate features. The ability of pseudocereal fibers to form complexes with water and lipid phases with great stability allows the development of stable emulsions in dressings, sauces, and mayonnaise. These fibers make food products that contain them tastier and have a better texture and feel in the mouth – thus with the help of such fibers, manufacturers can minimize the employment of such synthetic emulsifiers or fats as, for example, palm oil. We also mention the role of pseudocereal dietary fibers as emulsifying agents contributes to stability and prolonging the shelf life of processed foods where phase separation occurs over time (Sindhu & Khatkar, 2019). Therefore, pseudocereal dietary fibers are essential in fat binding, minimizing the absorption of oils, and also as natural emulsifying agents. Incorporation in food products serves to not only augment the nutritional content of foods but also the quality of the product, thereby making them useful in the food processing industry (Hitache et al., 2023; Sindhu & Khatkar, 2019).

2.3.3 Textural and Rheological Properties

DF from quinoa, amaranth, and buckwheat respectively and to a very large extent affect the texture, mouthfeel, and cohesiveness of food products particularly baked foods. These fibers positively impact to sensory characteristic of baked foods due to their ability to preserve moisture thereby reducing the rate of staling. Potent water retention properties help in retaining the soft and moist structures in products like breads, cakes, and muffin for a quite long shelf period and enhanced eating qualities. It helps in providing a good texture to the product structure and body to the food without using too much fat or sugar (Nandan et al., 2024; Sindhu & Khatkar, 2019).

In gluten-free baking system, which is always associated with a problem of right texture, pseudocereal fibers show significant functions. They assist in altering the characteristics of the dough and batter, mainly elasticity and cohesiveness, which are such scarce commodities in gluten free products. This leads to improved characteristics of the completed baked products such as reduced crumbliness or dryness akin to bread. These fibers help enhance the handling of the dough and contribute to enhancing the structure of what's formed, and hence, the final product resembles more of a wheat-based product (Torbica et al., 2022; Nandan et al., 2024).

In addition, pseudocereal fibers can affect the stability of batters and dough, improving this characteristic of the bakery products, providing a uniform texture. They also participate in the formation of a richer crumb structure that is generally more satisfactory to the palate. Hence, pseudocereal dietary fibers play a vital role in the texture, mouthfeel, and other related attributes of bakery products particularly in product which is marketed as "healthy" or "gluten free" (Nandan et al., 2024).

2.3.4 Gelling and Thickening Ability

Pseudocereal dietary fibers have a very important use as thickeners of soups, sauces, and dressings and present a natural advantage over substances of the cornstarch or modified types. Besides the increase of viscosity, pseudocereal fibers, when included in soups or sauces, swell, and thicken when in contact with water, so fat and chemical thickeners are not required. Besides amplifying the sensory characteristics of such products, it also reinforces stability as well as the shelf life of such products. In dressings, pseudocereal fibers function as the natural emulsifiers to facilitate the distribution of the oil and water phases in a formulation and thus prevent phase separation and undesirable textural qualities. Such fibers help to achieve the consistency and make the dressings thicker and more fulfilling without using the high fat (Ciudad-Mulero et al., 2019; Martínez-Villaluenga et al., 2020).

Moreover, pseudocereal dietary fibers find application in the formulation of low-fat or sugar-free foods. They are able to thicken or gel, making them suitable to replace fats or sugars in toppings such as puddings, ice creams, and low-calorie desserts. These fibers also afford structure and texture that allow the product to have a satisfactory mouthfeel associated with fats and/or sugars, but with the cutoff in calorie content. Due to these factors, pseudocereal fibers are incorporated commonly into the healthier formulations of foods that give the product good texture, stability yet creating a palatable mouthfeel (Nandan et al., 2024).

2.4 APPLICATIONS OF PSEUDOCEREAL DIETARY FIBER IN FOOD PRODUCT DEVELOPMENT

The details pertaining to various pseudocereal types including quinoa, amaranth, buckwheat, teff, fonio, chia seeds, kañiwa, sorghum, and wild rice, their fiber content, functional properties, and potential uses in food products are provided in Table 2.1.

2.4.1 Bakery Products

Over the past decade, pseudocereals, like quinoa, amaranth, and buckwheat, have gained interest as functional ingredients rich in dietary fiber in food matrices. Most of these ancient grains also have fiber and are also good sources of protein, vitamins, and minerals. Literature regarding the use of pseudocereal dietary fiber in bakery products has also been on the rise owing to the rising trend of health-conscious foods. The incorporation of pseudocereal dietary fiber with respect to bakery products necessitates alterations to its texture, storage life, and nutritional value of the final product (Sindhu & Khatkar, 2019).

2.4.1.1 Texture Properties

The sensory attributes of bakery products take texture into serious consideration. The use of pseudocereal dietary fiber can elicit both benefits and detriments on the texture of bakery products. On the favorable side, the dietary fiber fractions of pseudocereals can be in the improvement of the texture or mouthfeel of products such as breads, cakes, and cookies (Nandan et al., 2024). Because these fibers have high water-absorption capacity, they can cause the loaf to be moist and create a moist crumb in baked products. It is especially of considerable advantage in those formulations which have little or no gluten; these can lead to texture problems such as dryness or brittleness or crumbliness. Depending on the type of fiber used, the latter can replace the role of gluten by adding more structure and water absorption. Nonetheless, high incorporation of pseudocereal fibers may affect some unfavorable textural attributes such as hardness or poor elasticity. This is because fibers of course interfere with the gluten structure in wheat-based bakery products, hence making the structure denser or firmer. Hence, the fiber level should be optimized to obtain the required sensory characteristics wisely (Hitache et al., 2023).

2.4.1.2 Impact on Shelf Life

It also becomes apparent that pseudocereal fibers have their use in shelf life of bakery products. This sector experiences staling, which is mostly due to starch

TABLE 2.1
Various Pseudocereal Types, Their Fiber Content, Functional Properties, and Potential Uses in Food Products

Pseudocereal Type	DF Content (%)	DF Type	Techno-Functional Properties	Applications	Health Benefits	Reference
Quinoa	7%–15%	Soluble (fructans and pectins) and insoluble (cellulose and hemicellulose)	Water retention, gelling, emulsifying, thickening, texture enhancement	Gluten-free products (bread and pasta), breakfast cereals, and snacks	Improves digestion, gut health, and lowers cholesterol	Shahbaz et al. (2023)
Amaranth	7%–8%	Soluble (fructans and pectins) and insoluble (cellulose and hemicellulose)	High water absorption, fat absorption, and improves mouthfeel	Gluten-free bakery items, energy bars, and soups	Antioxidant-rich and anti-inflammatory gut health	Angel Huerta-Ocampo et al. (2011)
Buckwheat	10%–14%	Soluble (fructans) and insoluble (cellulose and hemicellulose)	Binds water, improves texture, stability in high-moisture products	Gluten-free pancakes, noodles, bread, and cookies	Regulates blood sugar and cardiovascular health	Sindhu and Khatkar (2019)
Teff	8%–10%	Insoluble (cellulose and hemicellulose) and soluble (pectins)	Thickening, enhances stability, and binding agent	Gluten-free bread, porridge, cookies, and crackers	Rich in iron, supports bone health and digestion	Satheesh and Fanta (2018)
Fonio	7%–8%	Soluble (pectins and fructans) and insoluble (cellulose)	Forms gel, thickens liquids, and enhances texture	Gluten-free baked goods, porridge, and snack bars	High in amino acids, supports digestion, energy boost	Zhu (2020)
Chia seeds	34% (on a dry weight basis)	Soluble (gums) and insoluble (cellulose)	High water retention, gel formation, and emulsifying	Hydrating beverages, energy gels, and gluten-free baking	Promotes hydration, heart health, and weight management	Czerwonka and Białek (2023)
Kañiwa	8%–10%	Soluble (fructans and pectins) and insoluble (cellulose)	High water absorption, good for texture and mouthfeel	Gluten-free flour, snack bars, and baked goods	High in antioxidants, helps manage cholesterol levels	Dávalos et al. (2023)
Sorghum	10%–13%	Insoluble (cellulose and hemicellulose) and soluble (fructans)	Thickening, moisture retention, emulsification	Gluten-free breads, pasta, snacks	Supports digestive health and reduces risk of heart disease	Taylor et al. (2014)
Wild rice	6%–8%	Insoluble (cellulose and hemicellulose) and soluble (fructans)	Stabilizes viscosity, enhances mouthfeel	Soups, pilafs, salads, and gluten-free bread	Supports heart health, rich in protein, and improves digestion	Mir et al. (2018)

retrogradation, as one of the main difficulties in bakery products. Due to the high fiber content especially soluble fibers, the occurrence of starch retrogradation can be controlled by binding water and the firming of the starch granules. This assists in preserving the new look, and soft nature of the bakery products for more days (Thakur et al., 2021).

Also, the presence of fiber possesses antimicrobial characteristic that helps to slow down the reproduction of spoilage microorganisms. Since fiber has a high water-absorption capacity its presence helps in denying moisture to germs and hence helps in enhancing the shelf life of the product. In gluten-free or low-moisture bread and bakery products addition of pseudocereal fibers can also add to more ripened and longer shelf life, ideal for bakery products compared to the use of synthetic preservatives (Sindhu & Khatkar, 2019).

2.4.2 Dairy Products

Pseudocereals including quinoa, amaranth, and buckwheat as food ingredients are gaining acceptance in developing better nutritive foods. All the above grains, although not cereal grains, give similar characteristics to the traditional cereal grains but are healthier foods because they are well endowed in dietary fiber (Bender & Schönlechner, 2021). Taking advantage of this, more applications of pseudocereal dietary fiber incorporated into dairy products are developing since consumers are demanding more of functional and healthier foods. It has important consequences for texture, stability, and nutrient content of dairy products; pseudocereal fibers are thus an interesting innovation for this sector (Thakur et al., 2021).

2.4.2.1 Texture Properties

This is well illustrated by texture since taste and feel of foods are very important for sensory motivations. Pseudocereal dietary fiber impacts mouthfeel and rheological properties of fluids and semifluid dairy products such as yogurt, cheese, and dairy beverages. They mentioned that one of the specific nutritional impacts of pseudocereal fiber towards texture is creaminess and thickness. The most important pseudocereal fibers are fibers with a high water-holding capacity, which also negatively affect viscosity and are most suitable for yogurt and other probiotically fermented dairy products. These fibers might increase the transparency and palatability which are the two basic indicators, which affect consumer acceptability (Nandan et al., 2024; Thakur et al., 2021).

In the case of low fat or reduced calorie dairy product, the use of pseudocereal fiber can be very effective since fat normally contributes to the smooth texture of dairy products. This fiber can have the license of fat, to provide the product with a thick texture of products containing fats and calories but with no additional fats added (Thakur et al., 2021). However, special caution must be taken while blending the fibers, lest they make the product too thick or viscous to be ingested or consumed. The level of fiber incorporation must be appropriately managed in order to achieve the right product texture as well as its appeal to customers (Bender & Schönlechner, 2021).

2.4.2.2 Impact on Shelf Life

Pseudocereal dietary fiber has shown the ability to increase the shelf life of dairy products which is a major factor in the dairy industry. In dairy products, shelf life is commonly affected by microbial activity, enzymatic activity, and changes in texture and compound formation. Thus, pseudocereal fibers, especially soluble ones, can be useful for the regulation of the TMI value that somewhat affects product stability. These fibers create a structure that retains moisture and better keeps the product firm and uniform in structure rather than allowing whey to drain off (Sindhu & Khatkar, 2019).

Apart from roles in water retention, dietary fiber can also prevent the growth of spoilage microorganisms. Due to the reformation of the microstructure of the dairy products, consumption of pseudocereal fibers might have an unfavorable effect on the microbes and may also enhance the shelf life of the product. This accentuates the already existing difficulty in producing food items such as yogurt, kefir, and other similar products which require the use of beneficial bacteria but are more susceptible to freshness spoilage by naturally occurring unwanted microorganisms, creating the need for smart packaging. Also, due to the gel forming of fiber and its moisture retention property, chances of syneresis in yogurt and such dairy products are minimized. This not only enhances the product quality but also retains the product capacity for a longer period of time in terms of texture quality (Bender & Schönlechner, 2021).

2.4.3 Confectionary, Snacks and Convenience Foods

Increasingly used in the development of better confectionery and snack foods, pseudocereals, including quinoa, amaranth, and buckwheat, contain high dietary fiber levels. These grains are botanical seeds, unlike true cereals, but they have nutritive values that can enhance the quality and comparability of numerous food items at once (De Angelis et al., 2024). Incorporation of pseudocereal dietary fiber in confectionery and snack items is particularly contributive in regards to consumer demands towards functional foods. These fibers affect several important factors in the formation of a final product like texture, durability or shelf life, and nutrient quality (Sindhu & Khatkar, 2019).

2.4.3.1 Texture Properties

Texture is one of the key factors in the sensory acceptance of confectionery and snack products. The presence of pseudocereal dietary fiber may influence mouthfeel, crispness, and final consistency of these products. High water-holding pseudocereal fibers might further improve the structure and retention of moisture in the case of snacks and confectionery items. For instance, pseudocereal fibers may improve the tenderness of baked snacks, such as crackers or cookies, and thus avoid a dry or crumbly texture in products (Sindhu & Khatkar, 2019). Moreover, water retention of the fiber could allow the desired chewy, or soft texture to be achieved in chewy candies or snack bars. This can be particularly useful in such products as energy bars, granola bars, and fruit chews, wherein consumers are looking for an acceptable mouthfeel. In puffed snacks, chips, and extruded items, the pseudocereal fibers also may contribute to the crispiness of these snack products. This makes them more desirable without having to rely on high-fat content. On the other hand, too much fiber may lead to undesirable textural changes that might be too hard or become gummy. For instance, fiber can be so high that it impedes the good structure a cookie or a bar will have, giving it over-density. As such, the optimal level of fiber must be known to adjust the texture; otherwise, the product ends up as either too stiff or as too chewy (Agregán et al., 2023).

2.4.3.2 Impact on Shelf Life

Shelf life represents one of the critical challenges to the product development that incorporates snack and confectionery items; such commodities are required to be both palatable and quality at some time points. Pseudocereal fibers do contribute significantly toward preventing moisture loss, mainly resulting in the staling process and other forms of texture degradation and thereby add long shelf life to a given product. Moisture is trapped by pseudocereal fibers, preventing products from becoming dry or hard with time, thereby ensuring a longer shelf life with the texture and freshness intact (Saha et al., 2024). Additionally, high fiber content in pseudocereals can slow down the oxidation rate, which is the primary cause of rancidity and off-flavors in snack foods that contain fats and oils. The antioxidant property of some pseudocereal fibers might extend the shelf life, as they would protect the fat content from oxidative damage (De Angelis et al., 2024).

Fiber-rich pseudocereal ingredients not only have textural and chemical benefits, but they can also inhibit the growth of spoilage microorganisms. This is particularly important in the case of snack bars, where the content of moisture can promote the development of mold as well as the bacterial proliferation. The binding of water and thus, the creation of a stronger structure within the product by fiber, thus reduces the microbial activity which improves the overall shelf stability of the product (Thakur et al., 2021).

2.4.4 Meat and Meat Alternatives

Rising consumer demands for the plant-based alternatives to traditional meat products have been the force behind the significant advancements in the development of plant-based meat analogs. These items are eco-conscious dead-on copies to the meat and vegan animal options as far as the taste, texture, and the nutritional profile are concerned and in addition being a more sustainable, ethical, and health-conscious choice. One of the difficulties of formulating a plant-based meat product is exactly how to get the right texture, moisture retention, and overall sensory appeal. Pseudocereal fibers created from grains like quinoa, amaranth, and buckwheat have come up as a new, innovative method to improve the quality of plant-based meat alternatives. They are the most important role in making a meat analog to be both nourishing and tasteful as best texture and moistures retention and also the overall nutritional profile are concerned.

2.4.4.1 Improving Texture of Meat Analogs

Since the consumer demand for plant-based alternatives to meat is growing, this has been a major factor in the development of plant-based meat analogues. All these products are made to behave and taste like animal-based meats, and yet, it exists as an environmental, ethical, and health option. One of the main challenges in creating plant-based meat is getting the right texture, moisture retention, and last but not least the overall sensory appeal. Pseudocereal fibers, the products of grains such as quinoa, amaranth, and buckwheat, are the new palm of the rising sun and have innately come up with an amazing way to improve the quality of plant-based meat substitutes. The most crucial role of pseudocereal fibers in the enhancement of the texture, moisture retention, and the overall nutritional profile of these products cannot be overstated as there is no way to think of a satisfying and healthy meat analog without them.

It is a proven fact that several people are influenced by the attraction of plant-based meat alternatives. In the context, it should be noted that the attainment of a masticable texture that can compare to other meat-based products is essential to human beings to appreciate a product. Pseudocereal fibers are very important in the meat analogs' structure, which in turn enhances the taste and mouthfeel of the final products by mimicking the chewiness and tenderness properties of the traditional meat (Bender & Schönlechner, 2021). Amongst the pseudocereal fibers, water-impenetrable fibers, in particular, form a jelly-like, stable structure in meat analogs. The ability to retain water does not only promote the moistness of the product but also improves its firmness during the cooking stage. When it comes to plant-based burgers or sausages, pseudocereal fibers can function as a direct and exhaust link in maintaining the structure of these products, thus preventing crumbling or breaking thereby resulting in a more uniform and enjoyable eating encounter.

Besides, DFs may be used for providing a fibrous texture to the plant-based meat products, which would be similar to the muscles' fibers found in the animal meats. In the products such as the plant-based chicken nuggets or steaks where the idea is to replicate the fibrous texture of meat, pseudocereal fibers can be included in the product to give a more realistic texture to it. This is the most important process in whole meat replacements, as they should have the same appearance and texture as meat (Langyan et al., 2024).

2.4.4.2 Moisture Retention in Meat Analogs

Maintaining moisture is another challenge in the production of plant-based meat alternatives. In nature, meat has lots of moisture, which is the real reason why it is so tender and juicy. For some conventional plant-based meats, the main water-controlling problem consists of not being able to maintain moisture, and thus the color and texture will suffer and become dry and unattractive. Water absorption is the property of pseudocereal fibers based on their relatively high water-binding ability, which is why they can be successful in removing the problem of sticking components (Agregán et al., 2023; Serna Saldívar & Hernández, 2020).

Through the integration of pseudocereal fibers in the meat analogue formulations, manufacturers can significantly improve moisture retention during the cooking process, and therefore, the final product will be more succulent with better texture. This is critical in technologies like the meatless burgers, sausages, and patties which demand the addition of the fibers where the loss of moisture will cause poor texture and appearance of the final product. The fibers create a network to catch the liquid within the product which makes the evaporation difficult in the cooking process and brings the product the texture and mouthfeel of the desired quality (Ciudad-Mulero et al., 2019). More than just increasing moisture retention, the fibers contribute to stability, ensuring an even spread of water or fat. Freezing and thawing can be the case for some products and are factors that the pseudocereal fibers can act against to have the retention of the original moisture and thus the texture is maintained (Kaur et al., 2023).

2.5 CONCLUSIONS

Pseudocereals originating from quinoa, amaranth, and buckwheat, have their own set of actions and qualities that give them an opportunity to perform in food product advancement. These fibers give very important values like water retention, and gelling abilities which are significant characteristics for enhancing the texture, consistency, and mouthfeel of food products. Their hygroscopic properties are important for the sustainability of bakery goods, dairy alternatives, and meat substitutes. While the soluble fibers in pseudocereals also increase viscosity, which lets smoothies and shakes taste better, and at the same time, the combination of the ingredients provides them with more thickness and creaminess without adding calories. The same fibers can also replace fats and help in the formulation of products such as low-calorie snacks and dairy items. Pseudocereal fibers also help in the improvement of the nutritional status of food by increasing the intake of fiber, natural support of digestion by beneficial bacteria, and the regulation of glucose in human blood. These fibers serve as prebiotics with the ability to bolster the growth of beneficial gut bacteria. High performance levels of them in these areas have allowed us to develop gluten-free products which are more stable and of better texture with the omission of gluten. To sum up, pseudocereal dietetic fibers can be called the basics of food processing that give the food texture, keep moisture, and provide nutrition, especially, in the case of production of healthier, functional foods.

REFERENCES

Agregán, R., Guzel, N., Guzel, M., Bangar, S. P., Zengin, G., Kumar, M., & Lorenzo, J. M. (2023). The Effects of processing technologies on nutritional and anti-nutritional properties of pseudocereals and minor cereal. *Food and Bioprocess Technology*, *16*(5), 961–986.

Angel Huerta-Ocampo, J., & Paulina Barba de la Rosa, A. (2011). Amaranth: A pseudo-cereal with nutraceutical properties. *Current Nutrition & Food Science*, *7*(1), 1–9.

Ameer, K., Shahbaz, H. M., & Kwon, J. H. (2017). Green extraction methods for polyphenols from plant matrices and their byproducts: A review. *Comprehensive Reviews in Food Science and Food Safety*, *16*(2), 295–315.

Bender, D., & Schönlechner, R. (2021). Recent developments and knowledge in pseudocereals including technological aspects. *Acta Alimentaria*, *50*(4), 583–609.

Ciudad-Mulero, M., Fernández-Ruiz, V., Matallana-González, M. C., & Morales, P. (2019). Dietary fiber sources and human benefits: The case study of cereal and pseudocereals. In Rosell, C. M. & Koksel, F. (eds.). *Advances in Food and Nutrition Research* (Vol. 90, pp. 83–134). Academic Press.

Czerwonka, M., & Białek, A. (2023). Fatty acid composition of pseudocereals and seeds used as functional food ingredients. *Life*, *13*(1), 217.

Dávalos, J. Z., Tirado, A., Romero, V., Cisneros, G., & Gamarra, F. (2023). Structural, thermal and energetic properties of Andean-pseudocereal flours with high nutritional values. *Journal of Thermal Analysis and Calorimetry*, *148*(14), 7207–7215.

De Angelis, D., Latrofa, V., Caponio, F., Pasqualone, A., & Summo, C. (2024). Techno-functional properties of dry-fractionated plant-based proteins and application in food product development: A review. *Journal of the Science of Food and Agriculture*, *104*(4), 1884–1896.

Hitache, Z., Al-Dalali, S., Pei, H., & Cao, X. (2023). Review of the health benefits of cereals and pseudocereals on human gut microbiota. *Food and Bioprocess Technology*, *16*(11), 2382–2399.

Kaur, H., Shams, R., Dash, K. K., & Dar, A. H. (2023). A comprehensive review of pseudocereals: Nutritional profile, phytochemicals constituents and potential health promoting benefits. *Applied Food Research*, *3*(2), 100351.

Langyan, S., Khan, F. N., & Kumar, A. (2024). Advancement in nutritional value, processing methods, and potential applications of Pseudocereals in dietary food: A review. *Food and Bioprocess Technology*, *17*(3), 571–590.

Martínez-Villaluenga, C., Peñas, E., & Hernández-Ledesma, B. (2020). Pseudocereal grains: Nutritional value, health benefits and current applications for the development of gluten-free foods. *Food and Chemical Toxicology*, *137*, 111178.

Mir, N. A., Riar, C. S., & Singh, S. (2018). Nutritional constituents of pseudo cereals and their potential use in food systems: A review. *Trends in Food Science & Technology*, *75*, 170–180.

Nandan, A., Koirala, P., Tripathi, A. D., Vikranta, U., Shah, K., Gupta, A. J., … & Nirmal, N. (2024). Nutritional and functional perspectives of pseudocereals. *Food Chemistry*, 139072.

Saha, D., Pradhan, A., Sarkar, A., & Tripathy, P. P. (2024). Recent trends in pseudocereal-based foods. In Dhull, S. B., Bains, A., Chawla, P., & Kaur, S. (eds.). *Pseudocereals* (pp. 264–291). CRC Press.

Sarkar, A., Saha, D., Pradhan, A., & Tripathy, P. P. (2024). Innovative processing technologies for pseudocereals. In Dhull, S. B., Bains, A., Chawla, P., & Kaur, S. (eds.). *Pseudocereals* (pp. 191–215). CRC Press.

Satheesh, N., & Fanta, S. W. (2018). Review on structural, nutritional and anti-nutritional composition of Teff (Eragrostis tef) in comparison with Quinoa (Chenopodium quinoa Willd.). *Cogent Food & Agriculture*, *4*(1), 1546942.

Serna Saldívar, S. O., & Hernández, D. S. (2020). Dietary fiber in cereals, legumes, pseudocereals and other seeds. In Welti-Chanes, J., Serna-Saldívar, S., Campanella, O., & Tejada-Ortigoza, V. (eds.). *Science and Technology of Fibers in Food Systems* (pp. 87–122).

Shahbaz, M., Raza, N., Islam, M., Imran, M., Ahmad, I., Meyyazhagan, A., … & Wan, C. (2023). The nutraceutical properties and health benefits of pseudocereals: A comprehensive treatise. *Critical Reviews in Food Science and Nutrition*, *63*(29), 10217–10229.

Sindhu, R., & Khatkar, B. S. (2019). Pseudocereals: Nutritional composition, functional properties, and food applications. In Chandra Deka, S., Seth, D., & Rachayya Swami Hulle, N. (eds.). *Food Bioactives* (pp. 129–147). Apple Academic Press.

Taylor, J. R., Belton, P. S., Beta, T., & Duodu, K. G. (2014). Increasing the utilisation of sorghum, millets and pseudocereals: Developments in the science of their phenolic phytochemicals, biofortification and protein functionality. *Journal of Cereal Science*, *59*(3), 257–275.

Thakur, P., Kumar, K., & Dhaliwal, H. S. (2021). Nutritional facts, bio-active components and processing aspects of pseudocereals: A comprehensive review. *Food Bioscience*, *42*, 101170.

Torbica, A., Radosavljević, M., Belović, M., Tamilselvan, T., & Prabhasankar, P. (2022). Biotechnological tools for cereal and pseudocereal dietary fibre modification in the bakery products creation–Advantages, disadvantages and challenges. *Trends in Food Science & Technology*, *129*, 194–209.

Zhu, F. (2020). Fonio grains: Physicochemical properties, nutritional potential, and food applications. *Comprehensive Reviews in Food Science and Food Safety*, *19*(6), 3365–3389.

3 Emerging Milling Techniques for Pseudocereals

Gitanjali Behera, Zahida Naseem,
Sajad Ahmad Mir, Jandrajupalli Sudhakara Rao,
and Deb Kumar Chakraborty

3.1 INTRODUCTION

"Pseudocereals" are non-grass plants which are generally consumed as cereals. Pseudocereals plant comes from broadleaf unlike the true cereals like rice, wheat, and corn, but they have similar nutrition profile (Schoenlechner et al., 2008). The grains such as quinoa (*Chenopodium quinoa*), buckwheat (*Fagopyrum esculentum*), and amaranth (*Amaranthus spp.*) are commonly known as pseudocereals. The aforementioned grain seeds resemble the functional and compositional characteristics of true cereals (Mir et al., 2018). Amaranth, quinoa, and buckwheat are primarily gluten-free starchy crops (Farooq et al., 2023), but they also contain some amount of protein, fiber, oils, flavonoids, phenolic compounds, minerals, vitamins, and fatty acids. All of these compounds have positive impact on human health, particularly in mitigating and preventing degenerative diseases. Recent research has shown that the presence of phenolic compounds in pseudocereals has several health benefits such as anticancer, preventing oxidative stress, antidiabetic, antihypertensive, anti-inflammatory, and preventing cardiovascular diseases. The versatility of these unique grains allows them to be mixed with a wide range of food products including snacks, gluten-free bread, and beverages. Because of the aforesaid characteristics of pseudocereals, these minor grains should be included in everyone's diet to increase the dietary diversity of human beings and to also maintain the food plant biodiversity (Mir et al., 2018).

Milling is a crucial step in processing and production of pseudocereals flour. It improves its usability, nutritional profile, and marketability. The outer layer of pseudocereals contains husk and bran which is inedible and should be thus removed during milling. Milling also ensures the separation of contaminants and impurities and improves the final product quality. The three main structures of pseudocereals are endosperm, embryo, and seed coat. The endosperm comprises of major starch and some amount of proteins. The embryo mainly stores oil, proteins, and some minerals. Lastly, the seed coat commonly known as bran, contains cellulose, hemicellulose, and lignin along with scarce amount of protein (Haros & Wronkowska,

 DOI: 10.1201/9781003584858-3

2017). The relative proportions of these three components differ among various pseudocereals. Typically, the cereal milling process involves separating the endosperm from the embryo and seed coat, followed by reducing the endosperm into flour or grits. Although whole-grain products made from pseudocereals do not include the term "whole" in their labeling, their flours are produced through the dry milling of whole grains.

The whole milling process is classified into dry milling and wet milling. The major difference lies in the amount of water used in the both processes. Dry milling techniques utilize water in very small quantity. Furthermore, the main aim of dry milling is to separate the various parts of grains like germ, pericarp, and endocarp but the major role of wet milling is to get the components like protein, starch, oil, and fiber from pseudocereals.

The milling of pseudocereals poses various challenges because of structure, shape, and small size of the grains. The outer coat being very tough adds further complication in milling of these cereals. Thus, it has become very important to explore the milling techniques for pseudocereals to achieve the potential utilization of these beneficial grains. There are various types of milling technologies being used presently and each has its impact on the various milling aspects like rate of milling, yield, quality of flour, energy consumption, and economic feasibility. Among various milling techniques some of the crucial milling techniques have been discussed in detail. However, very less information is available on the emerging milling techniques of the pseudocereals.

3.2 DRY MILLING TECHNIQUE

The dry milling method changes the size and shape of the pseudocereals. Dry milling is a process of grinding, separating, and again regrinding. The steps are always repeated to get particular parts of a pseudocereals. Generally, pseudocereals are cleaned to separate the foreign materials before milling. Sometimes, water is added to the grains and left undisturbed for tempering before the dry milling process. Application of mild heating and conditioning with water toughens the pericarp and germ, which helps in the milling process of pseudocereals (McKevith, 2004). The dry milling of amaranth was developed by Tosi *et al.* (2000), for getting protein (40%), dietary fiber and starch (79%) by employing conditioning of the grains. Quinoa kernels contains saponin in large amount which is present in the outer layer of seed. The dry milling can help in removing these bitter compounds, which is an alternative process to traditional washing (Farfan et al., 1978). Quinoa is generally small size grains and milled in the whole form after the saponins are separated. Buckwheat is made into flour from unroasted kernels, and the flours can be graded into light, medium, and dark depending on the amount of dark hull present. Completely dehulled buckwheat is used to make fancy buckwheat flour which mostly contains the central endosperm having starch (75%), protein (6%), lipid (1%), carbohydrates (1%), dietary fiber (3%), ash (1%) and remaining other components (13%) (Steadman et al., 2001). Dehulling of buckwheat is performed by precooking the grains in boiled water and drying under the sun. The cooking process helps in gelatinization of the starch and breaking of the hull. Furthermore, cooling helps in partly drying of the starch and husk which helps in efficient dehulling (Janeš et al., 2012). Emery grain milling was used by Takeuchi, (2001) for buckwheat for getting polished flour. The emery miller had strong peeling strength and it uses a pressure heating process for the milling. The dry milling methods for pseudocereals have been categorized into traditional milling and emerging milling techniques and explained further in the subsequent sections.

3.2.1 Traditional Methods of Milling

Traditional methods of milling pseudocereals involve techniques that are often manual or semi-mechanized. These methods are typically used in rural or small-scale setups where advanced milling technologies are unavailable. Some traditional milling methods for pseudocereals are hand pounding, stone grinding, roller crushing or roller milling, hulling and hammer milling, water power mills, etc., which are explained in the following sections.

3.2.1.1 Hand Pounding

The pseudocereals are milled by using the hand pounding methods from centuries either by using wood or stone pestles and mortar, a practice that still persists today (Figure 3.1a). The grains are further pounded before the sieving process to remove

FIGURE 3.1 (a–d) Traditional methods of milling.

the coarse material which results in flour and meal. Furthermore, the grains are winnowed for separation of un-husked and de-husked kernels. The un-husked kernels are pounded again using the above procedures. But these methods are laborious and take a lot of time (Haros & Wronkowska, 2017). Therefore, the hand pounding method has become obsolete nowadays, after the intervention of hullers.

3.2.1.2 Stone Grinding

Stone grinding is a traditional milling method used to process pseudocereals into flour or smaller particles for various culinary and industrial applications (Figure 3.1b). This technique involves using natural or artificial stone surfaces to crush and grind the grains (Kaushal & Kumar, 2022). The pseudocereals are fed between two large, rotating stone surfaces. The grains are ground into flour or smaller particles through friction and pressure. Stone milling has been used for getting 100% whole buckwheat flour followed by a subsequent sieving process for obtaining white flour (Izydorczyk et al., 2014). The flavor and taste of stone-milled buckwheat flour can be preserved by low storage temperature (Kawakami et al., 2008). Stone grinding protects the natural oils, vitamins, and minerals of pseudocereals and also prevents overheating of the grains. However, the stone grinding process is not efficient compared to modern roller mills because of the low throughput. Maintenance of the grinding stone becomes difficult due to the regular sharpening or replacement of the stone to maintain grinding quality.

3.2.1.3 Roller Crushing

Manual roller crushing involves using simple, hand-operated devices with rollers to crush pseudocereals (Figure 3.1c). This traditional approach is suitable for small-scale operations or household processing, offering an economical and accessible way to process these grains without the need for electricity or sophisticated machinery. The grains are cleaned to remove dust, stones, and debris and conditioned with water slightly to soften the outer husks for better crushing. Cylindrical rollers (two) are turned manually using a crank. Grains are crushed as they pass between the rollers, breaking them into smaller particles or flattening them (depending on the gap between the rollers) (Simon, 1889). One laboratory roller mill was used earlier which milled quinoa into bran and flour (Chauhan et al., 1992). This method of milling has low initial investment compared to electric mills or industrial rollers. Matsuhash et al. (1984) used a 12-roller milling system for the production of buckwheat flour and performed the compositional analysis of the milling fractions. The rollers are generally lightweight, thus can be portable and easy to use in rural or remote areas. The desired particle size of the grains can be achieved by setting the gap between rollers. However, the manual roller milling process can be labor intensive, have low efficiency, and getting uniform particle size can be challenging without precise adjustment of the rollers (Campbell, 2007).

3.2.1.4 Water Powered Milling

Water-powered mills are traditional mills that use the kinetic energy of flowing water to drive mill stones for grinding grains (Figure 3.3d). These mills are sustainable and have been used for centuries in various regions for grain processing. The water wheels convert the water flow into mechanical energy. Water wheel drives a system of gears and shafts to transfer energy to the millstone. Two circular millstones like a stationary bottom stone and a rotating top stone are present in this mill. Grains are fed through a central hole (hopper) in the top stone (Saltzman, 1990). The friction between the stones grinds the grains into flour or meal. Water-driven mills were first used by Janes et al. (2012) for buckwheat milling. They used two stones, an upper mill stone (rotating, 950 mm in diameter) and a lower mill stone (stationary) with 12 kg/h buckwheat milling capacity. This milling system is eco-friendly and sustainable and has low operating cost. However, seasonal variations in water availability can limit operation. In addition to this, the milling process is not efficient and achieving uniform particle size can be challenging.

3.2.1.5 Huller and Hammer Milling

Hulling is process of passing the grains through a huller machine where abrasion or impact removes the outer husk. Hulling involves frictional hulling and impact hulling. Frictional hulling uses abrasive surfaces to rub off the seed coat and impact hulling breaks the outer hull by striking the seeds against a hard surface (Dhankhar & Hissar, 2014). Huller milling is an essential step to improve the edibility and sensory properties of the pseudocereals by removing bitter or indigestible outer layers. After hulling, an aspirator separates the lighter husk from the heavier, hulled grain. Polishing of the grains might be an optional step after hulling to remove any remaining saponin or husk particles. Sinkovič et al. (2023), studied the effect of hulling, cleaning, and polishing on compositional qualities of common buckwheat and tartary buckwheat grains. They have reported that the implementation of threshing, cleaning, and polishing techniques on the harvested grains resulted in nutritious, ready-to-eat grain fractions, such as groats, which are appropriate for cooking or further transformation into flakes or whole meal flours. The milling characteristics of quinoa grains were investigated by Kumar et al. (2023). They have performed dehulling and pearling of quinoa for 5 minutes and evaluated the different milling fractions. The husk of 23.63% with the highest milling recovery of 63.58% and milling loss of 40.63% was also reported. Atalay et al. (2013), used a pearler and hammer mill to obtain buckwheat bran and flour. Huller milling can improve the taste of quinoa by removing the saponins and helps in increasing the market value of the product by improving its appearance and texture. However, huller machines can be expensive for small-scale farmers and the process can be time consuming if done on a small scale (Patil & Chawla, 2024).

3.2.2 Emerging Milling Techniques

3.2.2.1 Cryogenic Milling of Pseudocereals

Enhancing the structural and functional characteristics of the pseudocereals, the cryogenic milling has been adapted as one of the modern milling techniques. In this process to maintain the temperature at a desired value, the liquid nitrogen is being used. Mostly, the heat-sensitive compounds are being reduced to finer particle size. This milling technique has been carried out at pilot scale and large-scale level, involving various associated equipment and processing parameters. The physicochemical characteristics of pseudocereals are influenced both by the type of equipment and processing conditions. The most important step of this technique is to reduce the temperature to subzero with the help of liquid nitrogen (−196°C) (De Bondt et al., 2020). To achieve enhancement in the product quality and functionality, the degradation of heat-labile components is to be prevented and ultra-fine particle sizes should be obtained. Other properties like solubility and bioavailability of the nutrients in the pseudocereals are offered by the cryogenic milling (Chavarín-Martínez et al., 2019). The functional properties of the bran of pseudocereals like water-holding capacity and emulsification potential increase during this milling technique, thus making it the most beneficial technique for manufacturing of high-quality and health beneficial food products. Besides this, the growing demand of the consumers for minimally processed, nutrient rich, and novel food ingredients have been made possible by the application of cryogenic milling. Integration of an impact mill with the cryogenic milling to upscale this technique and to make it more impactful and efficient at industrial scale is being explored. It has been also observed that the ball mill in combination with cryogenic grinding was of less significance in terms of size reduction of the grains (Hemery et al., 2011).

3.2.2.1.1 Laboratory Scale Cryogenic Milling

Cryogenic ball mills are used for laboratory-scale cryogenic grinding (De Bondt et al., 2020). The sample size for milling typically ranges between 20 and 30 g. The sample is put into chamber that has 12 stainless steel beads (10 mm). After that, liquid nitrogen at −176°C is used for cooling to guarantee freezing prior to milling. However, precooling, a crucial step, is carried out at 300 rpm with oscillation before cooling. Then, the milling is done for 0.2, 15, 30, and 45 minutes at 1,800 rpm. Each milling cycle continues for 15 minutes, with a 5-minute cooling period in between at 300 rpm.

3.2.2.1.2 Industrial Scale Cryogenic Milling

Cryogenic milling uses extremely complex and advanced mills for industrial applications. A maximum throughput capacity of 10,000 kg/h is attained when the mill runs at 9,200 rpm. A cryogenic screw (15 rpm) is used to feed the pseudocereals that has been precooled with liquid nitrogen (−196°C). The inlet and outlet temperatures of the feed are generally maintained at -140°C and −80°C respectively. The substance is typically kept at an entering temperature of −140°C and an output temperature of −80°C. The dust particles are removed from milled bran using a cyclone separator.

3.2.2.1.3 Physicochemical Impact of Cryogenic Milling

Cryogenic milling improves the physicochemical characteristics and structural integrity of pseudocereals. Low temperatures in cryogenic milling enhance particle homogeneity, maintain functional qualities, and reduce heat-induced nutrient degradation. Furthermore, delicate bioactive substances like flavonoids and polyphenols that are frequently lost during traditional grinding procedures can be preserved by cryogenic milling. Additionally, by increasing surface area and improving textural qualities, this technique makes it easier to include into food compositions and improves its nutritional and functional qualities. Pseudocereals can be processed effectively using cryogenic milling, which has several benefits for maintaining nutritional value and producing tiny particle sizes. Since, this milling aids in the dispersion of phenolic chemicals resulting in variations in antioxidant capacity between the milling fractions. The potential of cryogenic milling to enhance the functionality of pseudocereal-based products is revealed by both laboratory- and industrial-scale operations (De Bondt et al., 2020). To increase its use in the food business, future studies should concentrate on improving operating factors.

3.2.2.2 Ultrafine Grinding

Ultrafine grinding has become a popular method for improving the structural and functional qualities of tartary buckwheat flour. The impact of airflow pulverization and other ultrafine grinding techniques on the physicochemical characteristics, shape, and particle size of the miracle grains will be emphasized.

Pseudocereals are first processed by sifting the grains and removing any external material. The mill is used to remove the bran and outer layer from the grains. The seeds are crushed into flour, sieved through a 120-mesh sieve, and then dried for an hour at 50°C in a blast oven set at a continuous temperature. With the aid of an airflow pulverization instrument, the air flow pulverization is subsequently carried out. To produce flour with varying particle sizes, grinding is done at pressures of 0.4, 0.5, 0.6, 0.7, and 0.8 MPa. Depending on how tiny the particles are, different classes of ultrafine milled flours can be made. The results of this grinding process show a notable improvement in quality characteristics, increasing the flour's suitability for a variety of culinary uses. A nutritionally dense pseudocereal (tartary buckwheat) is becoming more well-known for its practical uses and health advantages. By reducing particle size and changing morphology through varying grinding pressures, ultrafine grinding technology provides an effective way to improve its structural features.

This procedure improves the sensory and cooking characteristics of flour, in addition to its quality. There have been reports on the importance of using ultrafine grinding methods, such as airflow pulverization, and how they affect the different properties of tartary buckwheat flour. Buckwheat flour's color value, structure, solubility, chemical composition, water-holding capacity, and swelling power are impacted by these techniques (Xu et al., 2021).

The antioxidant qualities of tartary buckwheat are enhanced by ultrafine grinding, which dramatically raises the bioavailability of vital elements including rutin and quercetin. For a variety of food applications, including baked goods, beverages, and functional health products, the enhanced particle size distribution is perfect since it improves water absorption, gelation behavior, and compatibility with other ingredients.

3.2.2.2.1 Effects of Ultrafine Grinding

Ultrafine grinding greatly decreases the particle size, which enhances the texture and other qualities of various pseudocereal flours. The method changes the shape of the flour particles by applying varying grinding forces. While attaining a greater reduction rate, the ultrafine pulverization method improves energy efficiency. This makes it possible for pseudocereals, including tartary buckwheat, to be processed into finer powders with improved processing properties. The product quality and flavor are improved by the grinding process, which breaks up the starch granules into smaller pieces and narrower particle sizes. Additionally, the procedure preserves important nutritional elements, which increases its usefulness for products with a health focus. The ultrafine grinding procedure decreases the amount of protein, crude fat, and moisture. Customers appreciate these modifications since they are in line with the advantages of weight loss and wellness.

3.2.2.2.2 Implications for Food Quality and Applications

By changing the starch structure, ultrafine grinding processes improve the texture, quality, and nutritional value of flour made from various pseudocereals and increase its usage in culinary products. Because of these enhancements, the flour is ideal for use in health-conscious food applications, especially for customers who are concerned about nutritional advantages and weight control. The functional qualities of the flour, such as its ability to hold water and bind oil, are enhanced by ultrafine grinding and are crucial for creating novel food formulations. In addition to improving mouthfeel, the smaller particle size guarantees better incorporation into high-fiber and gluten-free foods, meeting the rising demand for specialized and functional diets.

Airflow ultrafine grinding has demonstrated substantial potential in enhancing the physicochemical properties of certain pseudocereals, particularly tartary buckwheat flour. It improves both functional and sensory qualities by creating finer and more consistent particles, opening the door for more widespread uses in the food and pharmaceutical sectors. The pace at which nutrients are released during digestion may be impacted by fine grinding, which increases the particles surface area and solubility. Furthermore, study is necessary to optimize process parameters and investigate their use in additional pseudocereals.

3.2.2.3 Impact Milling

Impact milling, a critical step in using pseudocereals, offers major benefits in improving their nutritional and functional profiles. Impact milling improves the swelling power and shear stability of pseudocereal flour by reducing particle size. The flour produced by this milling process has demonstrated moderate shear stability but significant swelling power in the case of amaranth. Quinoa flour has shown a variety of pasting characteristics, which can be ascribed to both cultivar composition and milling (De Bock et al., 2021). Buckwheat is appropriate for applications needing constant viscosity because of its great shear stability. Additionally, buckwheat exhibits the highest rate of retrogradation, which improves textural properties of processed foods.

3.2.2.3.1 Nutritional Enhancement through Impact Milling

Pseudocereals are abundant in unsaturated fatty acids, such as linoleic and oleic acids. Buckwheat has the largest phenolic content, which adds to its antioxidant qualities. Impact milling lessens oxidative damage during processing, thus nutrients are preserved. The bioavailability of vital minerals including calcium (Ca) and iron (Fe) is further increased by sprouting prior to milling (Manzoor et al., 2024). This procedure facilitates improved water absorption and mineral accessibility in flours (Paucar-Menacho et al., 2023).

3.2.2.3.2 Application and Functional Benefits

Addition of sprouted pseudocereal flours improves antioxidant activity and digestibility. Because milling produces smaller particle sizes, it also enhances the bioaccessibility of minerals in foods like pasta. Pseudocereal flours treated in impact mills aid in the absorption of phenolic components and gamma-aminobutyric acid, which reduce oxidative stress. Because of these advantages, pseudocereal flours provide great building blocks for functional food products. The functional and nutritional qualities of pseudocereal flours are significantly influenced by impact milling, which increases the possibility of using them in food applications with a health focus. One method that shows promise for creating nutrient-enriched food items is the use of impact-milled sprouting pseudocereals. To improve milling procedures and investigate novel uses in food systems, more study is required.

3.2.2.4 Hydrothermal Assisted Milling

Hydrothermal treatment is a preconditioning process which includes soaking, steaming, and drying before the milling and this process is also known as parboiling. This treatment softens the grain structure, breaks the cellular matrix, and reduces oxidation of lipids. It also results in higher flour yield due to starch gelatinization and finer particle size and improves the starch functional properties such as water absorption and oil retention. (Sharma et al., 2021). Hydrothermal treatments can improve the bioavailability of certain nutrients and reduces anti-nutritional factors like phytates and saponins (Agregán et al., 2023). Pseudocereals are submerged in water for preconditioning purpose which increases the moisture content. The hydration is followed by heating using either steam or hot water. Presently in the modern milling plants, super-heated steam with high pressure is being used for better heat penetration and higher nutrient retention. Also, the steaming step helps in starch expansion and gelatinization and splitting of the outer husk layers. The steamed pseudocereals are then dried to the optimum moisture content either by using sun drying or mechanical drying.

3.2.2.4.1 Impact of Hydrothermal Assisted Milling on Nutrition

The amount of macro and micro nutrients in pseudocereals get affected by the process of hydrothermal treatments. Hydrothermal treatments help in reducing the anti-nutritional factors like tannins, saponins, and phytates by either breaking them down or leaching process. Furthermore, this results in enhancement of bioavailability of essential nutrients (mainly minerals) like zinc, iron, and calcium which otherwise forms insoluble complex compound with phytates in case of untreated pseudocereals (Deka et al., 2023). Also, the pre-heating treatments before milling helps in protein structure alternation and improves the overall protein quality in the final flour, making it highly suitable for formulation of protein-based diets (Malik et al., 2021). Furthermore, hydrothermal treatments (involving both water and heat) modify the starch crystallinity and cause partial starch gelatinization which improves the functional properties like water absorption capacity, digestibility, and swelling power. Pseudocereals treated with these methods are more beneficial to be used as gluten-free food product and in the infant nutrition. If the process parameters are efficiently controlled during hydrothermal treatment, it will minimize the loss of valuable nutrients, especially heat sensitive vitamins (B-complex vitamins), and will also prevent the oxidation of polyunsaturated fatty acids of the pseudocereals (Malik et al., 2021).

3.2.2.5 Planetary Ball Milling

Planetary ball milling is mechanical milling techniques which uses high impact energy between the balls and the rotating jars to grind the cereal grains into very fine powder. The planetary designation comes from the fact that the milling jars rotate on their own axis while simultaneously rotating around a central axis (Burmeister & Kwade, 2013). Some research has been performed on the effect of planetary ball milling on rheological and thermal properties of pseudocereals (Roa et al., 2015; Roa et al., 2014). Roa et al. (2015), studied the effect of ball milling on thermal, rheological properties, and stability of amaranth flour. They have reported that the consistency index decreases and flow behavior index increases with an increase in the milling energy. Also, the ball milling decreased the gelatinization enthalpy and the highest storage stability was exhibited by the highest milled flours (Roa et al., 2015). Planetary methods of ball milling generate high energy due to a combination of both centrifugal and rotational forces which results in finer and faster grinding. However, high energy input can cause excessive heat, potentially affecting sensitive materials. It requires careful control of parameters to avoid undesirable effects like agglomeration or phase transformations (Burmeister & Kwade, 2013).

3.2.2.6 Air Classifier Milling

Air classifier milling is another promising technique, which can be highly efficient and versatile technique for processing pseudocereals. The air classifier mill operates by using airflow to separate fine and coarse particles after grinding, enabling the production of flours with targeted particle sizes and minimal overheating (Silventoinen et al., 2021). This method is suitable for removal of stones and other impurities from the grains, and the grains are separated based on their weight. Aspirator is used for removing light weight impurities. Grader separate the cleaned grains based on size.

3.2.2.7 Wet Milling Technique

Wet milling technique was first adopted for milling of maize (Inglett, 1970). Production of maize starch was performed by Thomas Kingsford, an American, on an industrial scale in 1842. Although wet milling produces a wide range of goods, it is a much more complicated compared to dry milling. While starch is the primary output of wet milling, the fiber-rich and protein-rich fractions are also valuable sub-products for the food sector and for technical applications. These fractions are utilized in the production of balanced foods in the case of maize, but they may have additional uses in the case of pseudocereals. These days, different types of starch are not only sold as such but are also utilized to formulate syrups, dextrin, and flammable alcohol. Additionally, the usage of starch as a raw material for the production of unique polymers for the plastics sector has increased in recent years.

Wet milling is nowadays adapted for milling of different pseudocereals. Wet milling is used to extract the grain's chemical components such as starch, fiber, protein, and lipids. First, in the wet milling procedure, different products such as starch, oil, protein and fibre are produced, which starts with the removal of germ from the grains in a wet environment with approximately 45% moisture content to increase the amount of germ produced. Second, it separates lighter germ that float to the top from heavy particles that settle to the bottom by using water as a medium. Third, it utilizes size to remove and wash fiber and germs using moist steeping. The details of wet milling process have been explained in the following sections.

3.2.2.7.1 Cleaning

Pseudocereals are cleaned similarly to how dry wheat is milled. The soaking duration and temperature must be properly regulated, since they are essential for the water to seep into the grain. Sulfur dioxide (SO_2) or alkali is also used to treat water, which speeds up the penetration of moisture (Haros & Wronkowska, 2017). The above treatment helps in the disintegration of the protein matrix around the starch granules and improves the proportion of refined starch production and further promotes the separation of starch and protein.

3.2.2.7.2 Steeping

Pseudocereal grains are soaked in tanks known as steeping. During steeping, soluble fractions like sugars, proteins, and some minerals from the grains are dissolved into the water. This makes the grain softer and prepares it for further processing. After soaking, the liquid containing the dissolved soluble fractions is separated from the grain and undergoes concentration by evaporation. The evaporation process removes water from the liquid, leaving behind a concentrated residue that can be used in feed formulations. The soaked grains are then ground in disk mills. The purpose of grinding is to release the germ, the part of the grain rich in nutrients and oils. The disk mills are designed with knobs on the grinding surface to ensure gentle grinding, preserving the germs size and preventing excessive crushing of the grain. The slurry of the ground grains is passed through a separating medium, and the germs are separated due to their light weight compared to the grains. The separated lighter fractions, which could include the germ, might be further processed or utilized, while the heavier fractions can be used in animal feed (Yuan et al., 1998).

3.2.2.7.3 Germ and Fiber Separation

There may still be germs in the material that starts to sink to the bottom, thus need to be extracted and recovered. To liberate the residual germ, this material is subsequently sent through a second set of grinding mills for comparatively close processing. The overflow is used for germ wet screening, which separates the discharged germ from the others. With the right screen size and type, wet screening enables the germ to be physically isolated from the others while being cleaned. After that, the germ is transferred for oil extraction and drying. The next stage of fiber removal is reached when the heavier material settles at the bottom. To separate the fiber from the starch and gluten, the remaining millet slurry is run through finer screens. Fiber and a few starch and fiber-attached particles make up the majority of the larger particles that cannot fit through the screen size.

3.2.2.7.4 Starch and Gluten Separation

The remaining slurry from the first fiber screening and the slurry left over from the fiber wash system are sent to the starch and gluten separation stage. The latter is made up of a number of centrifuges for thickening streams and clarifying process water. Also, a multistep starch-washing system for further concentration and refinement of the starch-rich slurry and another centrifuge for the gluten stream to further concentrate solids is operated. Gluten effectively captures water because it is insoluble in it and has a lower density than starch, whereas starch escapes with the water. After that, the gluten-rich material is transferred for filtering. Gluten is transferred for drying after filtering. For use in animal feed rations, this is subsequently marketed as gluten meal. The efficiency of the starch and gluten separation at this point is crucial to the wet milling process's economic feasibility. Profitability increases with the amount of high-protein gluten recovered with the least amount of starch lost. After the first separation, the starch-rich slurry may include just 3%–5% gluten.

3.2.2.7.5 Starch Refinement

After initial separation, the starch-rich slurry is transferred to a multi-step starch-washing procedure. This method involves repeatedly washing a diluted starch stream until all of the protein (gluten) is gone. The residual protein is separated from the starch using a battery of hydrocyclones in this procedure. Denser starch particles tend to sink to the bottom of a mixture of tiny starch and gluten particles, whereas lighter gluten particles prefer to float to the top. The force with which the water containing the gluten and starch particles hits the cyclone's interior determines how well this separation works. A number of pumps produce the force that is needed. As a result, the heavier starch keeps getting replenished as it moves from one set of hydrocyclones to the next, while the lighter, gluten-rich part would escape from the top of the hydrocyclone.

The starch that is produced in this manner is extremely refined and nearly 99.5% pure. The refined starch can be sold as a commercial starch and processed into syrup and dextrose. Starch can occasionally be altered and marketed as modified starch. First, starch is transformed into dextrose, which is subsequently fermented to produce alcohol.

The starch fractions obtained in the wet milling process can be calculated by using the following formula:

$$\text{Starch yield } (\%) = \frac{\text{Weight of dry starch obtained}}{\text{Weight of dry grain}} \times 100$$

The degree of recovery in the wet milling process can be calculated by using the following formula:

$$\text{Degree of recovery} = \frac{\text{Weight of dry strach obtained}}{\text{Weight of dry strach present in the grain}} \times 100$$

The yield values are influenced by the technique utilized and the type of grain selected, while the degree of recovery remains unaffected by the starch content of the grain. However, it is influenced by the milling method applied. The quality of the starch fraction is typically assessed based on the protein content present in that fraction (Haros & Wronkowska, 2017). The protein content of the grains in wet milling is affected by the methods of separation like lab scale or industrial scale.

3.2.2.7.6 Factors Affecting Wet Milling of Pseudocereals

The milling characteristics of a grain are determined by yield and quality of various fractions of grain. Wet milling process is a little complex compared to dry milling because wet milling is affected by the varieties of the grain, drying and storage methods, steeping conditions such as time, temperature, alkali, pH, and SO_2 concentrations. Other factors such as microbial load, fatty acids present in the germ, and amount of mycotoxin can affect the grain milling quality (Haros & Wronkowska, 2017). The factors affecting the milling characteristics of pseudocereals are discussed in the following sections.

3.2.2.7.6.1 Varieties of the Grains The variety of pseudocereals can significantly influence wet milling processes due to differences in their physicochemical properties, such as starch structure, protein content, lipid content, and fiber composition. Also, the structural differences in various pseudocereals in size, dimensions, outer husk covering, and endosperm structure behave differently during wet milling (Haros & Wronkowska, 2017). For example, amaranth which contains high protein and lysine content, may reduce its starch recovery due to protein-starch interaction. Quinoa contains high saponins, which requires prewashing and adds a preprocessing step before wet milling. Similarly, buckwheat is rich in fiber (Mir et al., 2025) which affects its separation efficiency during wet milling.

3.2.2.7.6.2 Treatment Given during Post-Harvest Processing Presence of excess moisture and temperature during the storage might cause insect infestation, microbial spoilage, and biochemical changes. Storage conditions are much more important for pseudocereals because of the higher oil content which turns rancid over longer period of time. Grain is susceptible to mechanical damage during storage and transportation due to handling and movement. To minimize such damage, it is essential to implement effective design strategies for drying and storage facilities. Additionally, optimizing the drying process can significantly reduce or eliminate the damage caused by these operations. Hot air drying of pseudocereals at higher temperature to obtain high dryer capacity affects the grain quality (grain becomes more fragile) and subsequently results in cracked or broken grains (Haros & Wronkowska, 2017). The use of higher drying temperatures leads to several adverse effects, including the denaturation of proteins and loss of vitamins. The decrease in grain quality not only impacts the grain itself but also directly affects the quality of the products. Therefore, optimizing the drying process is of utmost importance. A significant issue arising from the damage to grains intended for wet milling is the difficulty in separating components due to the alteration of the endosperm caused by high drying temperatures. Such temperatures can also modify the viscosity of the extracted starch, thereby influencing its ability to form uniform pastes. Additionally, high drying temperatures may result in pigment losses due to oxidation.

3.2.2.7.6.3 Steeping Conditions of the Grain Steeping involves immersion of grains in water while carefully regulating parameters like temperature, time period, sulfur dioxide or alkali concentration and lactic acid concentration. It is a complex method which is characterized by both chemical and biochemical reactions. It also involves physical transformations that lead to the degradation of the protein matrix within the endosperm. A significant physical alteration occurring during the steeping process is the absorption of water, which not only facilitates the softening of the grain but also serves as a medium for transporting sulfur dioxide and alkali (Ruan, 1992). Sulfur dioxide contributes to the weakening of the protein matrix by breaking disulfide bonds. Furthermore, the addition of sulfur dioxide in the steeping water suppresses microbial spoilage. The concentration of sulfur dioxide utilized during the steeping process typically ranges from 0.1% to 0.3%.

Starch from amaranth is difficult to extract by the wet milling method because of the small granule size, more protein content and strong association between protein and starch (Middlewood & Carson, 2012). Currently, there is no commercially available wet milling process for amaranth that facilitates the separation of starch from its other components. Consequently, various laboratory-scale wet milling techniques have been developed to extract starch from amaranth (Resio et al., 2009). The temperature and concentration of SO_2 significantly affect the water absorption. Alkaline soaking is necessary for the amaranth grain to extract the protein. However, this process can lead to the degradation of starch granules and a reduction in starch yield. An alternative method for isolating amaranth starch with enhanced recovery and purity was suggested by Radosavljevic et al. (1998), which involves the application of enzymes during the soaking process.

Similarly, like amaranth, one of the significant challenges associated with the wet milling of quinoa is the small size of its starch granules. Wet milling of quinoa has been performed by Atwell et al. (1983), using acetate buffer solution for the soaking. Similarly, Qian and Kuhn (1999) used sodium hydroxide for soaking using the similar wet milling techniques. Enzymes like cellulase, xylanase, and hemicellulose were utilized for the treatment of quinoa. Additionally, the diluted alkaline solution was utilized by Wright et al. (2002), for separation and characterization of starch obtained from bitter and sweet quinoa seeds.

Furthermore, wet milling techniques are being used for buckwheat milling to get 65% and 79% of protein and starch, respectively (Zheng et al., 1997). Wronkowska and Haros (2014) employed wet milling for buckwheat with and without hull to extract the starch. They reported a higher starch fraction from the buckwheat containing hull compared to the dehulled buckwheat.

3.3 CONCLUSION

Pseudocereals like buckwheat, quinoa, and amaranth have gained a lot of global potential due to their excellent nutrition profile, gluten-free properties, and potential health benefits. They are rich in fiber, protein, minerals, and polyphenolic compounds which make them highly suitable for the development of functional and health-promoting foods. Conventional milling techniques such as hand pounding, huller milling, stone grinding, and roller milling are not adequate for milling of pseudocereals because of their smaller grain size and higher oil content. However, on the other sides, emerging, milling technologies like cryogenic milling, ultrafine grinding, impact milling, planetary ball milling, and wet milling techniques improve the milling flour yields and quality and retention of nutrients in pseudocereals with less energy consumption. Therefore, emerging milling technology can be adopted for the milling of pseudocereals to enhance their better utilization in the food supply chain. The future milling techniques for pseudocereals can be improved by integrating with the smart sensors and monitoring and controlling the process with artificial intelligence. Such advancement will result in optimizing the whole milling process with respect to quality characteristics, less energy use, and little waste generations. However, new research is also needed to develop new solar-driven milling machineries for the pseudocereals and utilization of the milling by-products for contributing toward the eco-friendly food processing practices.

REFERENCES

Agregán, R., Guzel, N., Guzel, M., Bangar, S. P., Zengin, G., Kumar, M., & Lorenzo, J. M. (2023). The Effects of processing technologies on nutritional and anti-nutritional properties of pseudocereals and minor cereal. *Food and Bioprocess Technology*, *16*(5), 961–986.

Atalay, M. H., Bilgicli, N., Elgun, A., & Demir, M. K. (2013). Effects of buckwheat (fagopyrum esculentum moench) milling products, transglutaminase and sodium stearoyl-2-lactylate on bread properties. *Journal of Food Processing and Preservation*, *37*(1), 1–9.

Atwell, W. A., Patrick, B. M., Johnson, L. A., & Glass, R. W. (1983) Characterization of quinoa starch. *Cereal Chemistry*, *60*, 9–11.

Burmeister, C. F., & Kwade, A. (2013). Process engineering with planetary ball mills. *Chemical Society Reviews*, *42*(18), 7660–7667.

Campbell, G. M. (2007). Roller milling of wheat. *Handbook of Powder Technology*, *12*, 383–419.

Chauhan, G., Eskin, N., & Tkachuk, R. (1992). Nutrients and antinutrients in quinoa seed. *Cereal Chemistry*, *69*(1), 85–88.

Chavarín-Martínez, C. D., Reyes-Moreno, C., & Dorado, R. G (2019). Physicochemical Properties, Nutritional Profile, and Bioactive Compounds of Cereals and Pseudocereals. *Types, Process and Effects*, 119.

De Bock, P., Daelemans, L., Selis, L., Raes, K., Vermeir, P., Eeckhout, M., & Van Bockstaele, F. (2021). Comparison of the chemical and technological characteristics of wholemeal flours obtained from amaranth (Amaranthus sp.), quinoa (*Chenopodium quinoa*) and buckwheat (Fagopyrum sp.) seeds. *Foods*, *10*(3), 651.

De Bondt, Y., Liberloo, I., Roye, C., Windhab, E. J., Lamothe, L., King, R., & Courtin, C. M. (2020). The effect of wet milling and cryogenic milling on the structure and physicochemical properties of wheat bran. *Foods*, *9*(12), 1755.

Deka, G., Dutta, H., & Sharma, P. (2023). A review: Hydrothermal processing of millets and effect on techno-functional and nutritional properties. *Journal of Food Process Engineering*, *46*(9), e14408.

Dhankhar, P., & Hissar, T. (2014). Rice milling. *IOSR Journal of Engineering*, *4*(5), 34–42.

Farfan, A., Ciacco, C., Agapito, F., & Augusto-Ruiz, W. (1978). Removal of saponins from quinua (*Chenopodium quinoa* Wild), grain by milling. *5th International Congress. Food Science, Technology,* Kyoto, Japan *September*17–22.

Farooq, M., Wani, S. M., Mir, S. A., & Naseem, Z. (2023). An overview of buckwheat allergy: A rare allergenic food. *Journal of Food Composition and Analysis*, *123*, 105616.

Haros, C. M., & Wronkowska, M. (2017). Pseudocereal dry and wet milling: Processes, products and applications. In Claudia Monika Haros & Dr. Regine Schoenlechner (eds), Pseudocereals: *Chemistry and Technology* (pp. 140–162). Wiley.

Hemery, Y., Chaurand, M., Holopainen, U., Lampi, A.-M., Lehtinen, P., Piironen, V., Rouau, X. (2011). Potential of dry fractionation of wheat bran for the development of food ingredients, part I: Influence of ultra-fine grinding. *Journal of Cereal Science*, *53*(1), 1–8.

Inglett, G. E. (1970). *Corn, Culture, Processing,* Products: AVI.

Izydorczyk, M. S., McMillan, T., Bazin, S., Kletke, J., Dushnicky, L., & Dexter, J. (2014). Canadian buckwheat: A unique, useful and under-utilized crop. *Canadian Journal of Plant Science*, *94*(3), 509–524.

Janeš, D., Prosen, H., & Kreft, S. (2012). Identification and quantification of aroma compounds of tartary buckwheat (*Fagopyrum tataricum* Gaertn.) and some of its milling fractions. *Journal of Food Science*, *77*(7), C746–C751.

Kaushal, P., & Kumar, N. (2022). Processing of cereals. In Harish Kumar Sharma and Navneet Kumar (eds), *Agro-Processing and Food Engineering: Operational and Application Aspects* (pp. 415–454): Springer.

Kawakami, I., Murayama, N., Kawasaki, S., Igasaki, T., & Hayashida, Y. (2008). Effects of storage temperature on flavor of stone-milled buckwheat flour. *Journal of the Japanese Society for Food Science and Technology, 55(11),* 559–565.

Kumar, M., Singh, A., Upadhyay, A., & Thakur, R. (2023). Evaluation of Physical, Anti-nutritional and Optimisation of Milling Characteristics of Quinoa Grains for Sustainable Food Security. *International Journal Environment Climate Change*, *13*(11), 2490–2498.

Malik, A., Khamrui, K., & Prasad, W. (2021). Effect of hydrothermal treatment on physical properties of amaranth, an underutilized pseudocereal. *Future Foods*, *3*, 100027.

Manzoor, M. F., Ali, A., Ain, H. B. U., Kausar, S., Khalil, A. A., Aadil, R. M., & Zeng, X.-A. (2024). Bioaccessibility mechanisms, fortification strategies, processing impact on bioavailability, and therapeutic potentials of minerals in cereals. *Future Foods*, 100425.

Matsuhash, I. T., Kurokouchi, K., & Ohara, T. (1984). Composition and characteristics of buckwheat flour produced by a multiple stage milling system. *Research Report of the Nagano State Laboratory of Food Technology*, *12*, 138–147.

McKevith, B. (2004). Nutritional aspects of cereals. *Nutrition Bulletin*, *29*(2), 111–142.

Middlewood, P. G., & Carson, J. K. (2012). Extraction of amaranth starch from an aqueous medium using microfiltration: Membrane fouling and cleaning. *Journal of Membrane Science*, *411*, 22–29.

Mir, N. A., Riar, C. S., & Singh, S. (2018). Nutritional constituents of pseudo cereals and their potential use in food systems: A review. *Trends in Food Science & Technology*, *75*, 170–180.

Mir, S. A., Farooq, M., Naseem, Z., & Mugabi, R. (2025). Techno-functional, nutritional, antioxidant and prebiotic potential of underutilized buck wheat flour from Western Himalayas of Jammu and Kashmir. *Applied Food Research*, *5*(1), 100723.

Patil, N. D., & Chawla, P. (2024). Machines and equipment for pseudocereal processing. In Harish Kumar Sharma and Navneet Kumar (eds), *Pseudocereals* (pp. 216–247): CRC Press.

Paucar-Menacho, L. M., Vásquez Guzmán, J. C., Simpalo-Lopez, W. D., Castillo-Martínez, W. E., & Martínez-Villaluenga, C. (2023). Enhancing Nutritional profile of pasta: The impact of sprouted pseudocereals and cushuro on digestibility and health potential. *Foods*, *12*(24), 4395.

Qian, J. Y., & Kuhn, M. (1999) Characterization of *Amaranthus cruentus* and *Chenopodium quinoa* starch. *Starch/Stärke*, *51*, 116–120.

Radosavljevic, M., Jane, J., & Johnson, L. (1998). Isolation of amaranth starch by diluted alkaline-protease treatment. *Cereal Chemistry*, *75*(2), 212–216.
Resio, A. N. C., Tolaba, M. P., & Suárez, C. (2009). Correlations between wet-milling characteristics of amaranth grain. *Journal of Food Engineering*, *92*(3), 275–279.
Roa, D. F., Baeza, R. I., & Tolaba, M. P. (2015). Effect of ball milling energy on rheological and thermal properties of amaranth flour. *Journal of Food Science and Technology*, *52*, 8389–8394.
Roa, D. F., Santagapita, P. R., Buera, M. P., & Tolaba, M. P. (2014). Amaranth milling strategies and fraction characterization by FT-IR. *Food and Bioprocess Technology*, *7*, 711–718.
Ruan, R. (1992). Simultaneous and nondestructive measurement of transient moisture profiles and structural changes in corn kernels during steeping using microscopic nuclear magnetic resonance imaging. *Cereal Chemistry*, *69*, 600–606.
Saltzman, J. (1990). *Amazing Grains: Creating Vegetarian Main Dishes with Whole Grains*: HJ Kramer.
Schoenlechner, R., Siebenhandl, S., & Berghofer, E. (2008). Pseudocereals. In Elke Arendt, Fabio Dal Bello (eds), *Gluten-Free Cereal Products and Beverages* (pp. 149–VI): Elsevier.
Sharma, K. D., Sharma, B., & Saini, H. K. (2021). Processing, value addition and health benefits. In Mohar Singh, Salej Sood (eds), *Millets and Pseudo Cereals* (pp. 169–184): Elsevier.
Silventoinen, P., Kortekangas, A., Ercili-Cura, D., & Nordlund, E. (2021). Impact of ultra-fine milling and air classification on biochemical and techno-functional characteristics of wheat and rye bran. *Food Research International*, *139*, 109971.
Simon, H. (1889). On the Latest Development of Roller Flour Milling. *Proceedings of the Institution of Mechanical Engineers*, *40*(1), 148–192.
Sinkovič, L., Pipan, B., Neji, M., Rakszegi, M., & Meglič, V. (2023). Influence of hulling, cleaning and brushing/polishing of (Pseudo) cereal grains on compositional characteristics. *Foods*, *12*(13), 2452.
Steadman, K., Burgoon, M., Lewis, B., Edwardson, S., & Obendorf, R. (2001). Buckwheat seed milling fractions: Description, macronutrient composition and dietary fibre. *Journal of Cereal Science*, *33*(3), 271–278.
Takeuchi, Y. (2001). Buckwheat flour as ingredient for foodstuff preparation method, involves removing husk from buckwheat seed by thresh process, removing seed coat by emery grain milling, swelling by pressure heating and grinding. Patent JP 2001061421-A.
Tosi, E., Re, E., Lucero, H., & Masciarelli, R. (2000). Amaranth (Amaranthus spp.) grain conditioning to obtain hyperproteic flour by differential milling. *Food Science and Technology International*, *6*(6), 433–438.
Wright, K. H., Huber, K. C., Fairbanks, D. J., & Huber, C. S. (2002) Isolation and characterization of *Atriplex hortensis* and sweet *Chenopodium quinoa* starches. *Cereal Chemistry*, *79*(5), 715–719
Wronkowska, M., & Haros, M. (2014). Wet-milling of buckwheat with hull and dehulled–the properties of the obtained starch fraction. *Journal of Cereal Science*, *60*(3), 477–483.
Xu, Q., Zheng, F., Cao, X., Yang, P., Xing, Y., Zhang, P., & Bi, X. (2021). Effects of airflow ultrafine-grinding on the physicochemical characteristics of tartary buckwheat powder. *Molecules*, *26*(19), 5841.
Yuan, J., Chung, D., Seib, P., & Wang, Y. (1998). Effect of steeping conditions on wet-milling characteristics of hard red winter wheat. *Cereal Chemistry*, *75*(1), 145–148.
Zheng, G., Sosulski, F., & Tyler, R. (1997). Wet-milling, composition and functional properties of starch and protein isolated from buckwheat groats. *Food Research International*, *30*(7), 493–502.

4 Potential Utilization of Pseudocereals and Their By-Products in Food and Non-Food Industries

Dilek Dülger Altıner, Emine Aydın, and Gökçen Yıldız

4.1 INTRODUCTION

Pseudocereals are non-grass plants that produce seeds used similarly to grains and share physical traits with true cereals. They are commonly incorporated into flour-based products and are especially valued for their gluten-free nature, making them an important component of dietary nutrition. Examples of pseudocereals include quinoa, amaranth, kaniwa, teff, and sorghum. Among the most well-known pseudocereals are quinoa (*Chenopodium quinoa* Willd), amaranth (*Amaranthus* spp.), and buckwheat (*Fagopyrum esculentum*) (Graziano et al., 2022) (Figure 4.1). They are valued for their nutritional profiles, often containing higher protein levels and vital amino acids compared to traditional cereals (Malik & Singh, 2022). Pseudocereals are rich in fiber, vitamins, and minerals, making them a valuable addition to a well-rounded and nutritious diet (Kaur et al., 2023).

Pseudocereals, often regarded as "ancient grains" and included among the superfoods of the 21st century, have gained increased popularity due to consumers' growing awareness and focus on health. These pseudocereals stand out as alternative raw materials in the gluten-free product market due to their nutritional and functional properties, earning recognition as functional foods. Their low glycemic index (GI)

FIGURE 4.1 Image of selected pseudocereal grains.

DOI: 10.1201/9781003584858-4

further distinguishes them from other grains (Majaw et al., 2024). Compared to traditional cereals, pseudocereals have higher fatty acids (linoleic acid). Additionally, these pseudocereals exhibit resilience to pests, plant diseases, and harsh climatic conditions, offering environmental and economic benefits and promoting sustainable nutrition and agriculture (Pirzadah & Malik, 2020).

The utilization of diverse agricultural products in various sectors, the production of nutritionally valuable products, and the development of new products have become increasingly important in addressing global hunger and malnutrition-related health issues. This chapter discusses the general characteristics of pseudocereals such as quinoa, amaranth, and buckwheat, alongside their nutritional, functional, and health-related properties. Furthermore, it explores the applications of pseudocereals and their by-products in the food and non-food industries, providing insights into the products derived from these versatile crops.

4.2 QUINOA (*CHENOPODIUM QUINOA* WILLD.): CULTIVATION AND ITS IMPORTANCE IN GLOBAL AGRICULTURE

Quinoa, belonging to the Chenopodiaceae family (goosefoot family) and classified under the genus *Chenopodium*, is an annual plant species (Koyun, 2013). Known as an overlooked and underutilized crop, quinoa was widely cultivated in Latin America but faced production constraints due to Spanish colonization and conquests, which restricted its cultivation to local scales (Ruiz et al., 2014). Once a staple of the Inca civilization, quinoa has recently captured global interest. While it remains a major crop in Argentina, Bolivia, Ecuador, Peru, Chile, and Colombia, its cultivation has now extended across Europe, North America, and Africa (Vilcacundo et al., 2017).

While quinoa cultivation was initially concentrated in Andean countries, it has recently expanded worldwide. Globally, Bolivia, and Peru are recognized as the big producers and exporters of quinoa worldwide. Thanks to the genetic diversity of quinoa seeds, which have a 7,000-year cultivation history by farmers in the Andes Mountains, quinoa has spread to various other countries as well (Bazile, 2023). Recognized as a pseudocereal, quinoa is classified as a new functional food with a broad range of uses similar to cereals. It boasts high protein quality, bioactive phytochemicals, and both nutritional and technological properties, along with nutraceutical benefits, leading it to be identified as one of the "grains of the 21st century" (Abugoch, 2009).

With its broad adaptability and resilience to environmental stress, quinoa has been designated by the Food and Agriculture Organization (FAO) as a critical crop in the fight against hunger and in ensuring food security. In recognition of its value, the United Nations declared 2013 as the International Year of Quinoa (Miranda et al., 2012). Since this declaration, quinoa has been effectively incorporated into various food products to support global food security (FAO & CIRAD 2015). In this situation, climate change's growing threats to agricultural production, and the focus on salt- and stress-tolerant crops, such as quinoa, along with the development of resilient new genotypes, have become more critical than ever.

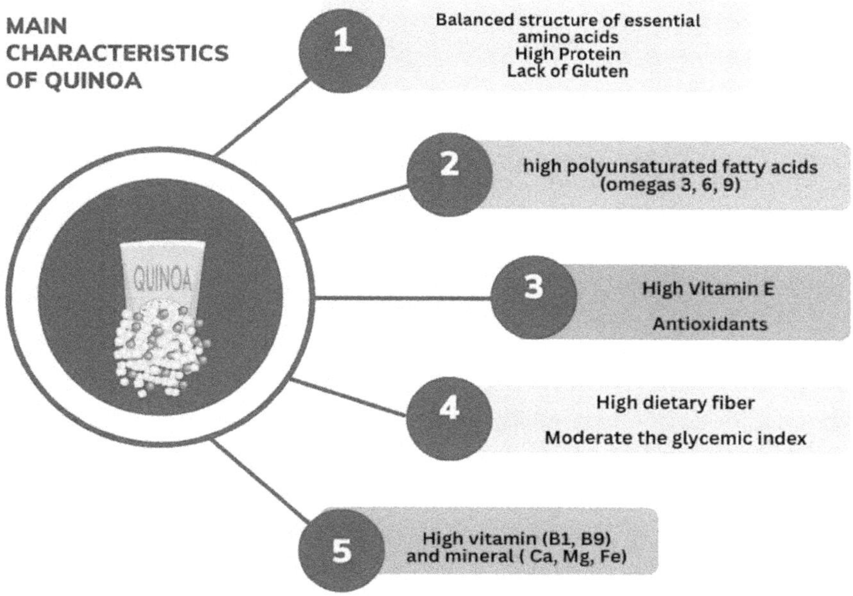

FIGURE 4.2 Simplified scheme summarizing the main characteristics of quinoa.

Figure 4.2 summarizes the key characteristics of quinoa, as highlighted in Bazile's (2023) research on global trends driving the expansion of quinoa cultivation and Ruiz et al.'s (2014) analysis. This chapter researches the chemical, nutritional, and functional attributes of quinoa, along with an exploration of potential applications for quinoa by-products in both the food industry and non-food sectors.

4.2.1 Chemical, Nutritional, Functional Properties and Health Benefits of Quinoa

An analysis of quinoa (*C. quinoa*) seeds reveals an average composition of approximately 15% protein, 476 mg/g starch, 46% carbohydrate, 7% fat, and 10% dietary fiber, establishing quinoa as a valuable source of antioxidants, vitamins, and minerals (Brakez et al., 2016; Abugoch, 2009). Dark-colored quinoa seeds, in particular, exhibit high antioxidant capacities due to their significant anthocyanin content (Pasko et al., 2009). Quinoa is recognized for its beneficial unsaturated fatty acid composition, with linoleic acid accounting for 52%, oleic acid 26%, and linolenic acid 4.8% of its unsaturated fats (Repo-Carrasco et al., 2003). The omega-6 to omega-3 ratio is approximately 6:1 (Tang et al., 2015). Additionally, quinoa stands out for its vitamin content, exceeding that of many common cereal grains (Graf et al., 2014). Its high levels of vitamin E not only enhance its antioxidant properties but also help maintain the stability of its fatty acid profile (Vega-Galvez et al., 2010). Nutritional details are provided in Figure 4.3.

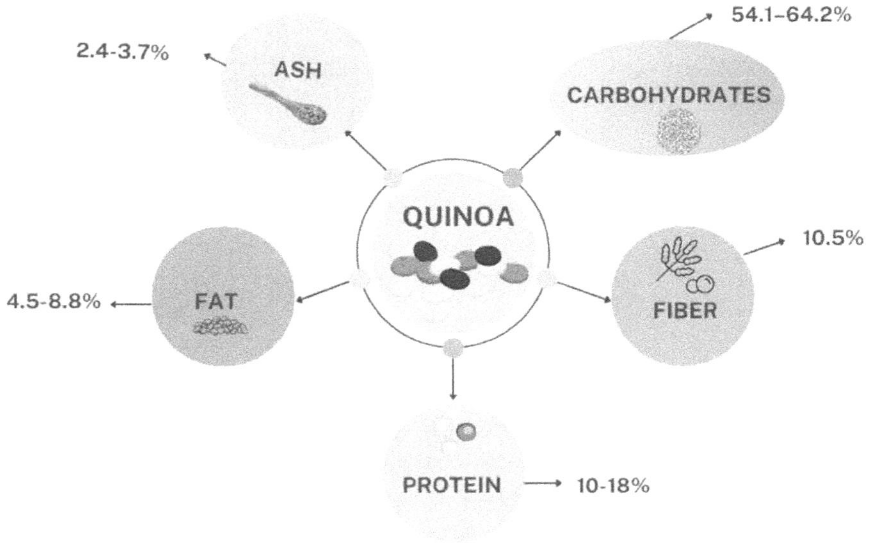

FIGURE 4.3 Nutritional composition of quinoa[1,2]. **Source**: The figure was created by the author. References: Olmos et al., (2022) and Angeli et al., (2020).

With the continuous increase in the global population, there is a growing need for plant-based protein sources to address rising food demands (Ghumman et al., 2021). Quinoa is notable for its protein richness, outperforming grains such as corn, wheat, and rice, and surpassing barley and rye in protein content, making it comparable to wheat. It features a well-balanced amino acid composition, including essential amino acids like threonine, lysine, and methionine, along with a total of 18 amino acids. This composition contributes to its high-quality protein content (12%–23%), with excellent digestibility (Abugoch, 2009; Valcarcel-Yamani & Lannes, 2012; Lan et al., 2023). Similar to milk protein, quinoa's protein content varies between 14% and 18% (Gordillo-Bastidas et al., 2016).

Quinoa seeds contain vitamins E, C, and complex B vitamins, along with essential minerals such as calcium, potassium, magnesium, iron, and phosphorus. Additionally, they contain anti-nutritional factors such as oxalates, phytic acid, saponins, nitrates, tannins, and protease inhibitors. While saponins, present in significant levels in quinoa seeds, contribute to a bitter taste, phytic acid – commonly found in grains and legumes – reduces the absorption and digestion of minerals and proteins (Demir & Bilgiçli, 2020; Filho et al., 2017). The sprouting process reduces the amount of these anti-nutritional factors (Demir & Bilgiçli, 2020). Processing techniques such as pearling can also reduce saponin content, diminishing quinoa's bitterness (Ren et al., 2023). Anti-nutritional factors in quinoa can be minimized or eliminated through common food processing methods, including atmospheric pressure treatment, roasting, pressurization, fermentation, sprouting, extrusion, and cooking, both industrially and at home (Nickel et al., 2016; Thakur et al., 2021; Jan et al., 2023).

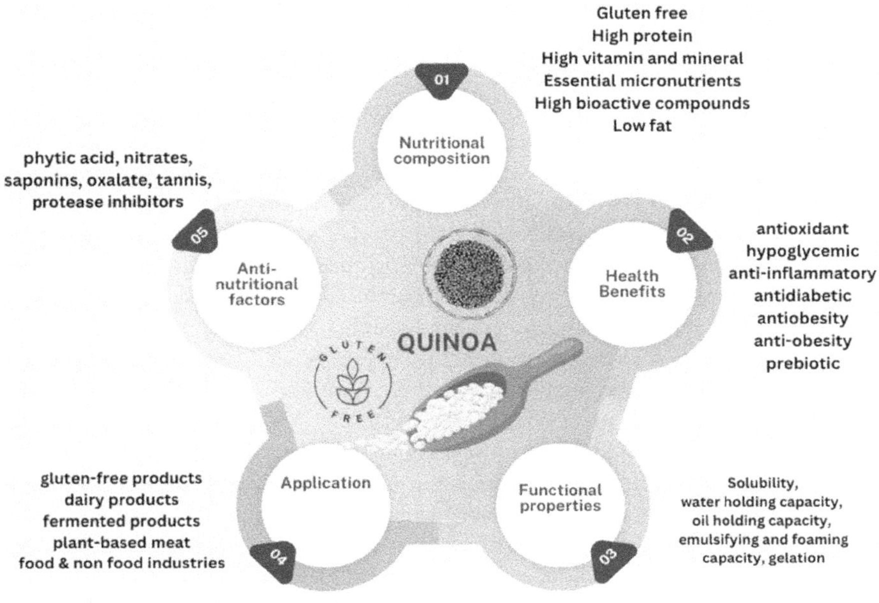

FIGURE 4.4 Nutritional value, functional properties, health benefits, and food applications of quinoa[1,2,3,4,5]. **Source**: Schematized by the author. References: [1]Graziano et al. (2022); Lan et al. (2023); [2]Jan et al. (2023); [3]Castro-Alba et al. (2019); [4]Abdelshafy et al. (2023); [5]Demir and Bilgiçli (2020); Filho et al. (2017).

According to Nascimento et al. (2013), a check of the nutritional properties and mineral content of Andean cereals – quinoa, amaranth, and purple corn – with rice revealed that rice contained significantly lower levels of minerals such as iron (Fe), copper (Cu), manganese (Mn), and zinc (Zn) in comparison to these Andean crops. Quinoa, due to its phosphorus and magnesium content, was reported to meet approximately 55% of the daily requirement, making it a valuable pseudocereal.

The general characteristics of quinoa, including its nutritional value, functional properties, health benefits, and food applications, are schematized in Figure 4.4.

4.2.2 Recent Advances in the Application of Quinoa and Its By-Products: Potential Utilization of Quinoa in Food and Other Industries

Quinoa has been identified by the National Aeronautics and Space Administration as a potential food source for long-duration space missions due to its resilience to environmental conditions and its high nutritional profile, making it suitable for sustaining crew members (Tang & Tsao, 2017). In recent studies, quinoa has been explored for its use in producing a variety of functional food crops, including bread, pasta, snacks, and gluten-free baked goods (Wang & Zhu, 2016).

The primary components of the quinoa seed are separated through dry and wet milling, with the purpose of dividing these essential parts according to milling methods for applications in the food industry and other sectors. Milling is typically categorized into two primary methods: dry milling and wet milling. Wet milling involves isolating major chemical ingredients of the seed, such as protein, crude oil, dietary fiber, and starch, through an industrial process combining physical, chemical, biochemical, and mechanical techniques. This process begins with soaking the grain, followed by mechanical separation. In contrast, dry milling produces components such as endosperm, bran, and germ, which are primarily utilized for flour production (Yalçın & Arslan, 2021). In a study where quinoa fiber, separated via wet milling, was incorporated into bologna-type sausage, results indicated an increase in dietary fiber content, an improved commercial quality of the sausage, extended shelf life, and a reduction in nitrite usage, resulting in a more natural product (Fernández-López et al., 2020).

Casalvara et al. (2024) identified several components of the quinoa plant as valuable by-products, including the husk, leaves, roots, seeds, and biomass residue. These components are utilized in various countries for purposes such as food ingredients, prebiotic sources, energy sources, animal feed, emulsifiers, food preservatives, and food colorants. The classification of quinoa and its by-products is schematized in Figure 4.5.

Quinoa flour (QF) is often mixed with or used as a substitute for other flours in the production of various foods, including bread, biscuits, pasta, cakes, and noodles

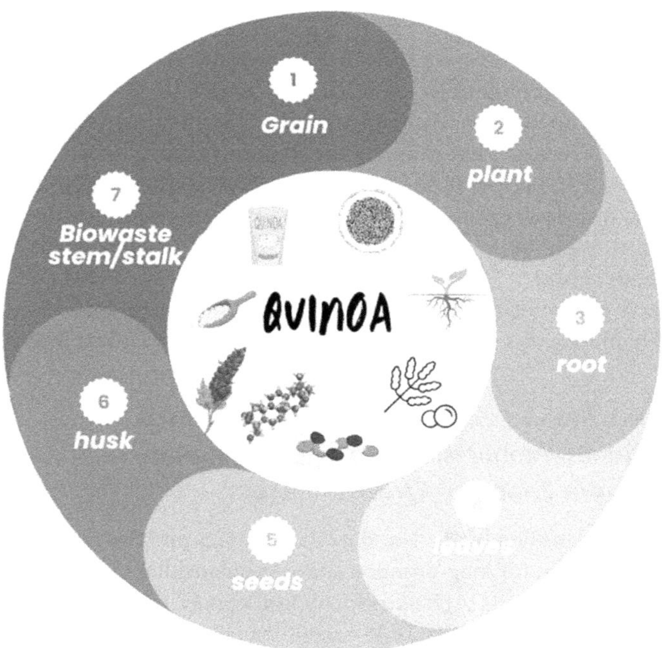

FIGURE 4.5 Quinoa and its by-products.

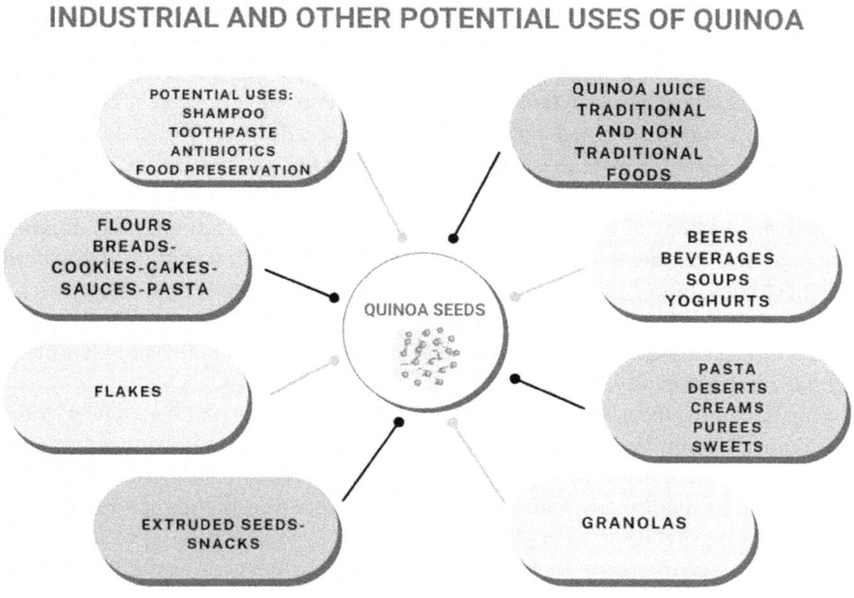

FIGURE 4.6 Industrial and other potential uses of quinoa and its by-products. **Source**: Schematized by the author. References: Montoya et al. (2005); Brito et al. (2022).

(Valencia-Chamorro, 2003). With the rising incidence of conditions such as celiac disease and gluten intolerance, as well as the increased popularity of gluten-free diets, the gluten-free product market has expanded (Aguiar et al., 2021). Quinoa, being naturally gluten-free, is recognized as a valuable ingredient for gluten-free products (Vicente et al., 2023). Additionally, it is being explored as a potential raw material for malt and beer production, alongside other grains and pseudocereals such as rice, millet, maize, buckwheat, and amaranth (Nic Phiarais & Arendt, 2008). Studies have shown that malting quinoa grain enhances its bioavailability, with sprouting reducing phytic acid content by an average of 36% and improving iron bioaccessibility by two to four times in vitro (Valencia et al., 1999). These showings underscore the significance of exploring quinoa's nutritional properties and its potential for innovative product development. The industrial and other applications of quinoa and its by-products are schematized in Figure 4.6 (Montoya et al., 2005; Brito et al., 2022).

4.2.2.1 Quinoa Flour

Ramos Diaz et al. (2015) developed a corn-based extruded snack product incorporating amaranth, quinoa, and kañiwa flours. Analysis of the sensory and physical properties indicated an increase in hardness with higher quinoa content. When QF was included at levels of 20%, 30%, and 50%, the product exhibited smaller particles, less stickiness, reduced crispness, and decreased crunchiness. These showings suggest that quinoa can be considered a potential component in producing cereal-based snacks with enhanced nutritional value.

A study evaluating the in vitro starch digestibility and estimated glycemic indices of five gluten-free breads made from buckwheat, oats, sorghum, quinoa, and teff flours found that quinoa bread had the highest estimated GI (GI = 95). This was followed by buckwheat bread (GI = 80), while oat bread had the lowest GI (GI = 71). Quinoa bread was also noted for higher starch digestibility (Wolter et al., 2013). Foods with a high GI release glucose rapidly, causing an increase in blood sugar levels (Brand-Miller et al., 2009).

Machado Alencar et al. (2015) evaluated the addition of quinoa and amaranth flours to gluten-free breads, finding that pseudocereal inclusion improved nutritional properties, particularly protein, lipid, and ash content, without statistically significant differences in taste. However, bitterness increased with higher quinoa content. Gluten-free breads made with a combination of QF and sweeteners were identified as potential competitors to starch-based gluten-free formulations.

In further study, the addition of QF (15%, 30%), amaranth, and buckwheat flour to wheat flour (WF) in bread production resulted in an increase in flavonoid substance, with antioxidant activity values rising as the proportion of quinoa increased. Sensory analysis showed that breads with buckwheat flour received higher scores in overall preference ratings (Chlopicka et al., 2012).

Baker et al. (2013) used QF as a gluten-free replacement for rice flour in muffin production. Comparing a control sample made with 100% rice flour to muffins with 30% and 50% quinoa substitution, sensory analysis by a panel of 75 untrained consumers reported that muffins with 50% QF had higher overall acceptability scores than the others. Additionally, muffins with QF were found to be significantly higher in protein, fiber, and mineral content (Mg, K, and Zn) compared to the rice flour control, suggesting that QF is a promising alternative gluten-free flour source.

Research has explored the quality attributes of noodles fortified with sprouted black quinoa powder and investigated the chemical and anti-nutritional property changes in pasta enriched with raw and sprouted QFs (Seol & Sim, 2017; Demir & Bilgiçli, 2020).

4.2.2.2 Quinoa Starch

Another study highlighted the favorable physicochemical properties of quinoa starch, such as water solubility, swelling power, and thermal and textural properties, establishing it as a high-quality food ingredient suitable for human consumption. The study noted significant differences among quinoa samples from various geographical origins. Quinoa starch exhibits a range of diverse physicochemical properties, such as swelling power, water solubility index, pasting behavior, gel texture, enzyme susceptibility, amylose leaching, and thermal characteristics. These attributes make it a versatile alternative for applications in both food and non-food industries. Additionally, due to its high enzyme susceptibility, careful processing of quinoa products is recommended (Li et al., 2016).

Quinoa seeds come in a variety of colors, including white, yellow, red, and black. High in starch content, they are commonly used as substitutes for rice and wheat in a range of products, including bread, crepes, baby food, and extruded breakfast cereals. The starch in quinoa seeds, located primarily in the perisperm, is characterized by low amylose content, a robust amylopectin structure, and freeze–thaw stability, comprising approximately 53%–69% of the seed's dry matter (Li et al., 2016; Wang & Zhu, 2016; Lindeboom et al., 2005).

4.2.2.3 Quinoa Protein

Quinoa seeds contain high protein levels, approximately 12%–17%, depending on species and variety (López et al., 2018). In a study examining the nutritional and functional properties of quinoa proteins, Lingiardi et al. (2022) reported that quinoa proteins had the potential to stabilize emulsion gel systems, which can serve as fat replacers and bioactive compound carriers in food development. In contrast to quinoa starch, quinoa protein isolates can be utilized directly without requiring any chemical modification, simplifying their application in various products. These proteins exhibit emulsifying abilities, emulsion stability, and gelation capabilities.

Compared to traditional cereal proteins, pseudocereal proteins have superior quality and higher nutritional value. Literature reviews discuss the nutritional, physicochemical, functional, and bioactive properties of quinoa protein (Brito et al., 2022). Wang et al. (2024) studied the alteration of quinoa protein's structural and techno-functional properties through extraction techniques and modification methods, finding that these techniques enhanced the structure and properties of quinoa proteins. As a result, improvements were observed in quinoa proteins' techno-functional properties, such as solubility, emulsification, and gelation, with noted advantages and disadvantages for each extraction method. In another study using quinoa seeds to obtain quinoa protein concentrate, these hydrolysates demonstrated antioxidant activity in zebrafish larvae (Vilcacundo et al., 2017).

Germination of seeds leads to the accumulation of secondary phytochemicals, vitamin C, and polyphenolic compounds. These germinated seeds, rich in lipids, proteins, carbohydrates, and exhibiting high digestibility, antimicrobial, and antioxidant effects, are gaining attention for use in new food products. Germinated seeds can serve as a source for producing protein concentrates (Carciochi et al., 2014). Piñuel et al. (2019) investigated the antioxidant capacity of protein isolates and hydrolysates derived from both germinated and non-germinated quinoa, reporting promising outcomes.

4.2.2.4 Quinoa Leaves

Quinoa leaves are also utilized for culinary purposes, as their flavor resembles that of spinach. They can be cooked or used in salads. With a composition of 3.3% protein, 1.8% fat, and 1.9% fiber, quinoa leaves contain higher levels of magnesium and sodium compared to spinach leaves (Valencia-Chamorro, 2003). In a study analyzing soluble dietary fibers isolated from different quinoa leaves, Hu et al. (2023) reported that quinoa dietary fibers exhibited antioxidant, prebiotic, and immune-regulating effects, highlighting their potential for use in functional food and health sectors.

4.2.2.5 Quinoa Husk (Biowaste)

Before quinoa seeds are used industrially, the outer layer is removed in a dehulling process, producing a by-product known as quinoa husk. The extracted husk biomass is rich in bitter compounds, particularly saponins, which are polyphenolic compounds and contain cellulose, hemicellulose, and lignin (Sun et al., 2019). Saponins, found in the outer husk, serve as anti-nutritional factors that protect quinoa seeds from pathogens and herbivores. These compounds are typically removed post-harvest through washing with cold water, abrasion, and dehulling. Saponins have applications in various industries, within cosmetics and pharmaceuticals (Angeli et al., 2020). Research on quinoa husk, an agro-industrial waste, is growing to explore its industrial and agricultural applications, aiming to prevent disposal through burning and instead convert it into value-added products. Quinoa husk offers significant potential for diverse applications, ranging from the creation of innovative food products to its use in packaging materials, biofuels, compost, commercial chemicals, and bioenergy production (de Cássia Spacki et al., 2022).

4.2.2.6 Edible Film

Quinoa starch has recently been adopted as a primary material in edible biofilm applications, a modern topic in food preservation (Kolata, 2009). In a study by Valenzuela et al. (2013), the mechanical, barrier, and structural features of edible films containing a quinoa–chitosan and sunflower oil blend were examined. The findings suggested that coating low-pH fresh fruits with these blended films could extend shelf life and enhance consumer satisfaction. In that study, after suspending QF in distilled water and adjusting the pH, the resultant supernatants from subsequent chemical treatments and centrifugation were termed quinoa protein extracts.

In another study focused on developing active biofilms containing gold nanoparticles within quinoa starch, quinoa starch was used to create biofilms with strong antimicrobial activity against foodborne pathogens such as *Escherichia coli and Staphylococcus aureus.* These active quinoa starch biofilms showed promise as active food packaging to pick up food safety and prolong shelf life in packaged products (Pagno et al., 2015).

A study on developing films based upon quinoa starch employed the dilute alkali solutions methodology with short contact time, as developed by Araujo-Farro (2008). The procedure for obtaining quinoa starch involved the following steps: soaking quinoa seeds in water, grinding, preliminary supernatant removal, centrifugation, chemical treatment, neutralization, washing with pure water, and removal of remaining proteins and salts (Araujo-Farro et al., 2010).

4.2.2.7 Applications in Cosmetics, Medicine, and Pharmaceuticals

Quinoa is utilized for various medicinal purposes. It has been informed to exhibit anti-inflammatory properties, function as a diuretic, act as an insect repellent, and be utilized in treatments for internal bleeding as well as a gastric protector. (Jan et al., 2023). Furthermore, quinoa seeds are highly recommended for inclusion in health-promoting diets due to their rich content of flavonoids, vitamin E, and protein, which contribute to the prevention of chronic diseases and overall well-being. The triterpenoid-rich saponins in quinoa husk extracts are noted for their strong affinity to cholesterol-containing lipid layers (Jurek et al., 2019).

Given its low glucose and fructose content, quinoa is suitable for use in developing alternative products for diabetic patients (Zheng et al., 2023) and, as a gluten-free choice, is beneficial for individuals with celiac disease, supporting its role as a functional food in promoting health.

In the cosmetics industry, quinoa also has diverse applications. Saponins, anti-nutritional compounds found in quinoa husk, are valued for their foaming properties and are used in the production of shampoos, detergents, and soaps. They also have potential as ingredients in mouthwash and toothpaste for reducing oral odor (Heliawati et al., 2002). Ethanol extracts from Brazilian quinoa have been identified as emulsifying surfactants that can be biotechnologically incorporated as additives in cosmetic products (Bezerra et al., 2020). Compounds from quinoa husk processed with ultrasound technology have shown potential as natural colorants and preservatives for both the cosmetics and food industries (Laqui-Vilca et al., 2018). Examples of functional food products made from quinoa and its by-products are shown in Table 4.1

Additionally, its contributions to food security, quinoa and its by-products stand out in packaging, cosmetics, medical, and biotechnology fields due to their superior qualities. With functional and nutritional properties varying by region, efforts to enhance quinoa's sustainability are ongoing.

4.3 AMARANTH (*AMARANTHUS* SPP.): CULTIVATION OF AMARANT AND ITS SIGNIFICANCE IN GLOBAL AGRICULTURE

Amaranth (*Amaranthus* spp.), a member of the Amaranthaceae family, is recognized as a cosmopolitan plant capable of thriving in both temperate and tropical regions due to its adaptability to various stress conditions and its high nutritional value (Netshimbupfe et al., 2023; Muyonga et al., 2014). The plant has garnered significant attention for its resilience and agricultural potential (Narwade & Pinto, 2018).

Etymologically derived from the Greek word meaning "eternal" or "immortal," amaranth is believed to have originated in the Americas, with evidence of its cultivation dating back to approximately 6700 BCE within the Aztec civilization (Mlakar et al., 2010). Historically, it served as an essential food source for the Aztec, Maya, and Inca civilizations; however, its production significantly declined following the collapse of Central American cultures (Jubete et al., 2010a; Venskutonis & Kraujalis, 2013).

Unlike typical monocot cereals, amaranth is a dicotyledonous plant, and is therefore categorized as a pseudocereal. Its seed structure comprises three distinct layers: the perisperm, the embryo, and the seed coat, which sets it apart from conventional grains (Malik et al., 2023; Balakrishnan & Schneider, 2022; Schoenlechner et al., 2008). Amaranth is an annual plant varying in height from 0.3 to 5 m, with seeds measuring between 1 and 1.5 mm in diameter (Rastogi & Shukla, 2013). The seeds are lens-shaped or spherical and exhibit colors ranging from cream to brown or yellowish beige (Malik et al., 2023).

Amaranth is distinguished by its inflorescences and leaves, which range in color from purple and red to golden yellow, and over 60 species have been identified (Narwade & Pinto, 2018). Of these, the leaves of at least 17 species are utilized as vegetables and fodder (Netshimbupfe et al., 2023), while the seeds of three species are consumed similarly to other grains (Burgos & Armada, 2015). Certain *Amaranthus* species, while cultivated for their grains, vegetables, or as ornamental plants, are also

TABLE 4.1
Examples of Functional Food Products Made from Quinoa and Its By-Products

Food Products	Product Features and Developed Aspects	References
Bread	• Quinoa flour (QF)-based flat bread (5%, 10%, 15%, 20%, 25%, and 30% of QF) • Enhanced physicochemical properties of the bread. • Received higher overall preference from panelists compared to the control • Indicated as a promising option for preventing malnutrition.	El-Sohaimy et al. (2019)
Cakes	• Pseudocereal flour- quinoa and amaranth (0%, 10%, 20%, and 30%) • Gluten-free cake formulation • Good protein, fat, ash, and fiber. • 20% and 30% quinoa-based cakes were the best	Hamzehpour and Dastgerdi (2023)
Cookies and biscuits	• Quinoa flour incorporation into Biscuits (12.5% or 25% QF) • Assessment of rheological and sensory attributes, alongside the impact of consuming 20% quinoa-based biscuits on diabetic rats • Enhanced nutritional and sensory qualities • Exhibited effective hypoglycemic properties	Alsuhaibani et al. (2022)
Beverages	• Fermented quinoa-based beverage • Proposed as an alternative to address food shortages • Pasankalla variety demonstrated higher protein levels, reduced saponin content, and minimal viscosity loss during fermentation • Maintains quality over a 28-day storage period • Contentful in protein, fiber, vitamins, and minerals, • Snack for individuals with celiac disease and lactose intolerance • Presents a novel and exotic option for consumers	Ludena Urquizo et al. (2016)
Pasta	• Quinoa germ–enriched pasta • Quinoa germ-with balanced amino acid profile and unsaturated fatty acid • Pasta supplemented with quinoa germ • Increased dough rheology and protein content • Increased firmness value of pasta • Decreased farinograph water absorption and dough stability	Ray et al. (2023)
Noodle	• Development of gluten-free Turkish noodle "erişte" using oat flour and QF. • Formulations include 100% oat flour, 100% QF, and a 50:50 blend of oat and QFs by weight • Positive feedback from panelists on the sensory qualities of erişte samples • Quinoa-based noodles exhibited higher protein, ash, and fat contents compared to oat-based noodles	Çalışkan Koç and Pandiselvam (2022)

(*Continued*)

TABLE 4.1 (*Continued*)
Examples of Functional Food Products Made from Quinoa and Its By-Products

Food Products	Product Features and Developed Aspects	References
Chinese steamed bread (CSB)	• Quinoa as a diverse food grain • Blends of QF with wheat flour (WF) in varying ratios: 85/15, 70/30, 55/45, 40/60, 25/75, and 10/90 • Increased shelf life of breads and physicochemical properties	Wang et al. (2015)
Traditional uses	• Quinoa soup: a nutritious dish featuring cooked quinoa combined with meat, tubers, and vegetables • Phisara: preparation of roasted and cooked quinoa grain • Andean soup-Juchacha: made from quinoa, katahui, served with roasted barley flour	Angeli et al. (2020) Bojanic (2011)
Infant foods	• Infant foods (ages 6–24 months) developed with QF • Comprehensive evaluation of physicochemical, rheological, molecular, thermal, and sensory properties • QF enhances nutritional qualities • Formulation with 8% QF received the highest sensory ratings	Ayseli et al. (2020)
Salad	• Traditional dish: *Chiwa* • Lliccha: young quinoa leaves • Utilized as a vegetable in soups and salads • High in vitamins and minerals, particularly calcium (Ca), potassium (K), and iron (Fe)	Bojanic (2011)
Sauce	• Canned anchovy (*Engraulis ringens*) in a sauce made from quinoa and kiwicha • Sensory evaluation conducted with 42 panelists • Canned anchovy with quinoa sauce provides valuable nutritional benefits	Castillo-Martínez et al. (2023)

considered weeds in various agricultural contexts, whereas *Amaranthus cruentus*, *Amaranthus hypochondriacus*, and *Amaranthus caudatus* are specifically grown for food purposes (Baraniak & Kania-Dobrowolska, 2022; Aderibigbe et al., 2022). These three species have been cultivated and widely enjoyed in Mexico, Peru, and Guatemala for many years (Caselato-Sousa & Amaya-Farfan, 2012). Currently, amaranth is also cultivated in countries such as the United States, Germany, Austria, the Czech Republic, Italy, Hungary, Poland, Russia, Slovakia, and Slovenia (Gönen et al., 2018).

Due to the significant diversity within the *Amaranthus* genus, taxonomic classification remains challenging (Svirskis, 2003). Commonly cultivated amaranth plants fall into four categories: leaf, grain, garden, and ornamental types (Svirskis, 2003; Kauffman & Weber, 1990). Factors such as high yield, seed color, stem height, early maturation, seasonal stability, seed shattering resistance, and satisfactory nutritional and utilitarian characteristics are considered in cultivation practices (Svirskis, 2003;

Weber, 1990). Optimal germination of amaranth seeds requires soil temperatures between 18°C and 25°C, along with ambient air temperatures exceeding 25°C. The plant's growth is further enhanced with increased light and temperature (Gönen, 2020). Seed yield can vary depending on factors such as soil moisture, structure, physical and chemical properties, climate, plant density, planting date, and fertilizer application (Gimplinger et al., 2008).

Amaranth exhibits a unique physiology as a C4-type CO_2-fixation plant (Breus, 1997; Malik et al., 2023), where the initial product of photosynthesis is a four-carbon compound (Malik et al., 2023). C4 plants are known for their highly efficient photosynthetic mechanisms, improved nitrogen metabolism, and distinctive physiological and metabolic characteristics (Breus, 1997). This means that, along with other anatomical features of amaranth, the use of CO_2 across a wide range of temperatures and under conditions of moisture stress leads to increased productivity, providing the plant with the capacity to adapt to diverse environmental conditions (Alemayehu et al., 2015; Cheng, 2018). In other words, amaranth demonstrates remarkable adaptation to arid conditions, poor soil quality, and extreme temperature fluctuations, and can also tolerate variations in soil pH and salinity (Caselato-Sousa & Amaya-Farfan, 2012). For these reasons, it is regarded as a sustainable and climate-resilient crop with the potential to enhance food security. (Alemayehu et al., 2015; Cheng, 2018). Given its ease of cultivation and harvesting, edible amaranth species are suggested as a low-cost alternative crop for agriculture (Ergun et al., 2014).

Amaranth provides a wide range of benefits due to its exceptional nutritional profile and the distinct properties of its components, making it a versatile resource for food production, the pharmaceutical industry, environmental sustainability, and various industrial applications (Rastogi & Shukla, 2013). These properties have earned amaranth the titles of "super grain," "miracle grain," and "grain of the future" (Ofitserov, 2001). The main characteristics of amaranth are schematized in Figure 4.7.

4.3.1 Chemical, Nutritional and Functional Properties of Amaranth

The nutritional composition of amaranth, as illustrated in Figure 4.8, is distinguished by its high-quality protein, well-balanced amino acid profile, and gluten-free characteristics. These attributes position amaranth as a highly appealing option for health-conscious consumers (Soriano-García & Aguirre-Díaz, 2019). Due to its diverse nutritional benefits, recent findings have classified amaranth not only as a food source but also as a pseudocereal with dual properties, possessing both nutritional and health-enhancing capabilities (Baraniak & Kania-Dobrowolska, 2022).

Amaranth, particularly popular in regions of Asia and Africa, is consumed as both leaves and seeds, which are highly nutritious (Netshimbupfe et al., 2023). As a rich source of vitamins, minerals, and proteins, amaranth surpasses the nutritional value of traditional grains and some legumes (Gorinstein et al., 2007).

Amaranth grains contain 65%–75% starch as the primary carbohydrate (Malik et al., 2023), along with 11.14%–20.6% dietary fiber (Caselato-Sousa & Amaya-Farfan, 2012). The main sugar present is sucrose, which is found at levels 2–3 times higher than in wheat (Venskutonis & Kraujalis, 2013). The starch, notable for its small granule size (average 1.38 μm), is composed of 97.9% amylopectin, with a low amylose

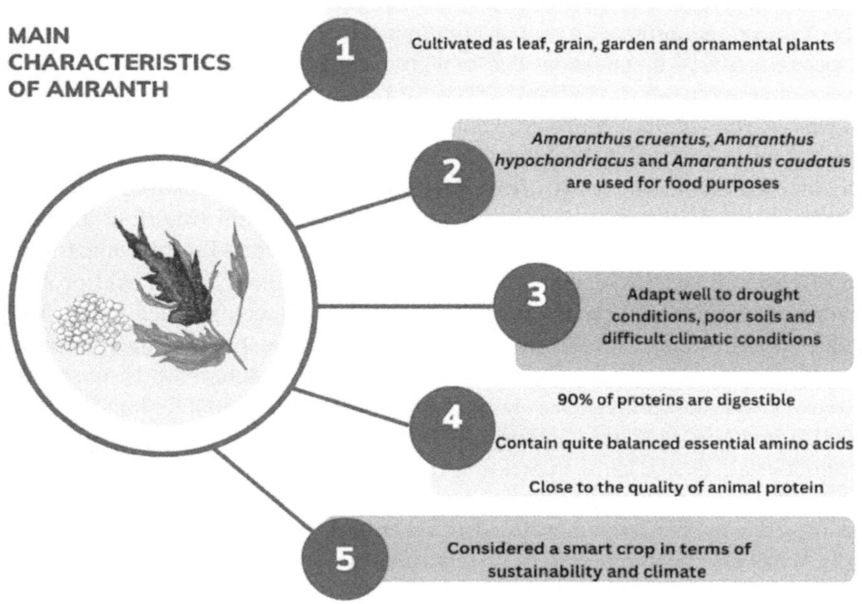

FIGURE 4.7 Simplified scheme summarizing the main characteristics of amaranth.

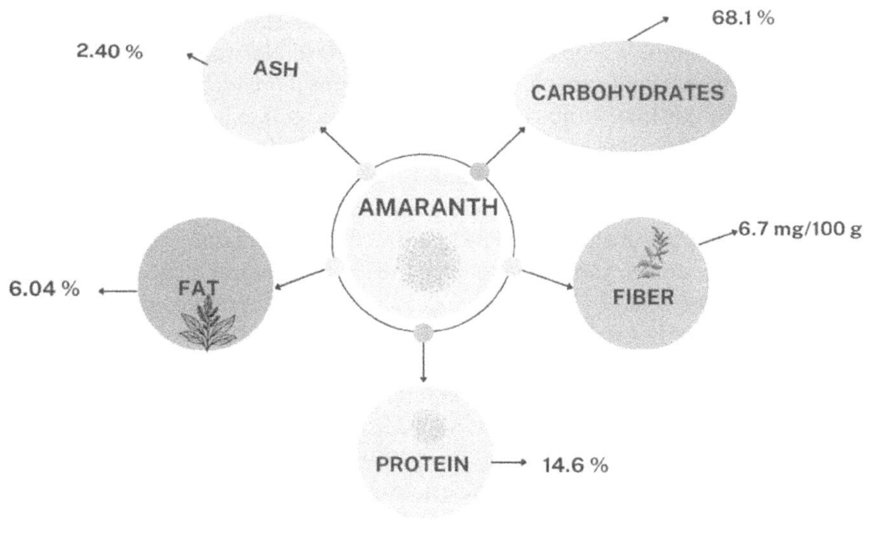

FIGURE 4.8 Nutritional composition of amaranth (Sattar et al., 2024).

content (Malik et al., 2023). The amylose content in amaranth flour and starch has been reported at 5.9% and 6.9%, respectively (Sindhu & Khatkar, 2016). This low amylose content is a significant factor influencing the textural, pasting, and thermal properties of amaranth-based products (Punia et al., 2020).

Amaranth boasts a higher protein content compared to other pseudocereals such as buckwheat and quinoa, as well as common cereals such as wheat, corn, and rice (Bock et al., 2021). It stands out due to its remarkably balanced amino acid profile when compared to wheat (13.5%–14.5%), corn (10.6%–13.8%), barley (10%–14.9%), and oats (12.4%–12.9%) (Malik et al., 2023), which contributes to its high nutritional value (Maurya & Arya, 2018; Marquez-Molina & Lopez-Martinez, 2020). The protein content in amaranth ranges from 15.4% to 16%, with reports indicating that it contains 47.6 g of essential amino acids per 100 g of protein (Malik et al., 2023). Notably, the protein composition of amaranth has been found to be comparable to that of cow's milk (Bekkering & Tian, 2019). The essential amino acid profile of amaranth protein aligns closely with the requirements recommended by the World Health Organization, approaching the quality of animal protein, thus surpassing that of conventional cereals (Rivero Meza et al., 2022). Amaranth is an excellent source of key amino acids such as lysine and tryptophan, essential for human growth, and also contains moderate amounts of threonine, leucine, valine, and isoleucine (Schmidt et al., 2021). Unlike cereals such as rice, oats, and corn, amaranth boasts high levels of lysine (Netshimbupfe et al., 2023). Additionally, its methionine and cysteine content surpasses that found in other cereals and legumes (Amare et al., 2015). It has also been reported to be relatively rich in sulfur-including amino acids (Maurya & Arya, 2018; Marquez-Molina & Lopez-Martinez, 2020). Cereals such as amaranth and quinoa, which are rich in essential amino acids, exhibit an optimal protein composition like that found in cow's milk (Bekkering & Tian, 2019). The primary protein fractions in amaranth are albumin and globulin (Malik et al., 2023). Unlike most cereals, amaranth is gluten-free (Sneha & Haripriya, 2018) and boasts high protein digestibility, reaching approximately 90% (Narwade & Pinto, 2018).

Amaranth, with a lipid content of 6%–9%, contains more fat than most other cereals, further distinguishing its nutritional profile (Emire & Arega, 2012). Its oil primarily comprises palmitic acid (19%), oleic acid (25%), and linoleic acid (47.5%–47.8%) (Jubete et al., 2010b). Although amaranth oil contains small amounts of linolenic acid, it is also rich in other lipid components such as phosphoric acid, sterols, and glycolipids (Srivastava et al., 2021). The lipid fraction of amaranth is a notable source of bioactive compounds, including squalene (8%), tocopherols (2%), phospholipids (10%), and phytosterols (2%) (Alegbejo, 2012). These compounds are rarely found together in commonly used oils (Shin et al., 2004). Due to the cholesterol-lowering effects of squalene and phytosterols (Jubete et al., 2010b), amaranth oil is considered an exceptionally valuable component (Shin et al., 2004). Although amaranth contains a low level of phytic acid (0.3%–0.6%), which can reduce the bioaccessibility of minerals (Repo-Carrasco-Valencia & Serna, 2011), it has been reported that phytic acid may also have cholesterol-lowering effects (Rastogi & Shukla, 2013).

Amaranth seeds are rich in macro and microelements (Malik et al., 2023), with the vitamin and mineral content varying according to the specific type of amaranth (Zhang et al., 2023a). The seeds contain 2.56 mg/100 g of zinc (Zn), 411 mg/100 g of potassium (K), 157 mg/100 g of calcium (Ca), 2.20 mg/100 g of manganese (Mn), 249 mg/100 g of magnesium (Mg), 443 mg/100 g of phosphorus (P), 0.51 mg/100 g of copper (Cu), 6.99 mg/100 g of iron (Fe), and 5.20 mg/100 g of sodium (Na) (Ramesh & Parakash, 2020).

Amaranth is a significant source of riboflavin (vitamin B2) and also contains abundant amounts of thiamine (vitamin B1) and ascorbic acid (vitamin C) (Repo-Carrasco-Valencia & Serna, 2011). Additionally, although amaranth seeds have a high oil content, they are resistant to oxidation due to the antioxidant properties of vitamin E (Valcarcel-Yamani & Lannes, 2012). Furthermore, amaranth grains are known for their significant antioxidant activity, which is attributed to their rich content of polyphenols, anthocyanins, flavonoids, and tocopherols (Escudero et al., 2011).

The small granule size of amaranth starch enhances its functional properties, such as high water-binding capacity, increased swelling power, lower gelatinization temperature, and notable resistance to amylase activity. These attributes make amaranth starch highly desirable in the food industry, particularly for its superior gelatinization behavior and freeze–thaw stability (Pal et al., 2001).

4.3.2 Health Benefits of Amaranth

Amaranth is widely recognized as a functional food due to its quantity of antioxidants, including phenolic compounds and flavonoids, which protect against oxidative stress and may lower the risk of chronic diseases such as heart disease and cancer (Schmidt et al., 2021; Bachar et al., 2020). Furthermore, amaranth provides numerous health benefits, such as reducing plasma cholesterol levels, enhancing immune function, exhibiting antitumor properties, lowering blood glucose levels, and alleviating conditions associated with hypertension and anemia. It is also noted for its anti-allergenic properties (Caselato-Sousa & Amaya-Farfan, 2012). The presence of vitamins A, C, E, and various B vitamins in amaranth positively impacts metabolism, skin health, and immune functions. Moreover, its mineral content, including iron (Fe), calcium (Ca), and zinc (Zn), supports bone and muscle health (Park et al., 2020; Alvarez-Jubete et al., 2009).

The general characteristics of amaranth, including its nutritional value, functional properties, health benefits, and food applications, are schematized in Figure 4.9.

4.3.3 Recent Advances in the Application of Amaranth and Its By-Products: Potential Utilization of Amaranth in Food and Other Industries

4.3.3.1 Amaranth Seeds

Amaranth species such as *Amaranthus caudatus*, *Amaranthus cruentus*, and *Amaranthus hypochondriacus* are utilized as functional ingredients in various food products across India, Africa, China, Japan, the Americas, and Europe. These species are valued for their high protein content, which enhances the essential amino acid profile of baked goods, increases dietary fiber, and enriches the product with calcium and iron (Gamel et al., 2006; Sanz-Penella et al., 2013).

Amaranth seeds can be processed in several ways, including popping, flaking, extrusion, and milling into flour, enabling their incorporation into diverse food products (Narwade & Pinto, 2018). Popped or roasted amaranth seeds can be used to

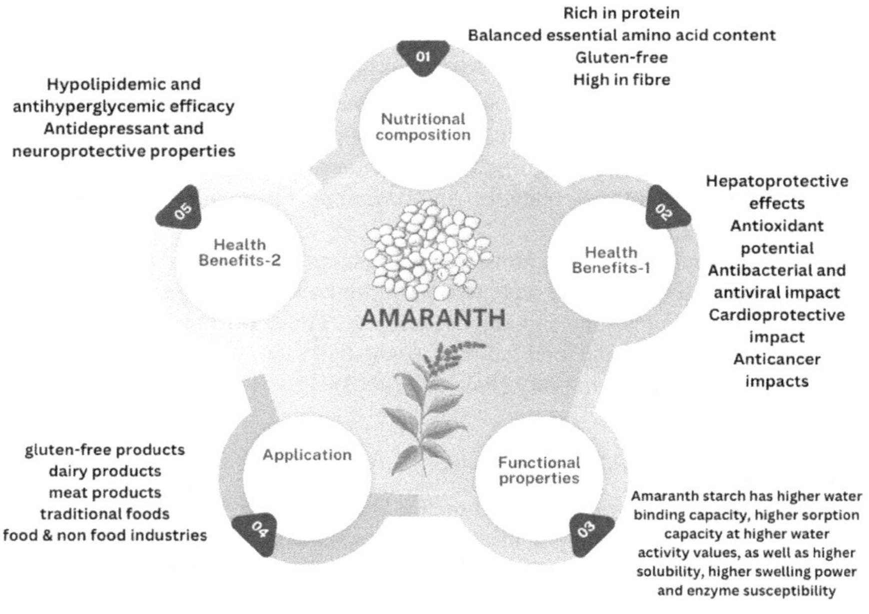

FIGURE 4.9 Nutritional value, functional properties, health benefits, and food applications of amaranth[1,2,3]. **Source**: Schematized by the authors. References: [1]Yadav and Yadav (2024); [2]Sattar et al. (2024); [3]Singhal and Kulkarni (1990).

produce amaranth flour, which serves as an ingredient in breakfast cereals, spaghetti, biscuits, crackers, bread, pancakes, and cakes, either on its own or blended with other components (Aderibigbe et al., 2022).

Buresova et al. (2017) reported that adding up to 35% amaranth flour to WF improved the organoleptic properties of the product, such as flavor, taste, color, and appearance, with the optimal sensory results observed at 25% substitution.

For home use, amaranth seeds can be prepared as porridge or malted for beer production (Yang & Gao, 2022). Amaranth flour has been found to partially or fully replace WF in the preparation of a wide array of foods, including pasta, cookies, bread, and porridge (Akande et al., 2017; Ramos Diaz et al., 2017).

Amaranth seeds are most used in flour-based products such as bread, pancakes, cereals, cakes, and snacks (Singh et al., 2024). They are also suitable for use in breakfast foods, gluten-free products, and extruded foods (Elizabeth, 2010). Furthermore, amaranth grain has been reported as an ingredient in soups, stews, sauces, porridge, and soufflés (Grobelnik Mlakar et al., 2009). Amaranth flour, specifically, is a valuable addition to baked goods due to its nutritional properties, making it a popular choice in baking and confectionery. The inclusion of amaranth flour in bread enhances its nutritional value and supports the intake of essential dietary components. It is recommended to add up to 10% amaranth flour or flakes to bread for optimal benefits (Nasir et al., 2020).

The addition of roasted amaranth seeds at 10% and 15% levels has been found to have a positive effect on the taste of food products, imparting a nutty and coffee-like flavor (Cacak-Pietrzak et al., 1995). Cakes made with amaranth seeds have demonstrated potential as successful substitutes for traditional products, offering consumer-acceptable taste comparable to conventional cookies, while providing higher nutritional value (Klebaniuk et al., 2018).

Amaranth seeds have also been reported to be used as a supplement in the production of snacks in chip form. Incorporating amaranth seeds into corn chips resulted in a richer cereal, sweet, and chocolate-like aroma (Gajewski et al., 2015). Due to their high nutritional value, amaranth seeds are also valuable as a supplement in yogurt (Sady et al., 2007).

4.3.3.2 Amaranth Leaves

Amaranth leaves and stems are rich in various vitamins, including riboflavin (vitamin B2), vitamin B6, folate, and niacin. They also contain significant levels of minerals such as calcium, iron, and magnesium (Jiménez-Aguilar & Grusak, 2017). Research indicates that baked products enriched with amaranth leaves contain significantly higher amounts of protein, fat, ash, and fiber compared to their non-enriched similar. Moreover, these fortified products demonstrate substantially increased levels of essential minerals such as calcium, magnesium, iron, and zinc (Singh et al., 2009). Due to their versatility, amaranth leaves are suitable for a variety of culinary uses, including soups, stews, and salads (Ruth et al., 2021).

4.3.3.3 Amaranth Flowers

Amaranth flowers are commonly used for decorative purposes in gardens and floral arrangements. Although they have potential health benefits, amaranth flowers have not been widely utilized as a food source. Traditionally, however, they have been employed as a remedy for ailments such as diarrhea, dysentery, cough, and bleeding (Baraniak & Kania-Dobrowolska, 2022).

4.3.3.4 Amaranth Flour

In recent years, amaranth flour has gained popularity in the production of gluten-free products, including pasta, bread, biscuits, and crackers. A study investigating the addition of amaranth flour to WF at 20%, 40%, 60%, 80%, and 100% ratios for cookie preparation found that incorporating up to 60% amaranth flour was sensory acceptable and preferable due to its nutritional benefits (Chauhan et al., 2016).

Gebreil et al. (2020) reported that crackers (25% corn flour: 75% amaranth flour) and tortillas (50% corn flour: 50% amaranth flour) made from corn and amaranth flour mixtures had superior physicochemical properties, essential amino acids, essential fatty acids, and mineral content compared to control samples made with 100% corn flour. However, the sensory analysis revealed that these products scored significantly lower in terms of color and overall acceptance, possibly due to the bitter taste imparted by amaranth flour, which may have reduced consumer preference.

In a study by Machado Alencar et al. (2015), gluten-free bread produced with 20% whole amaranth and QF showed specific volume, hardness, and water activity values similar to those of the control bread. These gluten-free breads were found to contain higher levels of protein, fat, and ash and exhibited a more open crumb structure with larger pores. Conversely, Turkut et al. (2016) found that adding 25% QF resulted in softer-textured bread with higher sensory scores, indicating better consumer acceptance.

4.3.3.5 Edible Films

Amaranth flour exhibits excellent film-forming capabilities, producing films with moderate solubility, high flexibility, and superior water vapor barrier properties. These edible films are composed of biopolymers suitable for food packaging applications (Diéguez et al., 2015). Amaranth flour has been evaluated as a raw material for the production of edible films and coatings (Tapia-Blácido et al., 2013). In a study by Colla et al. (2006), films created from a mixture of stearic acid, glycerol, and amaranth flour demonstrated better mechanical properties, lower solubility, and optimal barrier characteristics when compared to composite films prepared with other proteins and polysaccharides.

4.3.3.6 Emulsion-Type Meat Products

Amaranth meal has been explored as a plant-based protein isolate for the meat processing industry. Research has shown that amaranth protein isolate possesses significant functional and technological properties, including neutral organoleptic characteristics, making it a viable ingredient in the formulation of various meat products such as sausages and semi-finished meat products (Maksim et al., 2024).

4.3.3.7 Non-Food Use

One of the lesser-discussed historical uses of amaranth is as a source of natural dye. Civilizations such as the Aztecs, Incas, Hopi, and other Indigenous groups in the Americas used the vibrant leaves and flower clusters of amaranth to produce red dye for food, textiles, and religious ceremonies (Sauer, 1967; Howard et al., 2022).

Betalain pigments derived from amaranth species are already utilized as natural food colorants in China (Sauer, 1967). Compared to beetroot, plant-based amaranth as a food colorant provides several advantages, including a wider color spectrum, greater regional availability of raw materials, and a milder flavor profile that avoids the earthy taste often associated with beetroot, making it more appealing for a broader range of applications (Howard et al., 2022).

The antioxidant properties of amaranth have made it a valuable addition to dietary and nutritional plans for athletes. Athletes commonly use antioxidant supplementation to reduce oxidative stress, aid recovery, and enhance performance, positioning amaranth as a desirable component in sports nutrition (Neubauer & Yfanti, 2015).

Amaranth has gained attention as a feed ingredient for ruminant animals due to its high nutritional value and favorable fatty acid content (Peiretti, 2018). It can be used in various forms, including as a grain, or as fresh, dried, or ensiled forage (Abbasi et al., 2012).

4.3.3.8 Applications in Cosmetics, Medicine, and Pharmaceuticals

Due to its rich nutritional properties, certain amaranth preparations are used in the cosmetic industry (Baraniak & Kania-Dobrowolska, 2022). In some Indian villages, an infusion of *A. spinosus L.* shoots is applied for the treatment of eczema. Traditional healers also use amaranth leaves to treat skin allergies. In Africa, amaranth leaves are used as a nutraceutical or functional food for teething treatment in infants and as a diuretic to address worm infections, stomach pains, and constipation (Gandhi et al., 2021).

A recent study revealed that beyond its role as a nutritious food, amaranth possesses significant cosmetic potential. Extracts from *Amaranthus* spp. seeds have demonstrated potential skin-whitening properties due to the presence of myoinositol and squalene, which have shown a greater in vitro ability to reduce melanin production compared to common arbutin (Gandhi et al., 2021).

Oil derived from *A. cruentus* seeds has been found to provide pre- and post-UVA protection in skin fibroblasts, suggesting its utility in antiaging creams and as a natural sunscreen component (Wolosik et al., 2017). The leaf extract of *A. gangeticus*, rich in anthocyanins, can be used as a natural hair dye, providing a nontoxic alternative to synthetic hair dyes (Heliawati et al., 2022).

Amaranth oil is rich in unsaturated fatty acids, tocopherols, phytosterols, and squalene, which are highly beneficial for both skin and hair. Suitable for all skin types, amaranth oil serves as an excellent moisturizer, soothes irritation, accelerates wound healing, and exhibits antimicrobial and antioxidant properties. It also nourishes the skin and provides antiaging benefits. Due to these qualities, amaranth seed oil is commonly found in skin creams and lotions and is used as an ingredient in shampoos and shower gels. As a natural and abundant source of tocopherols, amaranth oil helps protect hair from the harmful effects of sunlight, strengthens hair, and prevents hair loss (Baraniak & Kania-Dobrowolska, 2022).

Amaranth and its by-products can be processed to enhance their value, enabling their use as functional foods with health-promoting properties. Some examples are listed in Table 4.2.

4.4 CULTIVATION OF BUCKWHEAT AND ITS SIGNIFICANCE IN GLOBAL AGRICULTURE

Buckwheat (*Fagopyrum esculentum*), though its name suggests otherwise, is unrelated to wheat. It is a pseudocereal that has been cultivated for thousands of years, particularly in the highlands of Southeast Asia, with the Yunnan region of China being a key area (Bvenura & Kambizi, 2022). Evidence indicates that buckwheat was domesticated approximately 6,000–7,000 years ago, marking it as one of humanity's oldest cultivated crops (Yao et al., 2023). It thrives in mountainous regions where traditional cereals struggle due to poor soil and challenging climates (Patel et al., 2024). As its cultivation spread along ancient trade routes, buckwheat became integral to diets in Central Asia, the Middle East, and Europe (Suvorova & Zhou, 2018). It gained particular popularity in Russia and Eastern Europe, where it remains a staple food, especially in Poland, Ukraine, and Slovenia (Pirzadah & Rehman, 2021). By the 15th century, buckwheat had made its way to Japan, where it is well-known for its role in soba noodle production (Tömösközi & Langó, 2017). In North America, European settlers introduced buckwheat cultivation in the 17th century, especially in northeastern regions, due to its resilience in less fertile soils (Cheng, 2018). However, the introduction of nitrogen fertilizers and higher-yielding crops in the 20th century led to a decline in its cultivation (Arduini & Mariotti, 2018).

TABLE 4.2
Examples of Functional Food Products Made from Amaranth and Its By-Products

Food Products	Product Features and Developed Aspects	References
Bread	• Fermented amaranth flour was incorporated at varying levels (0%, 15%, 30%, and 45%) into gluten-free bread. This showed in increased levels of ash, protein, phytic acid, total phenolic content, antioxidant activity, and essential minerals such as calcium (Ca), phosphorus (P), potassium (K), iron (Fe), magnesium (Mg), and zinc (Zn) in the bread. Bread containing 15%–30% pseudocereal flour received overall acceptability scores comparable to the control. Pseudocereals like amaranth offer a nutritious and healthy alternative for enhancing the quality of gluten-free breads.	Yeşil (2021)
	• The addition of amaranth flour at levels of 10%, 20%, and 30% to bread formulations resulted in a letup in specific volume and L values of the bread color. Additionally, it led to an increase in moisture content and crumb hardness across all formulations.	Alsaiqali (2018)
Cake	• Gluten-free cakes were developed using mixtures of raw and roasted amaranth flour at proportions of 0%, 25%, 50%, and 75%. Incorporating amaranth flour significantly enhanced the cakes' nutritional profile, increasing moisture, ash, protein, fat, total phenolic compounds, antioxidant activity, phytic acid, and total dietary fiber content.	Alp (2022)
Cookie	• Whole amaranth flour was added to cookie formulations at levels of 20%, 40%, 60%, 80%, and 100%, resulting in an increase in cookie diameter and spread, along with a reduction in hardness and overall acceptability scores. Sensory evaluations indicated that 60% supplementation represented the upper limit for producing cookies with acceptable sensory qualities.	Chauhan et al. (2016)
	• The health-promoting properties of gluten-free amaranth cookies were investigated, revealing that simulated gastrointestinal digestion released bioactive peptides with antithrombotic and antihypertensive activities.	Sabbione et al. (2019)
Beverage	• A stable and nutritious beverage was formulated using amaranth proteins, with a composition comparable to skimmed cow's milk. Thermal treatment led to protein denaturation and aggregation, while the inclusion of gums reduced the proteins' sensitivity to heat, enhancing the beverage's stability.	Manassero et al. (2020)

(Continued)

TABLE 4.2 (*Continued*)
Examples of Functional Food Products Made from Amaranth and Its By-Products

Food Products	Product Features and Developed Aspects	References
Pasta	• Various blend ratios of amaranth and wheat semolina were evaluated for pasta production. A higher proportion of amaranth in the mix resulted in increased solid loss during cooking, along with reduced weight gain and firmness of the pasta. However, the addition of amaranth significantly improved the nutritional profile of the pasta by increasing its dietary fiber content and enhancing the quality of its protein.	Bodroža-Solarov et al. (2022)
Noodle	• The incorporation of various levels and combinations of amaranth flour in the production of erişte (a traditional Turkish type of noodle) was studied. Substituting with this pseudocereal led to increases in mineral content, total phenolic compounds, phytic acid, crude protein, crude fat, and ash levels, enhancing the nutritional profile of the erişte.	Öncel (2017)
Mantı (Turkish type ravioli)	• Amaranth flour was substituted for wheat flour (WF) at levels of 0%, 5%, 10%, 15%, and 20%. The inclusion of amaranth flour resulted in higher protein and fat content in the manti samples, along with an increase in total phenolic compounds. Additionally, the cooking time extended as the proportion of substitute flour increased.	Aygün (2021)
Infant food	• Biscuits enriched with amaranth and inulin were developed to enhance the WF-based formula. While the total protein content of the amaranth biscuits was slightly lower than that of those made with WF, the digestible protein and bioaccessibility were 22% higher. Additionally, the soluble and insoluble dietary fiber content increased by 30.9%. Amaranth-enriched baby biscuits also exhibited higher mineral levels, including calcium (Ca), iron (Fe), magnesium (Mg), manganese (Mn), and copper (Cu).	Vitali et al. (2009)
Snack food	• A snack bar formulated with a blend of amaranth, acha, and pearl millet (90:5:5) demonstrated the highest protein content at 21.95%. Additionally, its essential amino acid composition surpassed that of the control bar, highlighting its superior nutritional quality. The snack bar also exhibited a low glycemic index (GI) and resulted in low postprandial blood glucose levels. Thus, it may be considered a healthful dietary snack due to its nutrient density, antioxidant properties, and hypoglycemic effects.	Ibidunni Olagunju et al. (2022)

(Continued)

TABLE 4.2 (*Continued*)
Examples of Functional Food Products Made from Amaranth and Its By-Products

Food Products	Product Features and Developed Aspects	References
Popped grain	• The formation of volatile aromatic compounds during a 10-second detonation of amaranth grains on a hot plate at 150°C–200°C was investigated. It was found that these volatile compounds in the popped kernels contributed to the nutty and corn-like flavor and roasted aroma of puffed amaranth.	Gamel and Linssen (2008)
Edible film	• Optimal conditions for producing amaranth-based edible films were determined to include 10 g of stearic acid, 26 g of glycerol per 100 g of flour, and a mixing speed of 12,000 rpm. Films created under these parameters exhibited superior mechanical strength, reduced solubility, and optimal barrier properties, outperforming other protein and polysaccharide composite films in these aspects.	Colla et al. (2006)
Traditional uses	• **Mexican sweets "Alegria"**: the most famous Mexican confectionery made by mixing popped amaranth with dark brown molasses syrup or honey.	Milan-Carrillo et al. (2012)
	• **Mexican drink "Atole"**: Natural amaranth flour is mixed with a certain amount of water and heat-treated at a certain temperature to evaporate the water and prepare a viscous drink.	González Acevedo et al. (2018)
	• **Indian sweets "Rajgeera" or "Ramdana"**: Prepared using puffed amaranth grains and flour and described as Indian savory	Spurthi et al. (2021)
	• **Nepalese traditional dish "Sattoo"**	Kauffman (1992)

Today, buckwheat is produced globally, with Russia being the largest producer, contributing over 50% of the world's supply (Small, 2017). This dominance is attributed to the crop's cultural significance and its adaptability to cold climates and marginal soils (Renzi et al., 2022). Additional leading producers include Ukraine, Poland, and China, the latter cultivating buckwheat primarily for domestic consumption and export (Ikeda, 2002). Japan also grows buckwheat, mainly for soba noodles, while France, Brazil, and Canada contribute to smaller niche markets (Chandra et al., 2021). In recent years, there has been a rise in buckwheat consumption, driven by growing interest in health foods and gluten-free diets (Giménez-Bastida et al., 2015). Once regarded as a "forgotten" crop, buckwheat is now recognized for its nutritional benefits and culinary adaptability, sparking renewed interest in its cultivation and consumption.

Botanically classified within the Polygonaceae family, which includes plants such as rhubarb and sorrel, buckwheat is often mistaken for a cereal grain due to its culinary applications (Thorssin, 2020). As a pseudocereal, buckwheat is not a true cereal grain from the grass family (Poaceae), but it shares many characteristics with them, including its use in various food products (Raungrusmee, 2023). Unlike true cereals, buckwheat is free from gluten, which makes it a great choice for people with celiac disease or gluten sensitivity (Rai et al., 2018). Nutritionally, buckwheat stands out for its amino acid profile, being particularly rich in lysine and arginine, which are often deficient in other plant-based foods (Kurek et al., 2024). This quality positions buckwheat as a high-quality protein source, especially beneficial for vegetarian and vegan diets. The seeds, known as groats, are typically consumed roasted or milled into flour. Groats can be boiled like rice, used in porridge, or processed into noodles such as soba (Nayak et al., 2024). Despite being gluten-free, buckwheat flour is versatile for baking and is often combined with other gluten-free flours to enhance texture (Schoenlechner, 2017).

From an agricultural perspective, buckwheat is distinguished by its adaptability. It flourishes in poor, acidic, and low-fertility soils where many cereal grains would fail (Umesh et al., 2019). With a limited growing season of only 10–12 weeks, buckwheat serves as an ideal cover crop, promoting soil health and controlling erosion between major cropping cycles (Small, 2017). Its resilience further enhances its ecological sustainability, requiring fewer inputs including fertilizers and pesticides (Liu et al., 2022). Jha et al. (2024) highlighted the significance of traditional cultivation techniques for buckwheat farming, emphasizing the importance of intercropping systems that can improve yield and soil health. Their findings suggest that the integration of buckwheat with legumes can significantly enhance nitrogen fixation, leading to improved soil fertility. In summary, buckwheat occupies a unique niche among food crops due to its exceptional nutritional properties, gluten-free status, and versatility. Its growing global production reflects increasing demand for health-conscious and sustainable food options, while its classification as a non-cereal highlights its botanical uniqueness and agricultural significance.

4.4.1 Nutritional Composition of Buckwheat

Macro- and Micronutrient Profile: Buckwheat is highly valued for its rich nutritional content, making it an excellent choice for health-conscious consumers. It provides a balanced composition of macronutrients – proteins, carbohydrates, and fats – along with a range of vital micronutrients, such as vitamins and minerals. Buckwheat is an excellent plant-based source of protein, comprising approximately 12%–15% protein by weight, depending on the variety (Appiani et al., 2021). What sets buckwheat protein apart from many other plant-based sources is its complete amino acid profile. Notably, it is rich in essential amino acids such as lysine, threonine, and tryptophan, which are often deficient in common cereal grains such as wheat, corn, and rice (Sá & House, 2024). The primary proteins in buckwheat consist of globulins and albumins, making up about 70%–80% of its total protein content (Janssen et al., 2017). These proteins exhibit greater solubility and digestibility compared to gluten-containing grains (Zhu, 2021).

In terms of carbohydrates, buckwheat serves as a good supplier of complex carbohydrates, having a carbohydrate composition of roughly 70%–80%. The carbohydrates in buckwheat are primarily in the form of starch, supplemented by a significant amount of dietary fiber, both soluble and insoluble (Christa & Soral-Śmietana, 2008). Soluble fiber is effective in regulating blood sugar levels and managing cholesterol by delaying digestion and increasing the sensation of fullness. Conversely, insoluble fiber is vital for maintaining digestive health, as it promotes regularity and prevents constipation by adding bulk to the stool and aiding in the movement of food through the digestive system (Guan et al., 2021). Buckwheat has a relatively low GI, indicating that it has a moderate impact on sugar levels in the blood, rendering it a suitable food option for those managing diabetes (Bączek et al., 2020). Furthermore, buckwheat contains approximately 10% dietary fiber, which contributes to its health-promoting effects, particularly in supporting cardiovascular and digestive health (Zhu, 2020). The high fiber content enhances gut motility and promotes satiety, making buckwheat a popular choice for weight management diets. Furthermore, the fiber content in buckwheat serves as a prebiotic, which provides essential sustenance for beneficial gut bacteria. This process not only promotes the growth of these positive microorganisms but also contributes significantly to overall gut health, improving digestion and potentially enhancing immune function (Rezende et al., 2021).

Buckwheat is recognized as a significant source of several essential B vitamins, including niacin (B3), riboflavin (B2), and pyridoxine (B6). These vitamins are integral to numerous critical physiological processes in the body. Niacin, for example, is crucial in energy metabolism as it assists in transforming carbohydrates, fats, and proteins into usable energy. Riboflavin is essential for the production of red blood cells and contributes to maintaining healthy skin, eyes, and nerve functions. Pyridoxine, on the other hand, is vital for synthesizing neurotransmitters, which are necessary for proper neurological health and function. Together, these B vitamins contribute to overall well-being, emphasizing the nutritional value of incorporating buckwheat into a balanced diet (Bojňanská et al., 2022). Buckwheat also includes small quantities of vitamin E, a potent antioxidant that aids in protecting cells from oxidative damage (Sonawane et al., 2024). When it comes to minerals, buckwheat is rich in magnesium, phosphorus, iron, copper, manganese, and zinc (Podolska et al., 2021). Magnesium, in particular, is present in significant amounts, contributing to muscle and nerve function, bone health, and blood pressure regulation. Notably, buckwheat contains a higher amount of iron compared to many cereals, which is critical for oxygen transport and preventing anemia.

4.4.2 Comparison of Buckwheat with Other Pseudocereals and Grains

Table 4.3 compares the nutritional attributes of buckwheat with other pseudocereals, such as quinoa and amaranth, and common cereals such as wheat, rice, and corn. It summarizes key nutritional aspects, including protein content, amino acid profile, carbohydrate content, dietary fiber, fat content, essential minerals, vitamins, gluten content, GI, digestibility, and associated health benefits. This comparison highlights the unique nutritional profile of buckwheat and its potential advantages for health-conscious consumers.

When comparing buckwheat to other pseudocereals such as quinoa and amaranth, as well as to common grains such as wheat, rice, and corn, several nutritional

TABLE 4.3
Comparative Nutritional Attributes of Buckwheat, Other Pseudocereals, and Cereals

Component/ Attribute	Buckwheat (*Fagopyrum esculentum*)	Quinoa (*Chenopodium quinoa*)	Amaranth (*Amaranthus* spp.)	Wheat (*Triticum* spp.)	Rice (*Oryza sativa*)	Corn (*Zea mays*)	References
Protein content	12%–15%	14%	13%–15%	12%–15%	7%–8%	9%–10%	Appiani et al. (2021)
Amino acid profile	Complete; high in lysine	Complete; high in lysine	Complete; high in lysine	Limited	Limited lysine	Limited lysine	Sá and House (2024)
Carbohydrate content	70%–80%	60%	65%–70%	70%–75%	80%–85%	74%–78%	Christa and Soral-Śmietana (2008)
Dietary fiber	~10%	7%	7%–8%	2%–3%	0.5%	7%	Guan et al. (2021)
Fat content	2%–3%	6%	5%–9%	1.5–2.5%	1%	4%–5%	Szakály and Kiss (2023)
Minerals	High in magnesium, iron and manganese	High in magnesium and phosphorus	High in iron and calcium	Moderate iron and zinc	Low iron and moderate magnesium	Low iron and moderate magnesium	Podolska et al. (2021)
Vitamins	B vitamins and vitamin E	B vitamins and vitamin E	B vitamins	B vitamins	B vitamins	B vitamins	Bojňanská et al. (2022)
Gluten content	Gluten-free	Gluten-free	Gluten-free	Not gluten-free	Gluten-free	Gluten-free	Mir et al. (2019)
Glycemic index	Low	Low	Low	Moderate	High	Moderate	Bączek et al. (2020)
Digestibility	High	High	High	Moderate	Moderate	High	Zhu (2021)
Use in gluten-free products	Excellent	Excellent	Good	Not suitable	Suitable	Suitable	Khairuddin and Lasekan (2021)
Health benefits	Supports blood sugar control, high in antioxidants	Anti-inflammatory, rich in antioxidants	High in antioxidants, promotes heart health	May increase blood sugar levels	Potentially high in arsenic	High in antioxidants	Nandan et al. (2024)

advantages of buckwheat emerge. First, buckwheat has a more complete amino acid profile than most cereals, featuring higher levels of lysine, an essential amino acid that is typically low in grains like wheat, corn, and rice (Kowalski et al., 2022). Quinoa and amaranth are also rich in lysine, making them similar to buckwheat in terms of protein quality; however, buckwheat is particularly valued for its high digestibility and solubility, which enhances its protein bioaccessibility (Zhu, 2021). Second, buckwheat provides more dietary fiber than most cereals, especially refined grains such as white rice and wheat (Kumar et al., 2024). Its carbohydrate content is comparable to other grains, but its low GI distinguishes it as a more favorable option for individuals with insulin sensitivity or diabetes. While quinoa and amaranth also offer a low GI and are rich in fiber, buckwheat further supports blood sugar stability (Zhu, 2020). Third, buckwheat is naturally low in fat, with most of its lipid content being unsaturated. It includes beneficial fatty acids such as linoleic acid, contributing to cardiovascular health (Ruan et al., 2020). In contrast, quinoa and amaranth contain slightly more fat due to their higher oil content, making buckwheat's low-fat profile advantageous for individuals seeking a low-calorie diet (Szakály & Kiss, 2023). Moreover, compared to grains such as wheat and corn, buckwheat generally offers higher levels of essential minerals such as magnesium, iron, and zinc (Sinkovič et al., 2022). While quinoa and amaranth also provide an excellent mineral profile, particularly in terms of Ca and Mg, buckwheat's higher manganese and copper content makes it stand out in supporting enzymatic and antioxidant activities (Mystkowska et al., 2024). Finally, a key distinction that sets buckwheat apart from conventional grains is its natural gluten-free status, which makes it a perfect substitute for those suffering from celiac disease or gluten sensitivity (Sofi et al., 2023). While other pseudocereals, such as quinoa and amaranth, also lack gluten, buckwheat is particularly celebrated for its superior texture and flavor, making it an ideal option for various gluten-free culinary uses. This includes its use in baked goods, noodles, and porridge, where it enhances not only the taste but also the overall quality of these products (Khairuddin & Lasekan, 2021). The growing recognition of buckwheat's versatility in gluten-free cooking has further solidified its position as a favored ingredient among those seeking healthy and delicious alternatives.

In summary, buckwheat's balanced macronutrient composition, complete amino acid profile, rich fiber content, and high levels of essential micronutrients make it a nutritionally superior alternative to many traditional cereals. When compared to other pseudocereals such as quinoa and amaranth, buckwheat stands out with a similarly high nutrient density while offering distinct advantages. These include its suitability for gluten-free diets and its beneficial effects on blood sugar regulation, making it a valuable option for individuals with specific dietary needs or metabolic concerns (Nandan et al., 2024).

4.4.3 Functional Components of Buckwheat

Buckwheat is abundant in bioactive compounds, particularly phenolics and flavonoids, which significantly contribute to its strong antioxidant activity (Zou et al., 2021). These compounds are essential in safeguarding the body from oxidative stress and lowering the risk of chronic diseases such as cardiovascular disease, cancer, and diabetes (Chaudhary et al., 2023). Buckwheat's phenolic content includes a variety of compounds, such as phenolic acids (e.g., ferulic acid, p-coumaric acid, and gallic acid), which have strong antioxidant capabilities (Huda et al., 2021). These phenolics enhance buckwheat's radical-scavenging ability, helping counteract harmful free radicals that can harm cells and tissues. Buckwheat is especially recognized for its high flavonoid content, particularly rutin, a flavanol glycoside extensively researched for its health benefits (Janovská et al., 2021; Kreft et al., 2022). Rutin is a potent antioxidant with anti-inflammatory, anticarcinogenic, and antidiabetic effects. It strengthens capillaries, reduces bleeding risk, and supports cardiovascular health by preventing platelet aggregation and lowering blood pressure (Ullah et al., 2020). Along with rutin, buckwheat contains other flavonoids such as quercetin and vitexin, which further enhance its antioxidant capacity and offer protection against metabolic and neurodegenerative disorders (Sonawane et al., 2024). The antioxidant capacity of buckwheat is one of its key functional traits, with studies showing that it possesses significantly higher antioxidant activity than many other grains and pseudocereals (Henrion et al., 2020; Ňorbová et al., 2022; Estivi et al., 2022). Its ability to neutralize free radicals, suppress lipid peroxidation, and reduce oxidative DNA damage is largely attributed to the synergistic benefits of its phenolic compounds and flavonoids (Chaudhary et al., 2023). These bioactive compounds make buckwheat a powerful food for promoting health and preventing oxidative stress-related diseases.

In addition to its bioactive compounds, buckwheat contains a unique set of proteins, predominantly albumins and globulins, which provide high-quality nutrition and possess functional properties that can benefit health and food processing. Albumins are water-soluble proteins that account for about 20% of buckwheat's total protein content (Ninomiya et al., 2022). These proteins have high digestibility and a favorable amino acid composition, particularly rich in lysine, an essential amino acid that is frequently deficient in other plant proteins. Albumins also exhibit bioactive properties, such as antioxidant and antihypertensive effects, due to their ability to bind to polyphenols like rutin (Li et al., 2022). The majority of buckwheat proteins (around 40%–50%) are globulins, which are salt-soluble proteins (Constantino & Garcia-Rojas, 2022). Globulins have excellent emulsifying and gelling properties, making them useful in food applications such as plant-based protein products. Beyond their functional roles in food, buckwheat globulins have been shown to exhibit health-promoting effects, including the ability to modulate immune responses and improve metabolic health (Chen et al., 2023). Along with being a source of complete protein, buckwheat proteins, especially albumins and globulins, are known to exert bioactive effects that can assist in regulating blood pressure, reducing cholesterol levels, and enhancing glucose metabolism (Zou et al., 2021). Several research indicate that buckwheat protein extracts may aid in lowering the risk of obesity and diabetes by improving insulin sensitivity and promoting lipid metabolism, making buckwheat an excellent source of both high-quality nutrition and functional food ingredients for disease prevention (Liu et al., 2021; Chiang et al., 2022; Shen et al., 2024). Tomotake et al. (2002) explored the functional characteristics of buckwheat proteins, revealing that they possess excellent emulsification and foaming capabilities, rendering them appropriate for a range of food uses. The study suggests that buckwheat proteins can improve the texture and stability of food products without gluten.

Buckwheat is additionally a notable source of fiber in the diet, comprising both soluble and insoluble fibers, as well as resistant starch, which provides various digestive and metabolic benefits (Chen et al., 2024). With about 10% dietary fiber, buckwheat offers a balanced mix of both fiber types (Zhu, 2020). Insoluble fiber aids digestive health by adding bulk to stools, encouraging regular bowel movements, and helping to prevent constipation. Soluble fiber assists in controlling blood glucose levels and lowering cholesterol (Barber et al., 2020). This fiber combination supports heart health by reducing low-density lipoprotein (LDL) cholesterol and improving cardiovascular function. Buckwheat's resistant starch acts like fiber, resisting digestion within the small intestinal tract and serving as a prebiotic in the colon, where it nourishes beneficial gut bacteria (Chen et al., 2024). The fermentation of resistant starch generates short-chain fatty acids, such as butyrate, which are linked to improved gut health, reduced inflammation, and lower colon cancer risks. Additionally, resistant starch enhances insulin sensitivity and blood sugar control, making buckwheat beneficial for individuals with, or prone to, type 2 diabetes (Chiang et al., 2022). The combined effects of dietary fiber and resistant starch in buckwheat support multiple aspects of metabolic health. These components aid in slowing the digestion and absorption of carbohydrates, which helps prevent rapid increases in blood sugar and enhances glycemic control. In addition, they foster a sense of fullness and satiety, which can play a significant role in supporting effective weight management by helping to reduce overall calorie intake and curb unnecessary snacking. Regular consumption of fiber and resistant starch from buckwheat has been linked to a lower risk of heart disease, obesity, and digestive disorders (Bojarczuk et al., 2022). In conclusion, buckwheat is rich in functional components such as phenolic compounds, flavonoids, and unique proteins, along with its high fiber and resistant starch content. These bioactive constituents provide numerous health benefits, including antioxidant activity, cardiovascular protection, digestive health support, and potential antidiabetic effects.

Table 4.4 presents a comprehensive overview of the key functional components found in buckwheat, along with their associated health benefits. It highlights various bioactive constituents, including phenolic compounds, flavonoids, dietary fiber, resistant starch, proteins, vitamins, and minerals. Every component contributes importantly to enhancing health and wellness, contributing to buckwheat's reputation as a nutritious pseudocereal.

4.4.4 Health Benefits of Buckwheat

Antioxidant Properties and Potential for Reducing Oxidative Stress: Buckwheat is a rich source of antioxidants, which are vital for neutralizing free radicals and minimizing oxidative stress in the body (Yang et al., 2020). Its high phenolic compound content, particularly flavonoids like rutin and quercetin, significantly enhances its antioxidant properties (García-Sánchez et al., 2020). Among these, rutin stands out for its potent ability to neutralize free radicals and prevent oxidative harm to lipids and proteins. Moreover, rutin has been found to boost the body's natural antioxidant defenses by increasing the activity of enzymes (Ali et al., 2023).

TABLE 4.4
Key Functional Components of Buckwheat and Their Health Benefits

Functional Component	Sources in Buckwheat	Description	Health Benefits	References
Phenolic compounds	Seeds and hulls	Contains phenolic acids such as ferulic acid, p-coumaric acid, and gallic acid.	Exhibits strong antioxidant properties; reduces oxidative damage and inflammatory responses may lower the risk of chronic diseases such as cardiovascular disorders and cancer.	García-Sánchez et al. (2020)
Flavonoids	Rutin (present in leaves and seeds)	Rich in rutin, quercetin, and vitexin.	Associated with vascular health; enhances blood circulation; reduces the risk of heart-related diseases; having potential anticancer properties owing to the ability to inhibit tumor growth and prevent metastasis.	Ullah et al. (2020)
Dietary fiber	Whole buckwheat	Contains both soluble and insoluble fibers.	Promoting digestive health by increasing stool bulk; regulating blood glucose levels; supporting weight control by increasing satiety; and beneficial for gut microbiota.	Barber et al. (2020)
Resistant starch	Whole buckwheat	Acts like fiber in the digestive system.	Contributes to lower glycemic responses; promotes feelings of fullness; acts as a prebiotic, and supports gut health by promoting the development of beneficial microorganisms.	Chen et al. (2024)
Proteins (albumins)	Seeds	Water-soluble proteins; about 20% of total protein content.	Supplies essential amino acids important for muscle recovery and growth; enhances the nutritional profile of meals, particularly for those with limited protein sources; and aids tissue repair and immune function.	Li et al. (2022)
Proteins (globulins)	Seeds	Salt-soluble proteins; around 40%–50% of total protein content.	Excellent emulsifying and gelling properties; modulates immune responses and improves metabolic health.	Chen et al. (2023)
Vitamins	Whole buckwheat	Rich in B vitamins (thiamine, riboflavin, and niacin).	Supports energy metabolism and neurological function; promotes overall health and wellness.	Tardy et al. (2020)
Minerals	Whole buckwheat	Contain essential minerals such as magnesium, zinc, and iron.	Vital for metabolic processes, immune function, and oxygen transport in the blood.	Alghamdi et al. (2023)
Antioxidant capacity	Seeds and hulls	Synergistic effects of phenolics and flavonoids.	Neutralizes free radicals, prevents lipid peroxidation, and safeguards DNA from oxidative damage.	Chaudhary et al. (2023)

Cardiovascular Health Benefits: Buckwheat has been linked to enhanced heart health, primarily due to its cholesterol-lowering effects and antihypertensive properties (Zou et al., 2021). Its dietary fiber, especially soluble fiber, is instrumental in reducing total and LDL (bad) cholesterol. Soluble fiber binds to cholesterol in the digestive system, inhibiting its absorption and facilitating its removal from the body (Barber et al., 2020). Oxidized LDL is a key factor in the development of atherosclerosis (hardening and narrowing of the arteries). By preventing this oxidation, buckwheat helps protect against the progression of atherosclerosis and supports overall heart health. Numerous studies have investigated the health-promoting properties of buckwheat, with a strong emphasis on its cardiovascular benefits. Sonawane et al. (2024) conducted a systematic review that highlighted buckwheat's ability to lower cholesterol levels, improve lipid profiles, and reduce blood pressure. The study pointed to the presence of rutin as a key factor in these health benefits.

Buckwheat also exhibits natural antihypertensive (blood pressure-lowering) properties, largely attributed to its high magnesium content and bioactive proteins like albumins and globulins (Sofi et al., 2023). Magnesium plays a vital role in relaxing blood vessels, improving circulation, and maintaining healthy blood pressure levels (Alghamdi et al., 2023). Studies have shown that regular consumption of magnesium-rich foods like buckwheat can help reduce the risk of hypertension (high blood pressure) (Ullah et al., 2020). Additionally, certain proteins in buckwheat have been found to inhibit the angiotensin-converting enzyme, which is involved in blood pressure regulation (Li et al., 2023). By inhibiting this enzyme, buckwheat can help lower blood pressure, reducing strain on the cardiovascular system and lowering the risk of heart attacks and strokes.

Anti-inflammatory and Anticancer Properties: Buckwheat contains several compounds that exhibit strong anti-inflammatory and anticancer properties, contributing to its role in disease prevention and health promotion (Yilmaz et al., 2020). The phenolic compounds in buckwheat, particularly flavonoids such as quercetin and rutin, have potent anti-inflammatory effects (García-Sánchez et al., 2020). These compounds help inhibit the production of pro-inflammatory cytokines and reduce the activation of inflammatory pathways in the body (Yu et al., 2022). Regular consumption of buckwheat has been linked to lower levels of inflammatory markers, such as C-reactive protein, which is associated with diminished risks of chronic inflammatory conditions (Zou et al., 2021). van den Berg et al. (2008) examined the antioxidant and anti-inflammatory features of buckwheat honey, reporting that its consumption is linked to reduced markers of inflammation in the body, which could have implications for chronic disease management. The antioxidant and anti-inflammatory qualities of buckwheat also contribute to its potential role in cancer prevention. Buckwheat's phenolic compounds, particularly rutin and quercetin, have been shown to induce apoptosis (programmed cell death) in cancer cells (Ullah et al., 2020). Furthermore, buckwheat contains D-chiro-inositol, a compound that has been linked to the suppression of certain cancer cell growth, particularly in hormone-dependent cancers such as breast and prostate cancer (Lahuta et al., 2023). Buckwheat's fiber content may also contribute to its anticancer properties by promoting a healthy gut microbiome, which is increasingly recognized as playing a role in reducing the risk of colorectal cancer (Barber et al., 2020). The fermentation of fiber in the colon produces short-chain fatty acids, such as butyrate, which have been shown to protect against the development of colon cancer (Vinelli et al., 2022).

Role in Managing Diabetes and Blood Sugar Regulation: Another important aspect is buckwheat's role in glycemic control. Buckwheat is considered a beneficial food for managing diabetes and regulating blood sugar levels due to its low GI, high fiber content, and the presence of unique bioactive compounds (Bączek et al., 2020). Buckwheat has a relatively low GI compared to many other grains and pseudocereals, meaning that it causes a slower and more gradual rise in blood sugar levels after consumption (Chiang et al., 2022). This makes buckwheat an ideal food for people with diabetes or those at risk of developing the condition. Its low GI helps prevent spikes in blood glucose levels and promotes better glycemic control over time (Vlachos et al., 2020). Chiang et al. (2022) demonstrated that incorporating buckwheat into the diet could significantly improve insulin sensitivity and glycemic responses in diabetic patients. The dietary fiber in buckwheat, particularly soluble fiber, plays a key role in slowing the digestion and absorption of carbohydrates (Barber et al., 2020). This helps moderate blood sugar levels and improve insulin sensitivity, making buckwheat a valuable addition to a diabetes-friendly diet. The fiber content also promotes satiety, which can aid in weight management – an important factor in preventing and managing type 2 diabetes (Akhlaghi, 2022). One of the unique bioactive compounds in buckwheat is D-chiro-inositol, which has been shown to improve glucose metabolism and insulin sensitivity (Duliński et al., 2020). This compound mimics insulin's effects by enhancing the uptake of glucose into cells, helping to lower blood sugar levels. Studies suggest that D-chiro-inositol may be particularly beneficial for people with insulin resistance, a hallmark of type 2 diabetes (Diamanti-Kandarakis et al., 2023; Nordio et al., 2023).

In summary, buckwheat offers a wide range of health benefits, particularly due to its antioxidant, anti-inflammatory, and cardioprotective properties. Its role in regulating blood sugar and supporting overall metabolic health makes it an excellent dietary choice for managing diabetes and preventing related complications (Figure 4.10).

4.4.5 Processing and Food Applications of Buckwheat

4.4.5.1 Traditional Uses

Buckwheat has been cultivated and consumed for centuries in various parts of the world, especially in Asia and Europe. Its unique nutritional profile and versatility have made it a staple food, often used in traditional dishes (Suvorova & Zhou, 2018). Buckwheat flour is one of the oldest and most widespread uses of buckwheat, prized for its dense, earthy flavor. In Eastern European and Russian cuisines, buckwheat flour is used in recipes like blinis, which are small, savory pancakes (Pirzadah & Rehman, 2021). In France, particularly in the Brittany region, buckwheat flour is the primary ingredient in galettes, a type of savory crepe often filled with ham, eggs, and cheese (Ikeda & Ikeda, 2016). Despite being gluten-free, buckwheat flour has excellent binding properties, making it a useful alternative in various baked goods, from muffins to bread. Because of its strong flavor, it is often mixed with other flours in modern recipes to balance the taste (Giménez-Bastida et al., 2015).

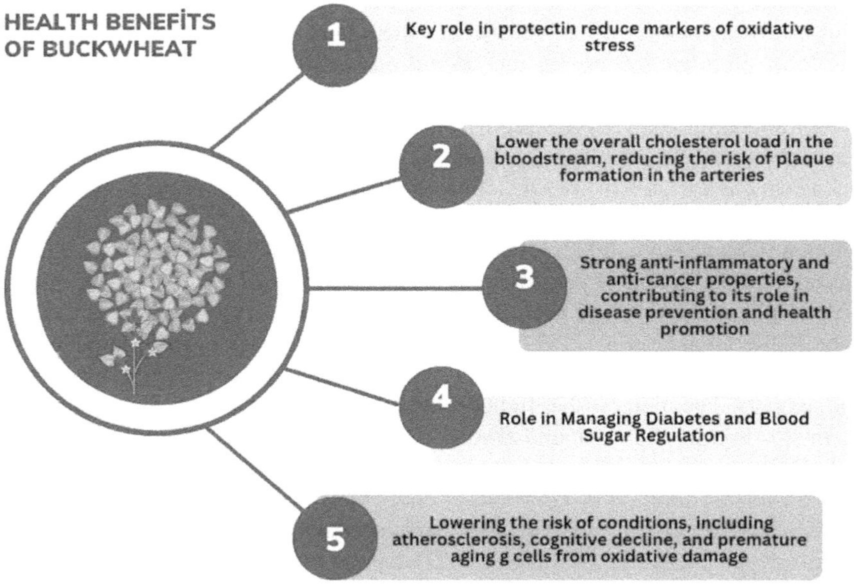

FIGURE 4.10 Health benefits of buckwheat. **Source**: Schematized by the authors. References: Ullah et al. (2020); Vlachos et al. (2020); Guan et al. (2021); Essa et al. (2023).

Soba, made from buckwheat flour, is an essential part of Japanese cuisine. These thin noodles can be served hot in a broth (as in soba soup) or chilled with a dipping sauce (zaru soba), and are typically consumed year-round (Jacob, 2021). The use of buckwheat flour gives soba noodles a slightly nutty flavor and a higher protein and fiber content compared to wheat-based noodles (Matías et al., 2024). Soba is also appreciated for its health benefits, including its low GI and ability to help regulate blood sugar levels (Baniwal et al., 2021). In Japanese culture, eating soba noodles on New Year's Eve is a tradition believed to bring longevity and good fortune (Komatsudaira, 2023).

Kasha refers to toasted buckwheat groats, a staple in Eastern European and Russian cuisines. The groats are boiled, toasted, and served as a porridge or side dish. Buckwheat kasha has a hearty and nutty flavor and is often mixed with sautéed onions, mushrooms, or even scrambled eggs to enhance its taste and nutritional value (Kreft et al., 2023). In Jewish cuisine, kasha varnishkes is a well-known dish, where the buckwheat groats are mixed with pasta (often bowtie-shaped) and flavored with onions and garlic (Stein & Isaacs, 2023). The toasting of buckwheat gives it a distinctive flavor, making it an essential dish during the cold seasons due to its warming and filling qualities.

4.4.5.2 Emerging Applications in Gluten-Free Products

In recent years, the demand for gluten-free products has surged, positioning buckwheat as a popular alternative due to its naturally gluten-free nature. Its increasing popularity is driven by its unique nutritional profile and functional properties in food applications. As awareness of gluten intolerance and celiac disease rises, buckwheat has become a key ingredient in gluten-free baking (Sofi et al., 2023). Unlike wheat, rye, or barley, buckwheat is free from gluten, making it suitable for individuals with celiac disease or non-celiac gluten sensitivity. Buckwheat flour is commonly used to make gluten-free bread, pancakes, and cookies, and its robust flavor complements other gluten-free flours such as rice or almond flour, enhancing both texture and taste in baked goods. However, since buckwheat lacks gluten, bakers often incorporate binding agents such as xanthan gum or eggs to achieve the necessary elasticity and structure in their products (Šmídová & Rysová, 2022).

In addition to traditional soba noodles, buckwheat is utilized to make other types of gluten-free pasta. Pasta made from buckwheat flour is dense and has a chewy texture, making it a hearty alternative to conventional wheat pasta (Dey & Kumar, 2024). These products are ideal for individuals seeking gluten-free options that are also nutrient-dense. Buckwheat pasta has a low GI, making it a suitable option for people with diabetes or those looking to control their blood sugar levels (Cui et al., 2024).

4.4.5.3 Fermented Products, Sprouts, and Functional Food Ingredients

As consumer preferences shift toward healthier and functional foods, buckwheat has found its place in a range of innovative products. Fermentation is a traditional process that enhances the nutritional quality of grains and seeds by increasing their digestibility and the bioaccessibility of nutrients (Adebo et al., 2022). In the case of buckwheat, fermentation improves its flavor profile and enhances its health benefits. Fermented buckwheat is used to make sourdough bread and beverages, such as buckwheat beer, and probiotic-rich foods (Zhang et al., 2023b). The fermentation process serves a dual purpose: it reduces anti-nutrients like phytic acid, which can hinder mineral absorption, while simultaneously enhancing the concentration of beneficial compounds, including phenolic acids and probiotics (Manzoor et al., 2021). Buckwheat-based fermented beverages have shown potential as functional drinks, with added health benefits such as improved digestion and gut health (Ziarno & Cichońska, 2021).

Sprouted buckwheat has gained popularity due to its increased nutritional value. The sprouting process activates enzymes in the seed, making the nutrients more bioavailable (Ikram et al., 2021). Sprouted buckwheat is rich in vitamins, especially B vitamins, antioxidants, and minerals like magnesium and manganese. Buckwheat sprouts are also high in rutin, a powerful flavonoid with anti-inflammatory and antioxidant properties (Sowdhanya et al., 2023). Sprouted buckwheat can be eaten raw in salads, added to smoothies, or dried and used in baking. This makes it a popular ingredient among health-conscious consumers and raw food enthusiasts who seek minimally processed and nutrient-dense foods.

In the realm of functional foods, buckwheat has gained attention as a health-promoting ingredient. Its high fiber content, along with its rich supply of antioxidants, makes it a sought-after ingredient for products targeting heart health, digestive health, and blood sugar management (Ahmed et al., 2014). Buckwheat protein concentrates and isolates are also becoming popular in protein bars, shakes, and meal replacements. These proteins are considered high quality, containing all nine essential amino acids, making buckwheat a valuable plant-based protein source for vegetarians and vegans (Jin et al., 2020). Buckwheat flour and groats are also being used in granola bars, cereals, and snack foods marketed as nutrient-dense and gluten-free options for health-conscious consumers (Otles & Nakilcioglu-Tas, 2022).

In summary, buckwheat's food applications are wide-ranging, from its traditional uses in soba noodles and kasha to its emerging role in gluten-free and functional foods. Its versatility, nutritional profile, and functional properties make it a valuable ingredient across both traditional and modern food systems. As demand for gluten-free, plant-based, and health-promoting foods continues to grow, buckwheat is well-positioned to become a staple in many diets worldwide.

4.4.5.4 Non-Food Applications of Buckwheat

Buckwheat is increasingly being used in cosmetic and skincare products due to its rich antioxidant content (Gonçalves et al., 2024). Its high levels of phenolic compounds, particularly rutin and quercetin, make buckwheat extracts ideal for creams and lotions aimed at enhancing skin health. These antioxidants shield the skin from oxidative stress resulting from environmental factors such as UV radiation and pollution, aiding in the prevention of skin damage and premature aging (Chen et al., 2021). Buckwheat extracts are highly valued for their antiaging properties, as they enhance skin elasticity and diminish the visibility of fine lines and wrinkles (Zhu, 2021). Moreover, the anti-inflammatory properties of buckwheat-derived flavonoids make them effective in soothing irritated skin, making it a popular choice for sensitive or inflamed skin formulations (Mondal, 2023).

Buckwheat's bioactive compounds have numerous pharmaceutical applications, particularly due to their health-promoting properties (Sofi et al., 2023). Rutin, a flavonoid that is notably present in buckwheat, is commonly used in supplements aimed at promoting vascular health. It strengthens blood vessels, reduces capillary fragility, and improves circulation, making it beneficial for individuals with conditions like varicose veins or chronic venous insufficiency (Ullah et al., 2020). Rutin is also known for its antioxidant and anti-inflammatory properties, contributing to its widespread use in pharmaceutical formulations (Choi et al., 2021). Furthermore, some of the phenolic compounds found in buckwheat have been investigated for their potential in anticancer therapies, as these compounds have been demonstrated to trigger apoptosis in cancer cells and suppress tumor growth (García-Sánchez et al., 2020). Their anti-inflammatory effects also make them promising candidates for drugs aimed at managing chronic inflammatory conditions, such as arthritis (Placha & Jampilek, 2021). In the pharmaceutical field, Ganeshpurkar & Saluja (2017) examined the health benefits of rutin derived from buckwheat, emphasizing its potential as a therapeutic agent for vascular health.

Buckwheat has nontraditional uses in the textile industry, particularly for eco-friendly and sustainable products. Buckwheat hulls, the outer shells of buckwheat seeds, are utilized as filling material in pillows, mattresses, and cushions (Nawirska-Olszańska et al., 2021). These hulls are hypoallergenic, breathable, and eco-friendly, making them an excellent alternative to synthetic fillings. Buckwheat-filled pillows are highly regarded for their ability to conform to the shape of the head and neck, offering orthopedic support and contributing to improved sleep quality (Lan, 2023). Zhang et al. (2020) discussed the environmental benefits of using buckwheat hulls in textiles, noting their hypoallergenic properties and sustainability as a natural filling material in bedding products.

Buckwheat residues are increasingly incorporated into animal feed, offering nutritional benefits to livestock. The leftover materials from buckwheat, such as the hulls and stems, can be used as animal feed, as they are abundant in fiber, protein, and essential minerals (Notup et al., 2024). These residues contribute to the overall health of livestock, improving digestion and nutrient absorption (Ge et al., 2021). Additionally, incorporating buckwheat into animal feed has been shown to enhance the immune system of livestock and improve growth performance, especially in poultry and ruminants (Sonawane et al., 2024). The presence of bioactive compounds like rutin may also have protective effects against oxidative stress in animals (Ullah et al., 2020).

Buckwheat's role extends to sustainable agricultural practices and bioenergy production. It is often used as a cover crop because of its capacity to improve soil health (Lalewicz et al., 2024). Buckwheat grows quickly and helps suppress weeds by outcompeting them for sunlight and nutrients, thus preventing soil erosion (Nath et al., 2024). It also adds organic matter back into the soil, enriching its nutrient profile and promoting better conditions for subsequent crops. Furthermore, buckwheat plants generate significant biomass, which has potential use in biofuel production. As an energy crop, buckwheat can be processed to produce bioethanol, contributing to renewable energy sources. Its rapid growth and adaptability to various environments make it an excellent candidate for bioenergy uses (Kulokas et al., 2021).

4.5 CONCLUSION

The three most well-known types of pseudocereals are amaranth, quinoa, and buckwheat. These grains are notable today for their high nutritional content, health benefits, and resilience to drought, poor soil, and adverse climate conditions. In addition to their use in food and nutrition, both the pseudocereals themselves and their by-products are also utilized in non-food industries.

Quinoa, a captivating pseudocereal consumed worldwide, is garnering increasing research interest for its unique properties, including those of its plant and by-products. With seeds resilient to various environmental conditions, quinoa continues to be a focus of breeding studies. Its bioactive compounds, great protein, mineral content, and fiber offer notable health benefits, making quinoa a valuable ingredient in the development of innovative food products, especially in gluten-free and diabetic-friendly foods.

Amaranth holds essential potential for use in both the food industry and medicine as a natural substance. With its great nutritional value and notable health features, amaranth seeds could be utilized to enrich foods with enhanced functional properties. To harness its pharmacological properties, which have been recognized in traditional medicine, it is essential to produce medical formulations that establish effective dosing and promote informed use for disease treatment. Given its status as the grain of the future, amaranth's ability to withstand various stress conditions makes it a promising candidate to produce high value-added products. This potential should be fully searched to support sustainability and enhance food security.

Buckwheat is an incredibly adaptable crop that provides a broad range of nutritional, functional, and health-promoting benefits. Its nutrient-dense profile, abundant in high-quality proteins, complex carbohydrates, vitamins, dietary fiber, and minerals, renders it a beneficial addition to a variety of diets, particularly for individuals seeking gluten-free alternatives. The unique functional components of buckwheat, such as phenolic compounds, flavonoids, and antioxidants, amplify its role as a functional food, offering health benefits through mechanisms like antioxidant activity, cardiovascular support, and anti-inflammatory effects. Its incorporation into cosmetics takes advantage of its antioxidant and anti-inflammatory benefits, while the potential of buckwheat-derived compounds, such as rutin, highlights its application in pharmaceuticals for supporting vascular health and anti-inflammatory treatments. Furthermore, buckwheat hulls can be utilized as eco-friendly materials in textiles, offering hypoallergenic filling options for pillows and mattresses. In agriculture, buckwheat contributes to soil health and biodiversity, promoting sustainable farming practices and serving as an effective cover crop.

The health benefits of pseudocereals are thoroughly researched, demonstrating their effectiveness in reducing oxidative stress, lowering cholesterol levels, managing blood sugar, and potentially providing anticancer properties. These attributes position the pseudocereals as a valuable food source, aligning with modern dietary trends that emphasize whole foods and natural ingredients. Pseudocereals are commonly included in the diets of individuals with allergies to traditional cereal products. Beyond their food applications, pseudocereals play a significant role in sustainable practices through various non-food uses. In summary, pseudocereals represent a holistic solution to the challenges faced by modern diets and agricultural systems. Their integration into our food supply emphasizes the need to renew focus on this ancient grain as a cornerstone of healthful eating and sustainable practices. By harnessing their full potential, we can promote a healthier population and a more sustainable planet.

In today's context, where issues such as climate change, food security, and accessibility are at the forefront, pseudocereals are emerging as a critical component of sustainable nutrition and a potential solution to future food crises. Recognized as functional foods, these grains hold significant potential; however, further investigations are needed to expand their practices in the global market and across various industrial sectors.

ACKNOWLEDGMENT

Figures were made using the Canva program. All authors have read and agreed to the published version of this book chapter.

REFERENCES

Abbasi, D., Rouzbehan, Y., & Rezaei, J. (2012). Effect of harvest date and nitrogen fertilization rate on the nutritive value of amaranth forage (*Amaranthus hypochondriacus*). *Animal Feed Science and Technology*, 171, 6–13.

Abdelshafy, A. M., El-Naggar, E. A., & Kenawi, M. N. (2023). Moringa leaves for improving the health benefits of quinoa fermented by probiotics. *Food Bioengineering*, 1, 264–275. https://doi.org/10.1002/fbe2.12035

Abugoch James, L. E. (2009). Chapter 1 Quinoa (Chenopodium quinoa Willd.): Composition, chemistry, nutritional, and functional properties. *Advances in Food and Nutrition Research*, 58, 1–31. https://doi.org/10.1016/S1043-4526(09)58001-1

Adebo, J. A., Njobeh, P. B., Gbashi, S., Oyedeji, A. B., Ogundele, O. M., Oyeyinka, S. A., & Adebo, O. A. (2022). Fermentation of cereals and legumes: Impact on nutritional constituents and nutrient bioavailability. *Fermentation*, 8(2), 63. https://doi.org/10.3390/fermentation8020063

Aderibigbe, O. R., Ezekiel, O. O., Owolade, S. O., Korese, J. K., Sturm, B., & Hensel, O. (2022). Exploring the potentials of underutilized grain amaranth (*Amaranthus* spp.) along the value chain for food and nutrition security: A review. *Critical Reviews in Food Science and Nutrition*, 62, 656–669.

Aguiar, E. V., Santos, F. G., Krupa-Kozak, U., & Capriles, V. D. (2021). Nutritional facts regarding commercially available gluten-free bread worldwide: Recent advances and future challenges. *Critical Reviews in Food Science and Nutrition*, 63(5), 693–705. https://doi.org/10.1080/10408398.2021.1952403

Ahmed, A., Khalid, N., Ahmad, A., Abbasi, N. A., Latif, M. S. Z., & Randhawa, M. A. (2014). Phytochemicals and biofunctional properties of buckwheat: A review. *The Journal of Agricultural Science*, 152(3), 349–369. https://doi.org/10.1017/S0021859613000166

Akande, O. A., Nakimbugwe, D., & Mukisa, I. M. (2017). Optimization of extrusion conditions for the production of instant grain amaranth-based porridge flour. *Food Science and Nutrition*, 5, 1205–1214.

Akhlaghi, M. (2022). The role of dietary fibers in regulating appetite: An overview of mechanisms and weight consequences. *Critical Reviews in Food Science and Nutrition*, 64(10), 3139–3150. https://doi.org/10.1080/10408398.2022.2130160

Alegbejo, J. O. (2012). Nutritional value and utilization of amaranthus (*Amaranthus* spp.)-a review. *Bayero Journal of Pure and Applied Sciences*, 6(1), 136–143. https://doi.org/10.4314/bajopas.v6i1.27

Alemayehu, F. R., Bendevis, M. A., & Jacobsen, S. E. (2015). The potential for utilizing the grain crop amaranth (*Amarathus* spp.) in East Africa as an alternative crop to support food security and climate change mitigation. *Journal of Agronomy and Crop Science*, 201, 321–329.

Alghamdi, M., Gutierrez, J., & Komarnytsky, S. (2023). Essential minerals and metabolic adaptation of immune cells. *Nutrients*, 15(1), 123. https://doi.org/10.3390/nu15010123

Ali, Y. A., Ahmed, O. M., Soliman, H. A., Abdel-Gabbar, M., Al-Dossari, M., El-Gawaad, N. S. A., El-Nahass, E.-S., & Ahmed, N. A. (2023). Rutin and hesperidin alleviate paclitaxel-induced nephrocardiotoxicity in Wistar rats via suppressing oxidative stress and enhancing the antioxidant defense mechanisms. *Evidence-Based Complementary and Alternative Medicine*, 2023, Article 5068304. https://doi.org/10.1155/2023/5068304

Alp, O. (2022). Using roasted amaranth flour and stevia in cake making. Harran University Graduate School of Natural and Applied Sciences Department of Food Engineering. *MSc Thesis*.

Alsaiqali, A. I. A. (2018). Effects of pseudocereals on the quality of gluten-free dough and bread (*Master's thesis*). Çukurova University, Institute of Natural and Applied Sciences, Department of Food Engineering, Adana, Turkey.

Alsuhaibani, A. M., Alkuraieef, A. N., Aljobair, M. O., & Alshawi, A. H. (2022). Technological, sensory, and hypoglycemic effects of quinoa flour incorporation into biscuits. *Journal of Food Quality*, 6484953. https://doi.org/10.1155/2022/6484953

Alvarez-Jubete, L., Arendt, E. K., & Gallagher, E. (2009). Nutritive value and chemical composition of pseudocereals as glutenfree ingredients. *International Journal of Food Science and Nutrition*, 60, 240–257.

Alvarez-Jubete, L., Auty, M., Arendt, E. K., & Gallagher, E. (2010a). Baking properties and microstructure of pseudocereal flours in gluten-free bread formulations. *European Food Research and Technology*, 230, 437–445.

Alvarez-Jubete, L., Arendt, E. K., & Gallagher, E. (2010b). Nutritive value of pseudocereals and their increasing use as functional gluten-free ingredients. *Trends in Food Science & Technology*, 21, 106–113. https://doi.org/10.1016/j.tifs.2009.10.014

Amare, E., Mouquet-Rivier, C., Servent, A., Morel, G., Adish, A., & Haki, G. D. (2015). Protein quality of amaranth grains cultivated in Ethiopia as affected by popping and fermentation. *Food and Nutrition Sciences*, 6, 38–48.

Angeli, V., Miguel Silva, P., Crispim Massuela, D., Khan, M. W., Hamar, A., Khajehei, F., Graeff-Hönninger, S., & Piatti, C. (2020). Quinoa (Chenopodium quinoa Willd.): An overview of the potentials of the "golden grain" and socio-economic and environmental aspects of its cultivation and marketization. *Foods*, 9(2), 216. https://doi.org/10.3390/foods9020216

Appiani, M., Rabitti, N. S., Proserpio, C., Pagliarini, E., & Laureati, M. (2021). Tartary buckwheat: A new plant-based ingredient to enrich corn-based gluten-free formulations. *Foods*, 10(11), 2613. https://doi.org/10.3390/foods10112613

Araujo-Farro, P. C. (2008). Development and optimization of biodegradables films made of products derivated from "Royal" variety quinoa (Chenopodium quinoa Willdenow) seeds. *PhD Thesis*. Unicamp. Brazil, p. 303.

Araujo-Farro, P. C., Podadera, G., Sobral, P. J. A., & Menegalli, F. C. (2010). Development of films based on quinoa (Chenopodium quinoa Willdenow) starch. *Carbohydrate Polymers*, 81(4), 839–848. https://doi.org/10.1016/j.carbpol.2010.03.051

Arduini, I., & Mariotti, M. (2018). Buckwheat cultivation in Mediterranean climates: Challenges and future outlook. In Frederick George (Ed.), *Buckwheat: Composition, Production and Uses* (pp. 43–92). Nova Science Publishers Inc.

Aygün, G. (2021). Determination of the effects of quinoa and amarant flour addition on the quality traits of Turkish Mantı. Tekirdağ Namık Kemal University, Graduate School of Natural and Applied Sciences, Department of Food Engineering.

Ayseli, M. T., Yilmaz, M. T., Cebi, N., Sagdic, O., Ozmen, D., & Capanoglu, E. (2020). Physicochemical, rheological, molecular, thermal and sensory evaluation of newly developed complementary infant (6–24 months old) foods prepared with quinoa (Chenopodium quinoa Willd.) flour. *Food Chemistry*, 315, 126208. https://doi.org/10.1016/J.FOODCHEM.2020.126208

Bachar, S. C., Bachar, R., Jannat, K., Jahan, R., & Rahmatullah, M. (2020). Hepatoprotective natural products. *Annual Reports in Medicinal Chemistry*, 55, 207–249.

Bączek, N., Jarmułowicz, A., Wronkowska, M., & Haros, C. M. (2020). Assessment of the glycemic index, content of bioactive compounds, and their in vitro bioaccessibility in oat-buckwheat breads. *Food Chemistry*, 330, 127199. https://doi.org/10.1016/j.foodchem.2020.127199

Baker, M. G., Hudson, H., Flores, L., Bhaduri, S., Ghatak, R., & Navder, K. P. (2013). Physical, textural and sensory properties of gluten-free muffins prepared using quinoa flour as a replacement for rice flour. *Journal of the Academy of Nutrition and Dietetics*, 113(9), A60. https://doi.org/10.1016/J.JAND.2013.06.210

Balakrishnan, G., & Schneider, R. G. (2022). The role of Amaranth, quinoa, and millets for the development of healthy, sustainable food products-a concise review, *Foods*, 11(16), 2442.

Baniwal, P., Mehra, R., Kumar, N., Sharma, S., & Kumar, S. (2021). Cereals: Functional constituents and their health benefits. *The Pharma Innovation Journal*, 10(2), 343–349. https://doi.org/10.22271/tpi.2021.v10.i2e.5681

Baraniak, J., & Kania-Dobrowolska, M. (2022). The dual nature of amaranth-functional food and potential medicine. *Foods*, 11(4), 618. https://doi.org/10.3390/foods11040618

Barber, T. M., Kabisch, S., Pfeiffer, A. F. H., & Weickert, M. O. (2020). The health benefits of dietary fibre. *Nutrients*, 12(10), 3209. https://doi.org/10.3390/nu12103209

Bazile, D. (2023). Global trends in the worldwide expansion of quinoa cultivation. *Biology and Life Sciences Forum*, 25(1), 13. https://doi.org/10.3390/blsf2023025013

Bekkering, C. S., & Tian, L. (2019). Thinking outside of the cereal box: Breeding underutilized (pseudo) cereals for improved human nutrition. *Frontiers in Genetics*, 10, 1289.

Bezerra, K. G., Durval, I. J., Silva, I. A., & Fabiola, C. G. (2020). Emulsifying capacity of biosurfactants from Chenopodium quinoa and Pseudomonas aeruginosa UCP 0992 with focus of application in the cosmetic industry. *Chemical Engineering Transactions*, 79, 211–216.

Bock, P., Daelemans, L., Selis, L., Raes, K., Vermeir, P., Eeckhout, M., & Bockstaele, F. V. (2021). Comparison of the chemical and technological characteristics of wholemeal flours obtained from amaranth (*Amaranthus* sp.), quinoa (*Chenopodium quinoa*) and buckwheat (*Fagopyrum* sp.) Seeds. *Foods*, 10(3), 651.

Bodroža-Solarov, M. I., Šimurina, O. D., Kojić, J. S., Krulj, J. A., Filipović, J. S., Cvetković, B. R., & Ilić, N. M. (2022). Utilization of *Amaranthus* Spp. grains in food. *Food and Feed Research*, 49(1), 37–52. https://doi.org/10.5937/ffr49-37163

Bojanic, A. (2011). Quinoa: An ancient crop to contribute to world food security. FAO Fiat Panis, Rome, Italy. Available online: https://www.fao.org/3/aq287e/aq287e.pdf (accessed on 26 October 2024).

Bojarczuk, A., Skąpska, S., Mousavi Khaneghah, A., & Marszałek, K. (2022). Health benefits of resistant starch: A review of the literature. *Journal of Functional Foods*, 93, 105094. https://doi.org/10.1016/j.jff.2022.105094

Bojňanská, T., Vollmannová, A., Lidiková, J., & Musilová, J. (2022). Use of common buckwheat in the production of baked and pasta products. In Viduranga Y. Waisundara (Ed.), *Advances in Food Production and Processing* (pp. 11–34). IntechOpen.

Brakez, M., Daoud, S., Harrouni, M. C., Tachbibi, N., & Brakez, Z. (2016). Nutritional Value of Chenopodium quinoa seeds obtained from an open field culture under saline conditions. In *Halophytes for Food Security in Dry Lands* (pp. 37–47). https://doi.org/10.1016/B978-0-12-801854-5.00003-0

Brand-Miller, J. C., Stockmann, K., Atkinson, F., Petocz, P., & Denyer, G. (2009). Glycemic index, postprandial glycemia, and the shape of the curve in healthy subjects: Analysis of a database of more than 1000 foods. *The American Journal of Clinical Nutrition*, 89(1), 97–105. https://doi.org/10.3945/AJCN.2008.26354

Breus, I. M. (1997). Productivity, chemical composition and fertilizing of amaranth growing for green matter yield. *Agricultural Chemistry*, 10, 52–74.

Brito, I. D., Chantelle, L., Magnani, M. & Cordeiro, A. M. (2022). Nutritional, therapeutic, and technological perspectives of quinoa (Chenopodium quinoa Willd.): A review. *Journal of Food Processing and Preservation*, 46, e16601. https://doi.org/10.1111/jfpp.16601

Buresova, I., Tokar, M., Marecek, J., Hrivna, L., Famera, O., & Sottnikova, V. (2017). The comparison of the effect of added amaranth, buckwheat, chickpea, corn, millet and quinoa flour on rice dough rheological characteristics, textural and sensory quality of bread. *Journal of Cereal Science*, 75, 158–164. https://doi.org/10.1016/j.jcs.2017.04.004

Burgos, V., & Armada, M. (2015). Characterization and nutritional value of precooked products of kiwicha grains (*Amaranthus caudatus*), *Food Science and Technology*, 35, 531–538.

Bvenura, C., & Kambizi, L. (2022). Future grain crops. In Rajeev Bhat (Ed.), *Future Foods* (pp. 81–105). Academic Press.

Cacak-Pietrzak, G., Dojczew, D., Haber, T., & Lewczuk, J. (1995). Wybrane wyroby cukiernicze z dodatkiem nasion amarantusa. *Żywność Technologia Jakość*, 2, 121.

Çalışkan Koç, G., & Pandiselvam, R. (2022). Evaluation of physicochemical, functional, and sensorial characteristics of gluten-free Turkish noodle "Erişte" formulated with oat and quinoa flours. *Journal of Food Quality*, 8622114, 7 pages. https://doi.org/10.1155/2022/8622114

Carciochi, R. A., Manrique, G. D., & Dimitrov K. (2014). Changes in phenolic composition and antioxidant activity during germination of quinoa seeds (Chenopodium quinoa Willd). *International Food Research Journal*, 21, 767–773.

Casalvara, R. F. A., Ferreira, B. M. R., Gonçalves, J. E., Yamaguchi, N. U., Bracht, A., Bracht, L., Comar, J. F., de Sá-Nakanishi, A. B., de Souza, C. G. M., Castoldi, R., Corrêa, R. C. G., & Peralta, R. M. (2024). Biotechnological, nutritional, and therapeutic applications of quinoa (Chenopodium quinoa Willd.) and its by-products: A review of the past five-year findings. *Nutrients*, 16(6), 840. https://doi.org/10.3390/nu16060840

Caselato-Sousa, V. M., & Amaya-Farfan, J. J. (2012). State of knowledge on amaranth grain: A comprehensive review. *Journal of Food Science*, 77, 93–104.

Castillo-Martínez, W. E., Símpalo-López, W. D., Galarreta-Oliveros, G. I., Miñan-Olivos, G. S., Esquivel Paredes, L. J., Quezada-Berrú, S., & Cordova Reyes, J. A. (2023). Evaluation of the acceptability of canned anchovy (Engraulis ringens) in Quinoa (Chenopodium quinoa Willd) and Kiwicha (Amaranthus caudatus L.) sauce. 21st LACCEI International Multi-Conference for Engineering, Education, and Technology: "Leadership in Education and Innovation in Engineering in the Framework of Global Transformations: Integration and Alliances for Integral Development", Hybrid Event, Buenos Aires – Argentina. https://www.researchgate.net/publication/373176513_Evaluation_of_the_acceptability_of_canned_anchovy_Engraulis_ringens_in_Quinoa_Chenopodium_quinoa_Willd_and_Kiwicha_Amaranthus_caudatus_L_sauce

Castro-Alba, V., Lazarte, C. E., Perez-Rea, D., Sandberg, A. S., Carlsson, N. G., Almgren, A., Bergenståhl, B., & Granfeldt, Y. (2019). Effect of fermentation and dry roasting on the nutritional quality and sensory attributes of quinoa. *Food Science & Nutrition*, 7(12), 3902–3911

Chandra, J. K., Chandora, R., Sood, S., & Malhotra, N. (2021). Chapter 2: Global production, demand, and supply. In M. Singh & S. Sood (Eds.), *Millets and Pseudo Cereals* (pp. 7–18). Woodhead Publishing. https://doi.org/10.1016/B978-0-12-820089-6.00002-1

Chaudhary, P., Janmeda, P., Docea, A. O., Yeskaliyeva, B., Abdull Razis, A. F., Modu, B., Calina, D., & Sharifi-Rad, J. (2023). Oxidative stress, free radicals and antioxidants: Potential crosstalk in the pathophysiology of human diseases. *Frontiers in Chemistry*, 11, Article 1158198. https://doi.org/10.3389/fchem.2023.1158198

Chauhan, A., Saxena, D. C., & Singh, S. (2016). Physical, textural, and sensory characteristics of wheat and amaranth flour blend cookies. *Cogent Food & Agriculture*, 2(1), 1125773. https://doi.org/10.1080/23311932.2015.1125773

Chen, J., Liu, Y., Zhao, Z., & Qiu, J. (2021). Oxidative stress in the skin: Impact and related protection. *International Journal of Cosmetic Science*, 43(6), 495–509. https://doi.org/10.1111/ics.12728

Chen, Q., Dong, L., Li, Y., Liu, Y., Xia, Q., Sang, S., Wu, Z., Xiao, J., Liu, L., & Liu, L. (2023). Research advance of non-thermal processing technologies on ovalbumin properties: The gelation, foaming, emulsification, allergenicity, immunoregulation, and its delivery system application. *Critical Reviews in Food Science and Nutrition*, 64(20), 7045–7066. https://doi.org/10.1080/10408398.2023.2179969

Chen, Z., Liang, N., Zhang, H., Li, H., Guo, J., Zhang, Y., Chen, Y., Wang, Y., & Shi, N. (2024). Resistant starch and the gut microbiome: Exploring beneficial interactions and dietary impacts. *Food Chemistry X*, 21, 101118. https://doi.org/10.1016/j.fochx.2024.101118

Cheng, A. (2018). Review: Shaping a sustainable food future by rediscovering long-forgotten ancient grains. *Plant Science*, 269, 136–142. https://doi.org/10.1016/j.plantsci.2018.01.018

Chiang, J. H., Hua, X. Y., Yu, A. H. M., Peh, E. W. Y., See, E., & Jeyakumar Henry, C. (2022). A review on buckwheat and its hypoglycemic bioactive components in food systems. *Food Reviews International*, 39(9), 6362–6386. https://doi.org/10.1080/87559129.2022.2103706

Chlopicka, J., Pasko, P., Gorinstein, S., Jedryas, A., & Zagrodzki, P. (2012). Total phenolic and total flavonoid content, antioxidant activity and sensory evaluation of pseudocereal breads. *LWT - Food Science and Technology*, 46(2), 548–555. https://doi.org/10.1016/J.LWT.2011.11.009

Choi, S.-S., Park, H.-R., & Lee, K.-A. (2021). A comparative study of rutin and rutin glycoside: Antioxidant activity, anti-inflammatory effect, effect on platelet aggregation and blood coagulation. *Antioxidants*, 10(11), 1696. https://doi.org/10.3390/antiox10111696

Christa, K., & Soral-Śmietana, M. (2008). Buckwheat grains and buckwheat products—Nutritional and prophylactic value of their components: A review. *Czech Journal of Food Science*, 26(3), 153–162.

Colla, E., do Amaral Sobral, P. J., & Menegalli, F. C. (2006). Amaranthus cruentus flour edible films: Influence of stearic acid addition, plasticizer concentration, and emulsion stirring speed on water vapor permeability and mechanical properties. *Journal of Agriculture and Food Chemistry*, 54, 6645–6653.

Constantino, A. B. T., & Garcia-Rojas, E. E. (2022). Proteins from pseudocereal seeds: Solubility, extraction, and modifications of the physicochemical and techno-functional properties. *Journal of the Science of Food and Agriculture*, 102(6), 2630–2639. https://doi.org/10.1002/jsfa.11750

Cui, C., Wang, Y., Ying, J., Zhou, W., Li, D., & Wang, L. J. (2024). Low glycemic index noodle and pasta: Cereal type, ingredient, and processing. *Food Chemistry*, 431, 137188. https://doi.org/10.1016/j.foodchem.2023.137188

de Cássia Spacki, K., Corrêa, R. C. G., Uber, T. M., Barros, L., Ferreira, I. C., Peralta, R. A., Moreira, de Fátima Peralta Muniz Moreira, R., Helm, C. V., de Lima, E. A., Bracht, A., & Peralta, R. M.. (2022). Full exploitation of peach palm (Bactris gasipaes Kunth): State of the art and perspectives. *Plants*, 11, 3175.

Demir, B., & Bilgiçli, N. (2020). Changes in chemical and anti-nutritional properties of pasta enriched with raw and germinated quinoa (Chenopodium quinoa Willd.) flours. *Journal of Food Science and Technology*, 57, 3884–3892. https://doi.org/10.1007/s13197-020-04420-7

Dey, M., & Kumar, P. (2024). Effect of nonconventional raw materials on the technological development of gluten-free pasta. In A. Gull, G. A. Nayik, & C. Brennan (Eds.), *Development of Gluten-Free Pasta* (pp. 45–64). Academic Press. https://doi.org/10.1016/B978-0-443-13238-4.00006-X

Diamanti-Kandarakis, E., Papalou, O., & Soulage, C. O. (2023). Effectiveness of myo- and d-chiro-inositol in the treatment of metabolic disorders. In V. Unfer & D. Dewailly (Eds.), *A Clinical Guide to Inositols* (pp. 31–51). Academic Press. https://doi.org/10.1016/B978-0-323-91673-8.00005-4

Diéguez, M. C. V., Pelissari, F. M., Sobral, P. J. A., & Menegalli, F. C. (2015). Effect of process conditions on the production of nanocomposite films based on amaranth flour and montmorillonite. *LWT - Food Science and Technology*, 61, 70–79.

Duliński, R., Zdaniewicz, M., Pater, A., Poniewska, D., & Żyła, K. (2020). The impact of phytases on the release of bioactive inositols, the profile of inositol phosphates, and the release of selected minerals in the technology of buckwheat beer production. *Biomolecules*, 10(2), 166. https://doi.org/10.3390/biom10020166

Elizabeth, A. A. (2010). Ancient grains: Opportunities and challenges for amaranth, quinoa, millet, sorghum and *tef* in gluten-free food products. *IFT Annual Meeting*, Chicago, IL.

El-Sohaimy, S. A., Shehata, M. G., Mehany, Taha, Zeitoun, M. A. (2019). Nutritional, physicochemical, and sensorial evaluation of flat bread supplemented with quinoa flour. *International Journal of Food Science*, 4686727, 15 pages. https://doi.org/10.1155/2019/4686727

Emire, S. A., & Arega, M. (2012). Value added product development and quality characterization of amaranth (*Amaranthus caudatus* L.) grown in East Africa. *African Journal of Food Science and Technology*, 3(6), 129–141.

Ergun, M., Özbay, N., Osmanoğlu, A., & Çalkır, A. (2014). Amaranth (*Amarant* spp) plant as vegetable and grain crop. *Iğdır University Journal of the Institute Science and Technology*, 4(3), 21–28.

Escudero, N. L., Alabarracin, G. J., Opez, L. L., & Gimenez, M. S. (2011). Antioxidant activity and phenolic content of flour and protein concentrate of *Amaranthus cruentus* seeds. *Journal of Food Biochemistry*, 35(4), 1327–1341.

Essa, M. M., Bishir, M., Bhat, A., Chidambaram, S. B., Al-Balushi, B., Hamdan, H., Govindarajan, N., Freidland, R. P., & Qoronfleh, M. W. (2023). Functional foods and their impact on health. *Journal of Food Science and Technology*, 60, 820–834. https://doi.org/10.1007/s13197-021-05193-3

Estivi, L., Pellegrino, L., Hogenboom, J. A., Brandolini, A., & Hidalgo, A. (2022). Antioxidants of amaranth, quinoa, and buckwheat wholemeals and heat-damage development in pseudocereal-enriched einkorn water biscuits. *Molecules*, 27(21), 7541. https://doi.org/10.3390/molecules27217541

FAO & CIRAD. (2015). State of the art report on quinoa around the world in 2013; D. Bazile, H. D. Bertero, C. Nieto (Eds.) Food and Agriculture Organization of the United Nations (FAO), Santiago, Chile.

Fernández-López, J., Lucas-González, R., Viuda-Martos, M., Sayas-Barberá, E., Ballester-Sánchez, J., Haros, C. M., Martínez-Mayoral, A., & Pérez-Álvarez, J. A. (2020). Chemical and technological properties of bologna-type sausages with added black quinoa wet-milling coproducts as binder replacer. *Food Chemistry*, 310, 125936. https://doi.org/10.1016/j.foodchem.2019.125936. Epub 2019 Dec 3. PMID: 31835224.

Filho, A. M. M., Pirozi, M. R., Borges, J. T. D. S., Pinheiro Sant'Ana, H. M., Chaves, J. B. P., & Coimbra, J. S. D. R. (2017). Quinoa: Nutritional, functional, and antinutritional aspects. *Critical Reviews in Food Science and Nutrition*, 57(8), 1618–1630. https://doi.org/10.1080/10408398.2014.1001811

Gajewski, M., Radzanowska, J., Jeznach, M., Jariene, E., & Danil-Cenko, H. (2015). Sensory quality evaluation of extruded corn snacks supplemented with pumpkin, Jerusalem artichoke and amaranthus flours, using profile analysis and consumers' test. *Postępy Techniki Przetwórstwa Spożywczego*, 2, 15–21.

Gamel, T. H., Linssen, J. P., Mesallam, A. S., Damir, A. A., & Shekib, L. A. (2006). Seedtreatments affect functional and antinutritional properties of amaranth flours. *Journal of the Science of Food and Agriculture*, 86, 1095–1102.

Gamel, T. H., & Linssen, J. P. H. (2008). Flavour compounds of popped amaranth seeds. *Journal of Food Processing and Preservation*, 32(4), 656–668. https://doi.org/10.1111/j.1745-4549.2008.00206.x

Gandhi, P., Samarth, R.M., Peter, K. (2021). Bioactive compounds of amaranth (Genus Amaranthus). In: Murthy, H.N., Paek, K.Y. (Eds.), *Bioactive Compounds in Underutilized Vegetables and Legumes*. Reference Series in Phytochemistry. Springer, Cham. https://doi.org/10.1007/978-3-030-57415-4_3

Ganeshpurkar, A., & Saluja, A. K. (2017). The pharmacological potential of rutin. *Saudi Pharmaceutical Journal*, 25(2), 149–164. https://doi.org/10.1016/j.jsps.2016.04.025

García-Sánchez, A., Miranda-Díaz, A. G., & Cardona-Muñoz, E. G. (2020). The role of oxidative stress in physiopathology and pharmacological treatment with pro- and antioxidant properties in chronic diseases. *Oxidative Medicine and Cellular Longevity*, 2020, 2082145. https://doi.org/10.1155/2020/2082145

Ge, X., Jing, L., Zhao, K., Su, C., Zhang, B., Zhang, Q., Han, L., Yu, X., & Li, W. (2021). The phenolic compounds profile, quantitative analysis, and antioxidant activity of four naked barley grains with different color. *Food Chemistry*, 335, 127655. https://doi.org/10.1016/j.foodchem.2020.127655

Gebreil, S., Ali, M., & Mohamed Mousa, E. A. (2020). Utilization of amaranth flour in preparation of high nutritional value bakery products. *Food and Nutrition Sciences*, 10(5), 336–354. https://doi.org/10.4236/fns.2020.105025

Ghumman, A., Mudgal, S., Singh, N., Ranjan, B., Kaur, A., & Rana, J. C. (2021). Physicochemical, functional and structural characteristics of grains, flour and protein isolates of Indian quinoa lines. *Food Research International*, 140, 109982. https://doi.org/10.1016/J.FOODRES.2020.109982

Giménez-Bastida, J. A., Piskuła, M., & Zieliński, H. (2015). Recent advances in the development of gluten-free buckwheat products. *Trends in Food Science & Technology*, 44(1), 58–65. https://doi.org/10.1016/j.tifs.2015.02.013

Gimplinger, D. M., Schulte, A., Erley, G., Dobos, G., & Kaul, H. P. (2008). Optimum crop densities for potential yield and harvestable yield of grain amaranth are conflicting. *European Journal of Agronomy*, 28(2), 119–125.

Gonçalves, M. A. A., Salvador, R., de Francisco, A. C., & Piekarski, C. M. (2024). Value recovery from waste in the processing of buckwheat: Opportunities for a circular bioeconomy. *Engineering Reports*, 6(7), e12757. https://doi.org/10.1002/eng2.12757

Gönen, E., Çolak, Y.B., & Yazar, A. (2018). Farklı Amarant Çeşitlerinin Çukurova Bölgesine Adaptasyonu. *Dünya Sağlık ve Tabiat Bilimleri Dergisi*, 1(2): 44–49.

Gönen, E. (2020). New approaches for adaptation to drought and global climate change: Effects of different deficit ırrıgatıon strategies applied with subsurface and surface drip systems on yield, yield components and water use efficiency of amaranth (*Amaranthus cruentus*). Kahramanmaraş Sütçü İmam University Institute for Graduate Studies in Science and Technology Department of Biosystem Engineering.

González Acevedo, O., Hernández Reyes, J., Gaytán Dario, H., Victoria Campos., C. I., & Palos Lucio, A. G. (2018). Amaranth drink and its potential effect on the body composition of older adults. *Nutricon Clínica y Dietetica Hospitalaria*, 38(2), 114–119. https://doi.org/10.12873/382ogonzalez

Gordillo-Bastidas, E., Díaz-Rizzolo, D. A., Roura, E., Massanés, T., & Gomis, R. (2016). Quinoa (Chenopodium quinoa Willd), from nutritional value to potential health benefits: An integrative review. *Journal of Nutrition and Food Sciences*, 6, 497. https://doi.org/10.4172/2155-9600.1000497

Gorinstein, S., Vargas, O. J. M., Jaramillo, N. O., Salas, I. A., Ayala, A. L. M., Arancibia-Avila, P., Toledo, F., Katrich, E., & Trakhtenberg, S. (2007). The total polyphenols and the antioxidant potentials of some selected cereals and pseudocereals, *European Food Research and Technology*, 225, 321–328.

Graf, B. L., Rojas-Silva, P., Rojo, L. E., Delatorre-Herrera, J., Baldeon, M. E., & Raskin, I. (2014). Innovations in health value and functional food development of quinoa (Chenopodium quinoa Willd.). *Comprehensive Reviews in Food Science and FoodSafety*. https://onlinelibrary.wiley.com/doi/10.1111/1541-4337.12135/epdf

Graziano, S., Agrimonti, C., Marmiroli, N., & Gullì, M. (2022). Utilisation and limitations of pseudocereals (quinoa, amaranth, and buckwheat) in food production: A review. *Trends in Food Science & Technology*, 125, 154–165. https://doi.org/10.1016/J.TIFS.2022.04.007

Guan, Z.-W., Yu, E.-Z., & Feng, Q. (2021). Soluble dietary fiber, one of the most important nutrients for the gut microbiota. *Molecules*, 26(22), 6802. https://doi.org/10.3390/molecules26226802

Hamzehpour, R., & Dastgerdi, A. A. (2023). The effects of quinoa and amaranth flour on the qualitative characteristics of gluten-free cakes. *International Journal of Food Science*, 6042636, 6 pages. https://doi.org/10.1155/2023/6042636

Heliawati, L., Lestari, S., Hasanah, U., Ajiati, D., & Kurnia, D. (2022). Phytochemical profile of antibacterial agents from red betel leaf (*Piper crocatum* Ruiz and Pav) against bacteria in dental caries. *Molecules*, 27(9), 2861. https://doi.org/10.3390/molecules27092861. PMID: 35566225; PMCID: PMC9101570.

Henrion, M., Labat, E., & Lamothe, L. (2020). Chapter 3: Pseudocereals as healthy grains. In M. Pojić & U. Tiwari (Eds.), *Innovative Processing Technologies for Healthy Grains*. https://doi.org/10.1002/9781119470182.ch3

Howard, J. E., Villamil, M. B., & Riggins, C. W. (2022) Amaranth as a natural food colorant source: Survey of germplasm and optimization of extraction methods for betalain pigments. *Frontiers in Plant Science*, 13, 932440. https://doi.org/10.3389/fpls.2022.932440

Hu, Y. C., Hu, J. L., Li, J., Wang, J., Zhang, X. Y., Wu, X. Y., Li, X., Guo, Z.-B., Zou, L., & Wu, D. T. (2023). Physicochemical characteristics and biological activities of soluble dietary fibers isolated from the leaves of different quinoa cultivars. *Food Research International*, 163, 112166.

Huda, M. N., Lu, S., Jahan, T., Ding, M., Jha, R., Zhang, K., Zhang, W., Georgiev, M. I., Park, S. U., & Zhou, M. (2021). Treasure from garden: Bioactive compounds of buckwheat. *Food Chemistry*, 335, 127653. https://doi.org/10.1016/j.foodchem.2020.127653

Ikeda, K. (2002). Buckwheat composition, chemistry, and processing. In *Advances in Food and Nutrition Research* (Vol. 44, pp. 395–434). Academic Press. https://doi.org/10.1016/S1043-4526(02)44008-9

Ikeda, K., & Ikeda, S. (2016). Factors important for structural properties and quality of buckwheat products. In M. Zhou, I. Kreft, S.- H. Woo, N. Chrungoo, & G. Wieslander (Eds.), *Molecular Breeding and Nutritional Aspects of Buckwheat* (pp. 193–202). Academic Press. https://doi.org/10.1016/B978-0-12-803692-1.00015-8

Ikram, A., Saeed, F., Afzaal, M., Imran, A., Niaz, B., Tufail, T., Hussain, M., & Anjum, F. M. (2021). Nutritional and end-use perspectives of sprouted grains: A comprehensive review. *Food Science & Nutrition*, 9, 4617–4628. https://doi.org/10.1002/fsn3.2408

Jacob, J. (2021). *Food Cultures of Japan (The Global Kitchen)*. Bloomsbury Publishing. ISBN: 9781440866845. https://digital.casalini.it/9781440866845

Jan, N., Hussain, S. Z., Naseer, B., & Bhat, T. A. (2023). Amaranth and quinoa as potential nutraceuticals: A review of anti-nutritional factors, health benefits and their applications in food, medicinal and cosmetic sectors. *Food Chemistry:* X, 18, 100687. https://doi.org/10.1016/J.FOCHX.2023.100687

Janovská, D., Jágr, M., Svoboda, P., Dvořáček, V., Meglič, V., & Hlásná Čepková, P. (2021). Breeding buckwheat for nutritional quality in the Czech Republic. *Plants*, 10(7), 1262. https://doi.org/10.3390/plants10071262

Janssen, F., Pauly, A., Rombouts, I., Jansens, K. J. A., Deleu, L. J., & Delcour, J. A. (2017). Proteins of amaranth (Amaranthus spp.), buckwheat (Fagopyrum spp.), and quinoa (Chenopodium spp.): A food science and technology perspective. *Comprehensive Reviews in Food Science and Food Safety*, 16, 39–58. https://doi.org/10.1111/1541-4337.12240

Jha, R., Zhang, K., He, Y., Mendler-Drienyovszki, N., Magyar-Tábori, K., Quinet, M., Germ, M., Kreft, I., Meglič, V., Ikeda, K., Chapman, M. A., Janovská, D., Podolska, G., Woo, S.-H., Studer, B., Georgiev, M. I., Chrungoo, N., & Zhou, M. (2024). Global nutritional challenges and opportunities: Buckwheat, a potential bridge between nutrient deficiency and food security. *Trends in Food Science & Technology*, 145, 104365. https://doi.org/10.1016/j.tifs.2024.104365

Jiménez-Aguilar, D. M., & Grusak, M. A. (2017). Minerals, vitamin C, phenolics, flavonoids and antioxidant activity of Amaranthus leafy vegetables. *Journal of Food Composition and Analysis*, 58, 33–39. https://doi.org/10.1016/j.jfca.2017.01.005

Jin, J., Ohanenye, I. C., & Udenigwe, C. C. (2020). Buckwheat proteins: Functionality, safety, bioactivity, and prospects as alternative plant-based proteins in the food industry. *Critical Reviews in Food Science and Nutrition*, 62(7), 1752–1764. https://doi.org/10.1080/10408398.2020.1847027

Jurek, I., Góral, I., Gęsiński, K., & Wojciechowski, K. (2019). Effect of saponins from quinoa on a skin-mimetic lipid monolayer containing cholesterol. *Steroids*, 147, 52–57.

Kauffman, C. S., & Weber, L. E. (1990). Grain amaranth. In J. Janick and J. E. Simon (Eds.), *Advances in New Crops* (pp. 127–139). Timmber Press.

Kauffman, C. S. (1992). Realizing the potential of grain amaranth. *Food Reviews International*, 8(1), 5–21. https://doi.org/10.1080/87559129209540927

Kaur, H., Shams, R., Dash, K. K., & Dar, A. H. (2023). A comprehensive review of pseudo-cereals: Nutritional profile, phytochemical constituents, and potential health-promoting benefits. *Applied Food Research*, 3(2), 100351. https://doi.org/10.1016/j.afres.2023.100351

Khairuddin, M. A. N., & Lasekan, O. (2021). Gluten-free cereal products and beverages: A review of their health benefits in the last five years. *Foods*, 10(11), 2523. https://doi.org/10.3390/foods10112523

Klebaniuk, R., Kowalczuk-Vasilev, E., Kiczorowska B., Grela, E. R., Winiarska-Mieczan, A., Samolińska W., Kwiecień M., Kwiatkowska, K., Rusinek-Prystupa, E., & Jarzyna, P. (2018). Nutritious value of selected food products with amaranth addition. *Problemy Higieny i Epidemiologii*, 99(1), 82–87.

Kolata, A. L., 2009. Quinoa: Production, consumption and social value in historical context, pp. 1–13.

Komatsudaira, Y. (2023). *Japanese Superfoods: Learn the Secrets of Healthy Eating and Longevity-the Japanese Way*. Tuttle Publishing.

Kowalski, S., Mikulec, A., Mickowska, B., & Buksa, K. (2022). Nutritional properties and amino acid profile of buckwheat bread. *Journal of Food Science and Technology*, 59(6), 3020–3030. https://doi.org/10.1007/s13197-022-05518-w

Koyun, S., 2013. Safe food: Quinoa (Chenopodium quinoa Willd.). *Mesleki Bilimler Dergisi MBD*, 2(2), 85–88.

Kreft, I., Germ, M., Golob, A., Vombergar, B., Bonafaccia, F., & Luthar, Z. (2022). Impact of rutin and other phenolic substances on the digestibility of buckwheat grain metabolites. *International Journal of Molecular Sciences*, 23(7), 3923. https://doi.org/10.3390/ijms23073923

Kreft, I., Golob, A., Vombergar, B., & Germ, M. (2023). Tartary buckwheat grain as a source of bioactive compounds in husked groats. *Plants*, 12(5), 1122. https://doi.org/10.3390/plants12051122

Kulokas, M., Praspaliauskas, M., & Pedišius, N. (2021). Investigation of buckwheat hulls as additives in the production of solid biomass fuel from straw. *Energies*, 14(2), 265. https://doi.org/10.3390/en14020265

Kumar, M., Kaushik, D., & Kasodhan, S. (2024). Buckwheat. In J. Singh, S. Kaur, P. Rasane, & J. Singh (Eds.), *Cereals and Nutraceuticals*. Springer. https://doi.org/10.1007/978-981-97-2542-7_9

Kurek, M. A., Napiórkowska, A., & Aktaş, H. (2024). Chapter 1: Plant sources for functional ingredients—proteins. In F. Boukid, C. M. Rosell, & N. Gasparre (Eds.), *Handbook of Plant-Based Food and Drinks Design* (pp. 3–18). Academic Press. https://doi.org/10.1016/B978-0-443-16017-2.00005-X

Lahuta, L. B., Górecki, R. J., Szablińska-Piernik, J., & Horbowicz, M. (2023). Changes in the carbohydrate profile in common buckwheat (Fagopyrum esculentum Moench) seedlings induced by cold stress and dehydration. *Metabolites*, 13(5), 672. https://doi.org/10.3390/metabo13050672

Lalewicz, P., Domagała-Świątkiewicz, I., & Siwek, P. (2024). Phacelia and buckwheat cover crops' effects on soil quality in organic vegetable production in a high tunnel system. *Agronomy*, 14(8), 1614. https://doi.org/10.3390/agronomy14081614

Lan, S. (2023). Assessing textile raw materials and exploring sustainable sources for future renewable textile materials for IKEA of Sweden. Swedish University of Agricultural Sciences, SLU Faculty of Horticulture and Crop Production Sciences, Agroecology - Master's Programme. Alnarp.

Lan, Y., Zhang, W., Liu, F., Wang, L., Yang, X., Ma, S., Wang, Y., & Liu, X. (2023). Recent advances in physiochemical changes, nutritional value, bioactivities, and food applications of germinated quinoa: A comprehensive review. *Food Chemistry*, 426, 136390. https://doi.org/10.1016/J.FOODCHEM.2023.136390

Li, D., Cakebread, J. A., & Loveday, S. M. (2022). Food proteins from animals and plants: Differences in the nutritional and functional properties. *Trends in Food Science & Technology*, 119, 428–442. https://doi.org/10.1016/j.tifs.2021.12.020

Li, G., Wang, S., & Zhu, F. (2016). Physicochemical properties of quinoa starch. *Carbohydrate Polymers*, 137, 328–338. https://doi.org/10.1016/J.CARBPOL.2015.10.064

Li, Y., Yang, N., Shi, F., Ye, F., & Huang, J. (2023). Isolation and identification of angiotensin-converting enzyme inhibitory peptides from Tartary buckwheat albumin. *Journal of the Science of Food and Agriculture*, 103(12), 5019–5027. https://doi.org/10.1002/jsfa.12573

Lindeboom, N., Chang, P. R., Falk, K. C., & Tyler, R. T. (2005), Characteristics of starch from eight quinoa lines. *Cereal Chemistry*, 82, 216–222. https://doi.org/10.1094/CC-82-0216

Lingiardi, N., Galante, M., de Sanctis, M., & Spelzini, D. (2022). Are quinoa proteins a promising alternative to be applied in plant-based emulsion gel formulation? *Food Chemistry*, 394, 133485. https://doi.org/10.1016/J.FOODCHEM.2022.133485

Liu, C., Plaza-Bonilla, D., Coulter, J. A., Kutcher, H. R., Beckie, H. J., Wang, L., Floc'h, J.-B., Hamel, C., Siddique, K. H. M., Li, L., & Gan, Y. (2022). Chapter six: Diversifying crop rotations enhances agroecosystem services and resilience. In D. L. Sparks (Ed.), *Advances in Agronomy* (Vol. 173, pp. 299–335). Academic Press. https://doi.org/10.1016/bs.agron.2022.02.007

Liu, S., Yin, X., Hou, C., Liu, X., Ma, H., Zhang, X., Xu, M., Xie, Y., Li, Y., & Wang, J. (2021). As a staple food substitute, oat and buckwheat compound has health-promoting effects for diabetic rats. *Frontiers in Nutrition*, 8, Article 762277. https://doi.org/10.3389/fnut.2021.762277

López, D. N., Galante, M., Robson, M., Boeris, V., & Spelzini, D. (2018). Amaranth, quinoa and chia protein isolates: Physicochemical and structural properties. *International Journal of Biological Macromolecules*, 109, 152–159. https://doi.org/10.1016/J.IJBIOMAC.2017.12.080

Ludena Urquizo, F. E., García Torres, S. M., Tolonen, T., Jaakkola, M., Pena-Niebuhr, M. G., von Wright, A., Repo-Carrasco-Valencia, R., Korhonen, H., & Plumed-Ferrer C. (2016). Development of a fermented quinoa-based beverage. *Food Science & Nutrition*, 5(3), 602–608. https://doi.org/10.1002/fsn3.436. PMID: 28572947; PMCID: PMC5448362.

Machado Alencar, N. M., Steel, C. J., Alvim, I. D., De Morais, E. C., & Andre Bolini, H. M., (2015). Addition of quinoa and amaranth flour in gluten-free breads: Temporal profile and instrumental analysis. *LWT - Food Science and Technology*, 62(2), 1011–1018. https://doi.org/10.1016/j.lwt.2015.02.029

Majaw, J., Bharath, J., & Prabhasankar, P. (2024). Effect of pseudocereals on gluten-free pasta: Rheological, microstructural, and cooking quality. In *Development of Gluten-Free Pasta* (pp. 135–156). https://doi.org/10.1016/B978-0-443-13238-4.00007-1

Maksim M. Danyliv, Olga A. Vasilenko, Olga N. Ozherelyeva, Alla E. Kutsova (2024). Amaranth cake as a source of food isolate of vegetable protein for the meat processing industry. *BIO Web of Conferences* 103, 00084. https://doi.org/10.1051/bioconf/202410300084 FIES 2023

Malik, A. M., & Singh, A. (2022). Pseudocereals proteins: A comprehensive review on its isolation, composition, and quality evaluation techniques. *Food Chemistry Advances*, 1, 100001. https://doi.org/10.1016/j.focha.2021.100001

Malik, M., Sindhu, R., Dhull, S. B., Bou-Mitri, C., Singh, Y., Panwar, S., & Khatkar, B. S. (2023). Nutritional composition, functionality, and processing technologies for amaranth. *Journal of Food Processing and Preservation*, 1753029, 24, https://doi.org/10.1155/2023/1753029

Manassero, C. A., Añón, M. C., & Speroni, F. (2020). Development of a high protein beverage based on amaranth. *Plant Foods for Human Nutrition*, 75(4), 599–607. https://doi.org/10.1007/s11130-020-00853-9

Manzoor, M., Singh, D., Aseri, G. K., Sohal, J. S., Vij, S., & Sharma, D. (2021). Role of lacto-fermentation in reduction of antinutrients in plant-based foods. *Journal of Applied Biology & Biotechnology*, 9, 7–16. https://doi.org/10.7324/JABB.2021.9302

Marquez-Molina, O., & Lopez-Martinez, L. X. (2020). Effect of various process conditions on the nutritional and bioactive compounds of amaranth. In V. Y. Waisundara (Ed.), *Nutritional Value of Amaranth* (pp. 1–11). IntechOpen.

Matías, J., Rodríguez, M. J., Carrillo-Vico, A., Casals, J., Fondevilla, S., Haros, C. M., Pedroche, J., Aparicio, N., Fernández-García, N., Aguiló-Aguayo, I., & Soler-Rivas, C. (2024). From 'Farm to Fork': Exploring the potential of nutrient-rich and stress-resilient emergent crops for sustainable and healthy food in the mediterranean region in the face of climate change challenges. *Plants*, 13(14), 1914. https://doi.org/10.3390/plants13141914

Maurya, N. K., & Arya, P. (2018). *Amaranthus* grain nutritional benefits: A review. *Journal of Pharmacognosy and Phytochemistry*, 7(2), 2258–2262.

Milan-Carrillo, J., Montoya-Rodríguez, A., Gutierrez- Dorado, R., Perales-Sanchez, X., & Reyes-Moreno, C. (2012). Optimization of extrusion process for producing high antioxidant instant amaranth (*Amaranthus hypochondriacus* L.) flour using response surface methodology. *Applied Mathematics*, 3, 1516–1525. https://doi.org/10.4236/am.2012.330211

Mir, N. A., Yousuf, B., Gul, K., Riar, C. S., & Singh, S. (2019). Cereals and pseudocereals: Genera introduction, classification, and nutritional properties. In Sankar Chandra Deka, Dibyakanta Seth, Nishant Rachayya Swami Hulle (Eds.), *Food Bioactives* (pp. 281–322). Apple Academic Press.

Miranda, M., Vega-Gálvez, A., Quispe-Fuentes, I., Rodriguez, M. J., Maureira, H., & Martínez, E. A. (2012). Nutritional aspects of six quinoa (Chenopodium quinoa Willd.) ecotypes from three geographical areas of chile. *Chilean Journal of Agricultural Research*, 72, 175–181.

Mlakar, S. G., Turinek, M., Jakop, M., Bavec, M., & Bavec, F. (2009). Nutrition value and use of grain amaranth: Potential future application in bread making. *Agricultura*, 6, 43–53.

Mlakar, S. G., Turinek, M., Jakop, M., Bavec, M., & Bavec, F. (2010). Grain amaranth as an alternative and perspective crop in temperate climate. *Journal for Geography*, 5(1), 135–145.

Mondal, S. (2023). Fagopyrum esculentum: A nutrient-dense part of nature. In T. Pullaiah (Ed.), *Phytochemical Composition and Pharmacy of Medicinal Plants: 2-Volume Set* (p. 325). Apple Academic Press.

Montoya Restrepo, L.; Martínez Vianchá, L., y Peralta Ballestero, J. (2005). Análisis de las variables estratégicas para la conformación de una cadena productiva de la quinua en Colombia. *Revista Innovar. Edit. Unibiblos*, 25, 103–120.

Muyonga, J. H., Andabati, B., & Ssepuuya, G. (2014). Effect of heat processing on selected grain amaranth physiochemical properties. *Food Science and Nutrition*, 2(1), 9–16. https://doi.org/10.1002/fsn3.75

Mystkowska, I., Plażuk, E., Szepeluk, A., Dmitrowicz, A. (2024). Gluten-containing flours and gluten-free flours as a source of calcium, magnesium, iron and zinc. *Scientific Reports*, 14, 14643. https://doi.org/10.1038/s41598-024-65530-2

Nandan, A., Koirala, P., Tripathi, A. D., Vikranta, U., Shah, K., Gupta, A. J., Agarwal, A., & Nirmal, N. (2024). Nutritional and functional perspectives of pseudocereals. *Food Chemistry*, 448, 139072. https://doi.org/10.1016/j.foodchem.2024.139072

Narwade, S., & Pinto, S. (2018). A functional food. *Concepts of Dairy & Veterinary Sciences*, 1(3), 72–77. https://doi.org/10.32474.CDVS.2018.01.000112

Nascimento, A. C., Mota, C., Coelho, I., Gueifão, S., Santos, M., Matos, A. S., Gimenez, A., Lobo, M., Samman, M., & Castanheira, I. (2013). Characterisation of nutrient profile of quinoa (Chenopodium quinoa), amaranth (Amaranthus caudatus), and purple corn (Zea mays L.) consumed in the North of Argentina: Proximates, minerals and trace elements. *Food Chemistry*, 148, 420–426.

Nasir, S., Allai, F. M., Gani, M., Ganaie, S., Gul, K., Jabeen, A., & Majeed, D. (2020). Physical, textural, rheological, and sensory characteristics of amaranth-based wheat flour bread. *International Journal of Food Science*, https://doi.org/10.1155/2020/8874872

Nath, C. P., Singh, R. G., Choudhary, V. K., Datta, D., Nandan, R., & Singh, S. S. (2024). Challenges and alternatives of herbicide-based weed management. *Agronomy*, 14(1), 126. https://doi.org/10.3390/agronomy14010126

Nawirska-Olszańska, A., Figiel, A., Pląskowska, E., Twardowski, J., Gębarowska, E., Kucharska, A. Z., Sokół-Łętowska, A., Spychaj, R., Lech, K., & Liszewski, M. (2021). Qualitative and quantitative assessment of buckwheat husks as a material for use in therapeutic mattresses. *International Journal of Environmental Research and Public Health*, 18(4), 1949. https://doi.org/10.3390/ijerph18041949

Nayak, P., Mishra, A. A., Shukla, R. N., & Biswas, R. (2024). Traditional foods based on pseudocereals. In Sanju Bala Dhull, Aarti Bains, Prince Chawla, Sawinder Kaur (Eds.), *Pseudocereals* (pp. 248–263). CRC Press.

Netshimbupfe, M. H., Berner, J., Kooy, F. V. D., Oladimeji, O., & Gouws, C. (2023). The importance and use of *Amaranthus* for crop diversification in the SADC region. *South African Journal of Botany*, 152, 192–202. https://doi.org/10.1016/j.sajb.2022.11.039

Neubauer, O., & Yfanti, C. (2015). Antioxidants in athlete's basic nutrition: Considerations towards a guideline for the intake of vitamin C and vitamin E. In M. Lamprecht (Ed.), *Antioxidants in Sport Nutrition*. CRC Press/Taylor & Francis. ISBN-13: 978-1-4665-6757-3

Nic Phiarais, B. P., & Arendt, E. K. (2008). Malting and brewing with gluten-free cereals. In *Gluten-Free Cereal Products and Beverages* (pp. 347–372). https://doi.org/10.1016/B978-012373739-7.50017-4

Nickel, J., Spanier, L. P., Botelho, F. T., Gularte, M. A., & Helbig, E. (2016). Effect of different types of processing on the total phenolic compound content, antioxidant capacity, and saponin content of Chenopodium quinoa Willd grains. *Food Chemistry*, 209, 139–143. https://doi.org/10.1016/J.FOODCHEM.2016.04.031

Ninomiya, K., Yamaguchi, Y., Kumagai, H., & Kumagai, H. (2022). Physicochemical and functional properties of buckwheat (Fagopyrum esculentum Moench) albumin. *Future Foods*, 6, 100178. https://doi.org/10.1016/j.fufo.2022.100178

Ňorbová, M., Vollmannová, A., Harangozo, Ľ., Franková, H., Čeryová, N., Jančo, I., & Fandrová, A. (2022). Risk elements, antioxidant activity and polyphenols in pseudocereal grains. *Agrobiodiversity for Improving Nutrition, Health and Life Quality*, 6(1). https://agrobiodiversity.uniag.sk/scientificpapers/article/view/418

Nordio, M., Bezerra Espinola, M. S., Bilotta, G., Capoccia, E., & Montanino Oliva, M. (2023). Long-lasting therapies with high doses of D-chiro-inositol: The downside. *Journal of Clinical Medicine*, 12(1), 390. https://doi.org/10.3390/jcm12010390

Notup, T., Vikram, K., Chaudhary, H., & Kumar, S. (2024). Nutritional potential and phytochemical screening of traditional buckwheat and barley crops of the Trans-Himalaya region. *Vegetos*, 37(3), 1071–1083. https://doi.org/10.1007/s42535-023-00645-x

Ofitserov, E. N. (2001). Amaranth: Perspective raw material for food processing and pharmaceutical industry. *Chemistry and Computational Simulation. Butlerov Communications*, 2(5), 1–4.

Olagunju, A. I., Arigbede, T. I., Makanjuola, S. A., & Oyebode, E. T. (2022). Nutritional compositions, bioactive properties, and in-vivo glycemic indices of amaranth-based optimized multigrain snack bar products. *Measurement: Food*, 7, 100039.

Öncel, E. (2017). The usage of different levels and combinations of amaranth, buckwheat and quinoa flour in eriste production. Necmettin Erbakan University the Graduate School of Natural and Applied Science Department in Food Engineering. *Ms Thesis*.

Otles, S., & Nakilcioglu-Tas, E. (2022). Chapter 3: Cereal-based functional foods. In N. Chhikara, A. Panghal, & G. Chaudhary (Eds.), *Functional Foods*. https://doi.org/10.1002/9781119776345.ch3

Pagno, C. H., Costa, T. M. H., de Menezes, E. W., Benvenutti, E. v., Hertz, P. F., Matte, C. R., Tosati, J. v., Monteiro, A. R., Rios, A. O., & Flôres, S. H. (2015). Development of active biofilms of quinoa (Chenopodium quinoa W.) starch containing gold nanoparticles and evaluation of antimicrobial activity. *Food Chemistry*, 173, 755–762. https://doi.org/10.1016/J.FOODCHEM.2014.10.068

Pal, J., Singhal, R. S., & Kulkarni, P. R. (2001). Physicochemical properties of hydroxypropyl derivative from corn and amaranth starch. *Carbohydrate Polymers*, 48, 19–53.

Park, S. J., Sharma, A., & Lee, H. J. A. (2020). Review of recent studies on the antioxidant activities of a third-millennium food: *Amaranthus* spp. *Antioxid*, 9, 1236.

Pasko, P., Barton´, H., Zagrodzki, P., Gorinstein, S., Fołta, M., Zachwieja, Z. 2009. Anthocyanins, total polyphenols and antioxidant activity in amaranth and quinoa seeds and sprouts during their growth. *Food Chemistry*, 115(3), 994– 998.

Patel, H., Sharma, M., Bains, A., & Chawla, P. (2024). Cultivation and production of pseudocereals. In Sanju Bala Dhull, Aarti Bains, Prince Chawla, Sawinder Kaur (Eds.), *Pseudocereals* (pp. 1–20). CRC Press.

Peiretti, P. G. (2018). Amaranth in animal nutrition: A review. *Livestock Research for Rural Development*, 30(5), 1–20. https://www.lrrd.org/lrrd30/5/peir30088.html

Piñuel, L., Boeri, P., Zubillaga, F., Barrio, D. A., Torreta, J., Cruz, A., Vásquez, G., Pinto, A., & Carrillo, W. (2019). Production of white, red and black quinoa (Chenopodium quinoa Willd Var. Real) protein isolates and its hydrolysates in germinated and non-germinated quinoa samples and antioxidant activity evaluation. *Plants (Basel)*, 8(8), 257. https://doi.org/10.3390/plants8080257.

Pirzadah, T. B., & Malik, B. (2020). Pseudocereals as super foods of 21st century: Recent technological interventions. *Journal of Agriculture and Food Research*, 2, 100052. https://doi.org/10.1016/J.JAFR.2020.100052

Pirzadah, T. B., & Rehman, R. U. (2021). *Buckwheat: Forgotten Crop for the Future: Issues and Challenges*. CRC Press.

Placha, D., & Jampilek, J. (2021). Chronic inflammatory diseases, anti-inflammatory agents and their delivery nanosystems. *Pharmaceutics*, 13(1), 64. https://doi.org/10.3390/pharmaceutics13010064

Podolska, G., Gujska, E., Klepacka, J., & Aleksandrowicz, E. (2021). Bioactive compounds in different buckwheat species. *Plants*, 10(5), 961. https://doi.org/10.3390/plants10050961

Punia, S., Dhull, S. B., Kunner, P., & Rohilla, S. (2020). Effect of γ-radiation on physico-chemical, morphological and thermal characteristics of lotus seed (*Nelumbo nucifera*) starch. *International Journal of Biological Macromolecules*, 157, 584–590.

Rai, S., Kaur, A., & Chopra, C. S. (2018). Gluten-free products for celiac susceptible people. *Frontiers in Nutrition*, 5. https://doi.org/10.3389/fnut.2018.00116

Ramesh, D., & Parakash, J. (2020). Nutritional and functional properties of amaranth grain flour fractions obtained by differential sieving. *Progress in Chemical and Biochemical Res*earch, 3(3), 272–286. https://doi.org/10.33945/SAMI/PCBR.2020.3.10

Ramos Diaz, J. M., Sundarrajan, L., Kariluoto, S., Lampi, A. M., Tenitz, S., & Jouppila, K. (2017). Effect of extrusion cooking on physical properties and chemical composition of corn-based snacks containing amaranth and quinoa: Application of partial least squares regression. *Journal of Food Processing and Engineering*, 40, e12320.

Ramos Diaz, J. M., Suuronen, J. P., Deegan, K. C., Serimaa, R., Tuorila, H., & Jouppila, K. (2015). Physical and sensory characteristics of corn-based extruded snacks containing amaranth, quinoa and kañiwa flour. *LWT - Food Science and Technology*, 64(2), 1047–1056. https://doi.org/10.1016/J.LWT.2015.07.011

Rastogi, A., & Shukla, S. (2013). Amaranth: A new millennium crop of nutraceutical values. *Critical Reviews in Food Science and Nutrition*, 53(2), 109–125. https://doi.org/10.1080/10408398.2010.517876

Raungrusmee, S. (2023). Underutilized cereals and pseudocereals' nutritional potential and health implications. In Anil Kumar Anal (Ed.), *Pandemics and Innovative Food Systems* (pp. 163–193). CRC Press.

Ray, A., Srivastava, A. K., & Sakhare, S. D. (2023). Quinoa germ–enriched pasta: Technological, nutritional, textural, and morphological properties. *Journal of Food Science*, 88, 4907–4917. https://doi.org/10.1111/1750-3841.16813

Ren, G., Teng, C., Fan, X., Guo, S., Zhao, G., Zhang, L., Liang, Z., & Qin, P. (2023). Nutrient composition, functional activity and industrial applications of quinoa (Chenopodium quinoa Willd.). *Food Chemistry*, 410, 135290. https://doi.org/10.1016/J.FOODCHEM.2022.135290

Renzi, J. P., Coyne, C. J., Berger, J., von Wettberg, E., Nelson, M., Ureta, S., Hernández, F., Smýkal, P., & Brus, J. (2022). How could the use of crop wild relatives in breeding increase the adaptation of crops to marginal environments? *Frontiers in Plant Science*, 13. https://doi.org/10.3389/fpls.2022.886162

Repo-Carrasco, R., Espinoza, C., & Jacobsen S. E. (2003). Nutritional value and use of the Andean crops quinoa (Chenopodium quinoa) and kañiwa (Chenopodium pallidicaule). *Food Reviews Internatinoal*, 19, 179–189.

Repo-Carrasco-Valencia, R. A. M., & Serna, L. A. (2011). Quinoa (*Chenopodium quinoa*, Willd.) as a source of dietary fiber and other functional components. *Food Science and Technology*, 31(1), 225–230. https://doi.org/10.1590/S0101-20612011000100035

Rezende, E. S. V., Lima, G. C., & Naves, M. M. V. (2021). Dietary fibers as beneficial microbiota modulators: A proposed classification by prebiotic categories. *Nutrition*, 89, 111217. https://doi.org/10.1016/j.nut.2021.111217

Rivero Meza, S. L., Hirsch Ramos, A., Cañizares, L., Raphaelli, C. de O., Bueno Peres, B., Gaioso, C. A., Egea, I., Estrada, Y., Flores, F. B., & Oliveira, M. de. (2022). A review on amaranth protein: Composition, digestibility, health benefits and food industry utilisation. *International Journal of Food Science and Technology*, 58, 1564–74.

Ruan, J., Zhou, Y., Yan, J., Zhou, M., Woo, S. H., Weng, W., Cheng, J., & Zhang, K. (2020). Tartary buckwheat: An under-utilized edible and medicinal herb for food and nutritional security. *Food Reviews International*, 38(4), 440–454. https://doi.org/10.1080/87559129.2020.1734610

Ruiz, K. B., Biondi, S., Oses, R., Acuña-Rodríguez, I. S., Antognoni, F., Martinez-Mosqueira, E. A., Coulibaly, A., Canahua-Murillo, A., Pinto, M., Zurita-Silva, A., Bazile, D., Jacobsen, S.-E., & Molina-Montenegro, M. A. (2014). Quinoa biodiversity and sustainability for food security under climate change: A review. *Agronomy for Sustainable Development*, 34(2), 349–359. https://doi.org/10.1007/s13593-013-0195-0

Ruth, O. N., Unathi, K., Nomali, N., & Chinsamy, M. (2021). Underutilization versus nutritional-nutraceutical potential of the Amaranthus food plant: A mini-review. *Applied Sciences*, 11, 6879. https://doi.org/10.3390/app11156879

Sá, A. G. A., & House, J. D. (2024). Protein quality of cereals: Digestibility determination and processing impacts. *Journal of Cereal Science*, 117, 103892. https://doi.org/10.1016/j.jcs.2024.103892

Sabbione, A. C., Suarez, S., Añón, M. C., & Scilingo, A. (2019). Amaranth functional cookies exert potential antithrombotic and antihypertensive activities. *International Journal of Food Science and Technology*, 54(5), 1506–1513. https://doi.org/10.1111/ijfs.13930

Sady, M., Domagala, J., Grega, T., & Kalicka, D. (2007). Effect of the storing period on micro-flora in yoghurts containing amaranth seeds and oat grains added. *Żywnośċ Nauka Technolo-gia Jakosċ*, 14(6), 242–250.

Sanz-Penella, J. M., Wronkowska, M., Soral-Smietana, M., & Haros, M. (2013). Effect of whole amaranth flour on bread properties and nutritive value. *LWT - Food Science and Technology*, 50, 679–685.

Sattar, M., Saeed, F., Afzaal, M., Rasheeda, A., Asifa, A., Sharifa, S., Hussain, M., Asad Ur Rehman, H., Raza, M. A., Munira, H., & Jbawi, E. A. (2024). An overview of the nutritional and therapeutic properties of amaranth. *International Journal of Food Properties*, 27(1), 263–272. https://doi.org/10.1080/10942912.2024.2304266

Sauer, J. D. (1967). The grain amaranths and their relatives: A revised taxonomic and geographic survey. *Annals of the Missouri Botanical Garden*, 54, 103–137. https://doi.org/10.2307/2394998

Schmidt, D., Verruma-Bernardi, M. R., Forti, V. A., & Borges, M. T. M. R. (2021). Quinoa and amaranth as functional foods: A review. *Food Reviews International*, 39, 2277–2296.

Schoenlechner, R. (2017). Pseudocereals in gluten-free products. In C. M. Haros & R. Schoenlechner (Eds.), *Pseudocereals*. https://doi.org/10.1002/9781118938256.ch9

Schoenlechner, R., Siebenhandl, S., & E. Berghofer, E. (2008). Pseudo-cereals. In E. K. Arendt & F. Dal Bello (Eds.), *Gluten-Free Cereal Products and Beverages* (pp. 149-VI). Academic Press.

Seol, H., & Sim, K. H. (2017). Quality characteristics of noodles with added germinated black quinoa powder. *Korean Journal of Food and Nutrition*, 30, 19–30.

Shen, L., Li, C., Wang, W., Wang, X., Tang, D., Xiao, F., & Xia, T. (2024). Buckwheat extracts rich in flavonoid aglycones and flavonoid glycosides significantly reduced blood glucose in diabetes mice. *Journal of Functional Foods*, 113, 106029. https://doi.org/10.1016/j.jff.2024.106029

Shin, D. H., Heo, H. J., Lee, Y. J., & Kim, H. K. (2004). Amaranth squalene reduces serum and liver lipid levels in rats fed a cholesterol diet. *British Journal of Biomedical Science*, 61(1), 11–14. https://doi.org/10.1080/09674845.2004.11732639

Sindhu, R., & Khatkar, B. S. (2016). Thermal, microstructural and pasting properties of starch and flour of grain amaranthus. *Annals of Biology*, 32, 269–273.

Singh, N., Samarthb, R. M., Vashishtha, A., & Pareek, A. (2024). Amaranthus as a potential dietary supplement in sports nutrition. *CYTA–Journal of Food*, 22(1), 2375253 https://doi.org/10.1080/19476337.2024.2375253

Singh, S., Punia, D., & Khetarpaul, N. (2009). Nutrient composition of products prepared by incorporating amaranth (*Amaranthus tricolour*) leaf powder. *Nutrition & Food Science*, 39, 218–226. https://doi.org/10.1108/00346650910957465

Singhal, R. S., & Kulkarni, P. R. (1990). Effect of puffing on oil characteristics of amaranth (rajgeera) seeds. *Journal of the American Oil Chemists' Society*, 67, 952–954.

Sinkovič, L., Deželak, M., Kopinč, R., & Meglič, V. (2022). Macro/microelements, nutrients and bioactive components in common and Tartary buckwheat (Fagopyrum spp.) grain and stone-milling fractions. *LWT*, 161, 113422. https://doi.org/10.1016/j.lwt.2022.113422

Small, E. (2017). Buckwheat – the world's most biodiversity-friendly crop? *Biodiversity*, 18(2–3), 108–123. https://doi.org/10.1080/14888386.2017.1332529

Šmídová, Z., & Rysová, J. (2022). Gluten-free bread and bakery products technology. *Foods*, 11(3), 480. https://doi.org/10.3390/foods11030480

Sneha, A., & Haripriya, A. (2018). Development of amaranth grain (*Amaranthus cruentus*) based instant Dosa mix and its quality characteristics. *International Journal of Food Science and Nutrition*, 3(1), 06–11.

Sofi, S. A., Ahmed, N., Farooq, A., Rafiq, S., Zargar, S. M., Kamran, F., Dar, T. A., Mir, S. A., Dar, B. N., & Mousavi Khaneghah, A. (2023). Nutritional and bioactive characteristics of buckwheat, and its potential for developing gluten-free products: An updated overview. *Food Science & Nutrition*, 11, 2256–2276. https://doi.org/10.1002/fsn3.3166

Sonawane, S., Shams, R., Dash, K. K., Patil, V., Pandey, V. K., & Dar, A. H. (2024). Nutritional profile, bioactive properties and potential health benefits of buckwheat: A review. *eFood*, 5(4), e171. https://doi.org/10.1002/efd2.171

Soriano-García, M., & Aguirre-Díaz, I. S. (2019). Nutritional functional value and therapeutic utilization of amaranth. In V. Y. Waisundara (Ed.), In *Nutritional Value of Amaranth* (pp. 101–118). https://www.intechopen.com/books/8949

Sowdhanya, D., Singh, J., Rasane, P., Kaur, S., Kaur, J., & Kumar, M. (2023). Buckwheat sprouts. In I. Majid, B. A. Kehinde, B. Dar, & V. Nanda (Eds.), *Advances in Plant Sprouts* (pp. 35–47). Springer. https://doi.org/10.1007/978-3-031-40916-5_3

Spurthi, N., Lakshmi, K., Lakshmi, J., Raghavendra, M., & Devi, G. N. (2021). Acceptability of traditional savouries made using puffed amaranth seeds. *The Pharma Innovation Journal*, 8, 101–104.

Srivastava, S., Sreerama, Y. N., & Dharmaraj, U. (2021). Effect of processing on squalene content of grain amaranth fractions. *Journal of Cereal Science*, 100, 103218.

Stein, L., & Isaacs, R. H. (2023). *Let's Eat: Jewish Food and Faith*. Rowman & Littlefield.

Sun, X., Yang, X., Xue, P., Zhang, Z., & Ren, G. (2019). Improved antibacterial effects of alkali-transformed saponin from quinoa husks against halitosis-related bacteria. *BMC Complement Alternative Mediine*, 19, 1–10.

Suvorova, G., & Zhou, M. (2018). Chapter 3: Distribution of cultivated buckwheat resources in the world. In M. Zhou, I. Kreft, G. Suvorova, Y. Tang, & S. H. Woo (Eds.), *Buckwheat Germplasm in the World* (pp. 21–35). Academic Press. https://doi.org/10.1016/B978-0-12-811006-5.00003-3

Svirskis, Š. (2003). Investigation of amaranth cultivation and utilisation in Lithuania. *Agronomy Research*, 1(2), 253–264.

Szakály, Z., & Kiss, M. (2023). Consumer acceptance of different cereal-based "healthy foods." In M. Rakszegi, M. Papageorgiou, & J. M. Rocha (Eds.), *Developing Sustainable and Health Promoting Cereals and Pseudocereals* (pp. 467–488). Academic Press. https://doi.org/10.1016/B978-0-323-90566-4.00002-3

Tang, Y., Li, X., Zhang, B., Chen, P. X., Liu, R., & Tsao, R. (2015). Characterisation of phenolics, betanins and antioxidant activities in seeds of three Chenopodium quinoa Willd. genotypes. *Food Chemistry*, 166, 380–388. https://doi.org/10.1016/j.foodchem.2014.06.018. Epub 2014 Jun 14.

Tang, Y., & Tsao, R. (2017). Phytochemicals in quinoa and amaranth grains and their antioxidant, anti-inflammatory, and potential health beneficial effects: A review. *Molecular Nutrition & Food Research*, 61(7). https://doi.org/10.1002/mnfr.201600767

Tapia-Blácido, D.R., Sobral, P. J. A., & Menegalli, F.C. (2013). Effect of drying conditions and plasticizer type on some physical and mechanical properties of amaranth flour films. *LWT - Food Science and Technology*, 50, 392–400.

Tardy, A.-L., Pouteau, E., Marquez, D., Yilmaz, C., & Scholey, A. (2020). Vitamins and minerals for energy, fatigue, and cognition: A narrative review of the biochemical and clinical evidence. *Nutrients*, 12(1), 228. https://doi.org/10.3390/nu12010228

Thakur, P., Kumar, K., & Dhaliwal, H. S. (2021). Nutritional facts, bio-active components and processing aspects of pseudocereals: A comprehensive review. *Food Bioscience*, 42, Article 101170.

Thorssin, H. A. (2020). Paleoethnobotany in the Central Aleutian Islands: A case study from ADK-00237, an upland archaeological site on Adak Island, Alaska. University of Alaska Anchorage. *Master's Thesis*.

Turkut, G. M., Cakmak, H., Kumcuoglu, S. & Tavman, S. (2016). Effect of quinoa flour on gluten-free bread batter rheology and bread quality. *Journal of Cereal Science*, 69, 174–181. https://doi.org/10.1016/j.jcs.2016.03.005

Tömösközi, S., & Langó, B. (2017). Chapter 7: Buckwheat: Its unique nutritional and health-promoting attributes. In J. R. N. Taylor & J. M. Awika (Eds.), *Gluten-Free Ancient Grains* (pp. 161–177). Woodhead Publishing. https://doi.org/10.1016/B978-0-08-100866-9.00007-8

Tomotake, H., Shimaoka, I., Kayashita, J., Nakajoh, M., & Kato, N. (2002). Physicochemical and functional properties of buckwheat protein product. *Journal of Agricultural and Food Chemistry*, 50(7), 2125–2129. https://doi.org/10.1021/jf011248q

Ullah, A., Munir, S., Badshah, S. L., Khan, N., Ghani, L., Poulson, B. G., Emwas, A. H., & Jaremko, M. (2020). Important flavonoids and their role as a therapeutic agent. *Molecules*, 25(22), 5243. https://doi.org/10.3390/molecules25225243

Umesh, M. R., Angadi, S., Gowda, P., Ghimire, R., & Begna, S. (2019). Climate-resilient minor crops for food security. In M. Hasanuzzaman (Ed.), *Agronomic Crops*. Springer. https://doi.org/10.1007/978-981-32-9151-5_2

Valcarcel-Yamani, B., & Lannes, S. D. S. (2012). Applications of quinoa (*Chenopodium quinoa* Willd.) and amaranth (*Amaranthus* spp.) and their influence in the nutritional value of cereal based foods. *Food and Public Health*, 2(6), 265–275.

Valencia, Svanberg Ulf, Ann-Sofie Sandberg, Jenny Ruales, S. (1999). Processing of quinoa (Chenopodium quinoa, Willd): Effects on in vitro iron availability and phytate hydrolysis. *International Journal of Food Sciences and Nutrition*, 50(3), 203–211. https://doi.org/10.1080/096374899101247

Valencia-Chamorro, S. A. (2003). Quinoa. In B. Caballero (Ed.), *Encyclopedia of Food Sciences and Nutrition* (pp. 4895–4902). Academic Press. https://doi.org/10.1016/B0-12-227055-X/00995-0

Valenzuela, C., Abugoch, L., & Tapia, C. (2013). Quinoa protein–chitosan–sunflower oil edible film: Mechanical, barrier and structural properties. *LWT - Food Science and Technology*, 50(2), 531–537. https://doi.org/10.1016/J.LWT.2012.08.010

van den Berg, A. J., van den Worm, E., van Ufford, H. C., Halkes, S. B., Hoekstra, M. J., & Beukelman, C. J. (2008). An in vitro examination of the antioxidant and anti-inflammatory properties of buckwheat honey. *Journal of Wound Care*, 17(4), 172–174, 176–178. https://doi.org/10.12968/jowc.2008.17.4.28839

Vega-Galvez, A., Miranda, M., Vergara, J., Uribe, A., Puente, L., Martinez, E., A., 2010. Nutrition facts and functional potential of quinoa (Chenopodium quinoa willd.), an ancient Andean grain: A review. Published online in Wiley Online Library. https://doi.org/10.1002/jsfa.4158, from https://onlinelibrary.wiley.com/doi/10.1002/jsfa.4158/pdf

Venskutonis, P. R., & Kraujalis, P. (2013). Nutritional components of amaranth grains and vegetables: A review on composition, properties and uses. *Comprehensive Reviews in Food Science and Food Safety*, 12, 381–412.

Vicente, A., Villanueva, M., Caballero, P. A., Muñoz, J. M., & Ronda, F. (2023). Microwave modification of quinoa grains at constant and varying water content modulates changes in structural and physico-chemical properties of the resulting flours. *Foods (Basel, Switzerland)*, 12(7), 1421. https://doi.org/10.3390/foods12071421

Vilcacundo, R., Barrio, D., Carpio, C., García-Ruiz, A., Rúales, J., Hernández-Ledesma, B., & Carrillo, W. (2017). Digestibility of quinoa (Chenopodium quinoa Willd) protein concentrate and its potential to inhibit lipid peroxidation in the Zebrafish larvae model. *Plant Foods for Human Nutrition*, 72, 294–300. https://doi.org/10.1007/s11130-017-0626-1

Vilcacundo, R., & Hernández-Ledesma, B. (2017). Nutritional and biological value of quinoa (Chenopodium quinoa Willd.). *Current Opinion in Food Science*, 14, 1–6. https://doi.org/10.1016/J.COFS.2016.11.007

Vinelli, V., Biscotti, P., Martini, D., Del Bo', C., Marino, M., Meroño, T., Nikoloudaki, O., Calabrese, F. M., Turroni, S., Taverniti, V., & Unión Caballero, A. (2022). Effects of dietary fibers on short-chain fatty acids and gut microbiota composition in healthy adults: A systematic review. *Nutrients*, 14(13), 2559. https://doi.org/10.3390/nu14132559

Vitali, D., Vedrina Dragojević, I., & Šebečić, B. (2009). Effects of incorporation of integral raw materials and dietary fibre on the selected nutritional and functional properties of biscuits. *Food Chemistry*, 114, 1462–1469.

Vlachos, D., Malisova, S., Lindberg, F. A., & Karaniki, G. (2020). Glycemic index (GI) or glycemic load (GL) and dietary interventions for optimizing postprandial hyperglycemia in patients with T2 diabetes: A review. *Nutrients*, 12(6), 1561. https://doi.org/10.3390/nu12061561

Wang, S., Miao, S., & Sun, D. W. (2024). Modifying structural and techno-functional properties of quinoa proteins through extraction techniques and modification methods. *Trends in Food Science & Technology*, 143, 104285. https://doi.org/10.1016/J.TIFS.2023.104285

Wang, S., Opassathavorn, A., & Zhu, F. (2015), Quinoa bakery products. *Journal of Texture Studies*, 46, 281–292. https://doi.org/10.1111/jtxs.12128

Wang, S., & Zhu, F. (2016). Formulation and quality attributes of quinoa food products. *Food Bioprocess Technology*, 9, 49–68. https://doi.org/10.1007/s11947-015-1584-y

Weber, L. E. 1990. *Amaranth Grain Production Guide* (28 p). Rodale Press.

Wolosik, K., Zareba, I., Surazynski, A., & Markowska, A. (2017). The possible pre- and post-UVA radiation protective effect of Amaranth oil on human skin fibroblast cells. *Pharmacognosy Magazine*, 13(2), 339–343.

Wolter, A., Hager, A. S., Zannini, E., & Arendt E. K., 2013. In vitro starch digestibility and predicted glycaemic indexes of buckwheat, oat, quinoa, sorghum, teff and commercial gluten-free bread. *Journal of Cereal Science*, 58, 431–436.

Wright, K. H., Pike, O. A., Fairbanks, D. J., & Huber, C. S. 2002. Composition of Atriplex hortensis, sweet and bitter Chenopodium quinoa seeds. *Journal of Food Science*, 67(4), 1383–1385. http://dx.doi.org/10.1111/j.1365-2621.2002.tb10294.x

Yadav, A., & Yadav, K. (2024). From humble beginnings to nutritional powerhouse: The rise of amaranth as a climate-resilient superfood. *Tropical Plants*. https://doi.org/10.48130/tp-0024-0037

Yalçın, E., & Arslan, A. (2021). Wet milling technology of cereal and pseudocereals. *GIDA*, 46(2), 463–473. https://doi.org/10.15237/gida.GD20147

Yang, D., & Gao, X. (2022). Progress of the use of alternatives to malt in the production of gluten-free beer. *Critical Reviews in Food Science and Nutrition*, 62(10), 2820–2835.

Yang, Q., Luo, C., Zhang, X., Liu, Y., Wang, Z., Cacciamani, P., Shi, J., Cui, Y., Wang, C., Sinha, B., Peng, B., Tong, G., Das, G., Shah, E., Gao, Y., Li, W., Tu, Y., Qian, D., Shah, K., Akbar, M., Zhou, S., & Song, B. J. (2020). Tartary buckwheat extract alleviates alcohol-induced acute and chronic liver injuries through the inhibition of oxidative stress and mitochondrial cell death pathway. *American Journal of Translational Research*, 12(1), 70–89. https://pubmed.ncbi.nlm.nih.gov/32051738/

Yao, Y.-F., Song, X.-Y., Xie, G., Tang, Y.-N., Wortley, A. H., Qin, F., Blackmore, S., Li, C.-S., & Wang, Y.-F. (2023). New insights into the origin of buckwheat cultivation in southwestern China from pollen data. *New Phytologist*, 237(6), 2467–2477. https://doi.org/10.1111/nph.18659

Yeşil, S. (2021). The effects of fermented buckwheat, quinoa and amaranth flours and different emulsifiers on gluten-free bread quality. Karamanoğlu Mehmetbey University Graduate School of Natural and Applied Sciences Department of Food Engineering. *MsThesis*.

Yilmaz, H. Ö., Ayhan, N. Y., & Meriç, Ç. S. (2020). Current trends in nutrition and food science: A review. *Current Nutrition & Food Science*, 16(1), 29–34. https://doi.org/10.2174/1573401314666180910140021

Yu, C., Wang, D., Yang, Z., & Wang, T. (2022). Pharmacological effects of polyphenol phytochemicals on intestinal inflammation via targeting TLR4/NF-κB signaling pathway. *International Journal of Molecular Sciences*, 23(13), 6939. https://doi.org/10.3390/ijms23136939

Zhang, J., Yao, Y., Li, J., Ju, X., & Wang, L. (2023b). Impact of exopolysaccharides-producing lactic acid bacteria on the chemical, rheological properties of buckwheat sourdough and the quality of buckwheat bread. *Food Chemistry*, 425,136369. https://doi.org/10.1016/j.foodchem.2023.136369

Zhang, W., Yao, J., Huang, P., & Xing, S. (2020). Aqueous extraction of buckwheat hull and its functional application in eco-friendly dyeing for wool fabric. *Textile Research Journal*, 90(5–6), 641–654. https://doi.org/10.1177/0040517519877465

Zhang, X., Shi, J., Fu, Y., Zhang, T., Jiang, L., & Sui, X. (2023a). Structural, nutritional, and functional properties of amaranth protein and its application in the food industry: A review. *Sustainable Food Protins*, 1(1), 45–55. https://doi.org/10.1002/sfp2.1002

Zheng, C. Y., An, T., Liang, Z. T., Lv, B. H., Liu, Y. T., Hu, X. H., Zhang, Y. L., Liu, N. N., Tao, S. Y., Deng, R. X., Liu, J. X., & Jiang, G. J. (2023). Revealing the mechanism of quinoa on type 2 diabetes based on intestinal flora and taste pathways. *Food Science & Nutrition*, 11(12), 7930–7945. https://doi.org/10.1002/fsn3.3710

Zhu, F. (2020). Dietary fiber polysaccharides of amaranth, buckwheat and quinoa grains: A review of chemical structure, biological functions and food uses. *Carbohydrate Polymers*, 248, 116819. https://doi.org/10.1016/j.carbpol.2020.116819

Zhu, F. (2021). Buckwheat proteins and peptides: Biological functions and food applications. *Trends in Food Science & Technology*, 110, 155–167. https://doi.org/10.1016/j.tifs.2021.01.081

Ziarno, M., & Cichońska, P. (2021). Lactic acid bacteria-fermentable cereal- and pseudocereal-based beverages. *Microorganisms*, 9(12), 2532. https://doi.org/10.3390/microorganisms9122532

Zou, L., Wu, D., Ren, G., Hu, Y., Peng, L., Zhao, J., García-Pérez, P., Carpena, M., Prieto, M. A., Cao, H., Cheng, K.-W., Wang, M., Simal-Gandara, J., John, O. D., Rengasamy, K. R. R., Zhao, G., & Xiao, J. (2021). Bioactive compounds, health benefits, and industrial applications of Tartary buckwheat (Fagopyrum tataricum). *Critical Reviews in Food Science and Nutrition*, 63(5), 657–673. https://doi.org/10.1080/10408398.2021.1952161

5 Applicability of Pseudocereals in Novel Food Products, Their Formulations, and Consumer Perception and Market Strategy

Nighat Raza, Huma Qayyum, Muhammad Shahbaz, Umar Farooq, Ambreen Naz, Nida Firdous, and Bisma Shabbir

5.1 INTRODUCTION

Pseudocereals are gluten-free dicotyledonous plants – and a rich source of saponins, high biological value proteins, and essential nutrients – different from cereals in some aspects. Due to a presence of high biological value essential amino acids and high protein efficiency ratio (PER), these grains are superior to other cereals in nutritional value. Buckwheat, chia seeds, amaranth, quinoa, and chenopods are included in pseudocereals. Their demand in market is increasing because of their use in the diet of celiac patients as they contain no gluten. These pseudocereals reduce the risk of chronic and noncontagious diseases such as cancer, high blood pressure, and cardiovascular diseases due to the presence of phytochemicals. These are rich in antioxidants which help in combating the oxidative stress through neutralizing the free radicals and breaking oxidative chain reactions (Kaur et al., 2023).

Most common foods that are included in pseudocereals are buckwheat (*Fagopyrum* sp.), amaranth (*Amaranth* sp.), chia (*Salvia hispanica*), and quinoa (*Chenopodium quinoa* Wild) as in Figure 5.1 (Shahbaz et al., 2023). The family of *Fagopyrum* is Polygonaceae. It has three species whose origin is Western and Central China, which includes *F. cymosum* tall buckwheat, *F. sculentum* common buckwheat, and *F. tataricum* tartary buckwheat. The family of amaranth and qunioa is Chenopodiaceae which is native of the Andean region in South America (Martínez-Villaluenga et al., 2020).

Amaranth is a Greek word derived from "Anthos" which means perpetual or eternal. It is belonged to the family of Amaranthaceae. Its most common species are

DOI: 10.1201/9781003584858- 5

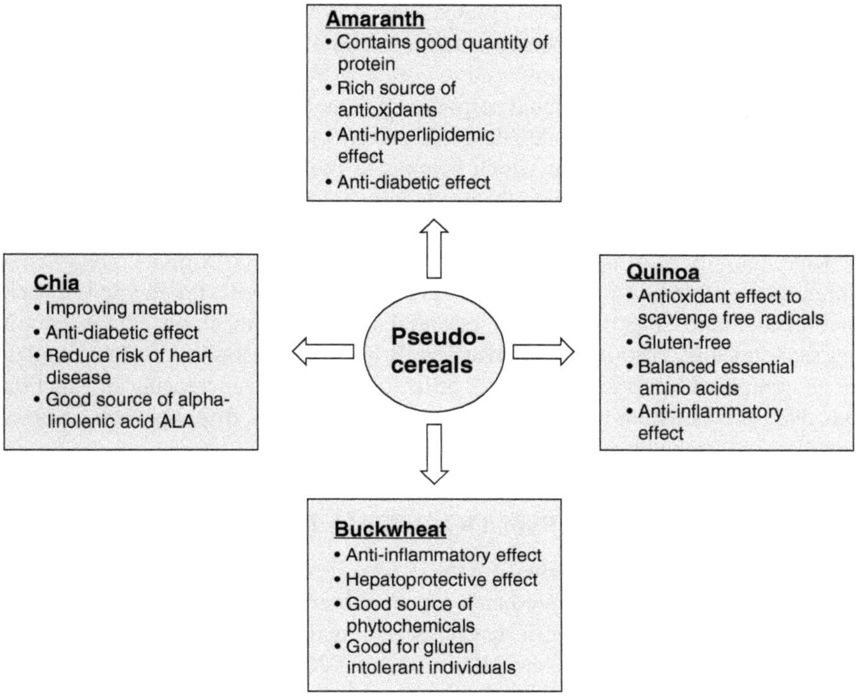

FIGURE 5.1 Types of pseudocereals.

Amaranthus crutentus, *Amaranthus caudatus*, and *Amaranthus hypochondriacus*. It is a functional food because of the good nutritional value such as rich source of omega-6 and omega-3 fatty acids, essential amino acids, good quality protein, minerals, vitamins, phenolic contents, flavonoids, dietary fiber, phytates, squalene, and tocopherols. Maize, rice, and wheat are deficient in amino acids such as tryptophan and lysine while amaranth has higher content of these amino acids. Sulfur-containing amino acids are also present in rich amount in amaranth but these are limited in the pulses (Yadav, 2022).

Quinoa is also a functional food due to its good nutritional value which depends on agronomic and environmental conditions or variety. It helps in reducing the levels of hypercholesterolemia and improving the metabolism because of the higher content of dietary fiber. It contains phenolic contents which act as metal chelators and antioxidants to scavenge free radicals. It has more protein content as compared to other cereals. Most of the protein is present in endosperm, in which 44%–77% of the protein consists of globulins and albumin. Perisperm contains carbohydrates which shows 64.2% of the dry weight of quinoa. Its lipid content is 2.0%–9.5% of the dry weight. It has essential fatty acids such as linolenic acid (8.7%–11.7%), oleic acid (19.7%–29.5%), which are the cellular components of plasma membrane and act as anti-inflammatory (Schmidt et al., 2023).

Buckwheat is a nutritionally important food as it contains vitamins, polyphenols, phenolic contents, phytosterols, d-fagomine, fagopyrins, and balanced amino acids with more protein content which is gluten-free. It prevents from chronic diseases. It has two types: tartary buckwheat, and common buckwheat. Tartary buckwheat has more important nutrients as compared to common buckwheat but it causes anaphylactic reactions with IgE-mediated response such as dermatitis, anaphylactic shock, asthma, and eczema due to some allergenic proteins (Kurćubić et al., 2024).

Chia is an herbaceous plant which belongs to the *Lamiaceae* mint family. Chia is a Latin word which means oily. Its product includes seed oil, mucilage, flour, and whole seeds. Chia seeds can be added to other dishes such as yoghurt, fruits, and salads. Rich quantity of minerals, fat, protein, dietary fiber, antioxidants, polyphenolic compounds, and omega-3 fatty acids are present in chia seeds. It helps in improving metabolism, increasing the level of essential fatty acids in blood, and preventing from diabetes and cardiovascular diseases. It has a great source of alpha-linolenic acid as compared to other sources. Its utilization has been increasing in functional foods due to its nutrition benefits that are related to chronic diseases such as cancer, heart diseases, diabetes, and obesity (Shahbaz et al., 2023).

5.2 APPLICABILITY OF PSEUDOCEREALS IN FOOD INDUSTRY

Cereals support healthy life as they are the good source of providing calories to body. Different cereals and other ingredients are combined through semi-processing and processing for consumption. Different types of cereals are rice, wheat, barley, buckwheat, quinoa, amaranth, chia, etc. (Baniwal et al., 2021).

Wheat (family: Poaceae) is a staple diet in all over the world. Worldwide production of wheat is 763.1 million tons in 2019 according to the statistical report of Food and Agriculture Organization in 2020 (Liu et al., 2020). It is used in bakery products such as bread, biscuit, cakes, pastry, and other products due to their soft character. Wheat serves as a chief ingredient in the manufacturing of flour, semolina, pasta, noodles, and fermented foods like kefir, biscuits, and desserts (Baniwal et al., 2021).

Pseudocereals have been used in the formation of biscuits, bread, noodles, baked goods, and pasta. Wheat bread is a staple diet in many counties. Although, it is a best source of carbohydrates, micronutrients, copper, and phosphorus but deficient in threonine, tryptophan, and lysine (Stamatovska et al., 2018). The lipid content in quinoa and amaranth is 2–3 times higher than wheat (Kaur et al., 2023).

Large number of populations especially Asia produced and consumed rice (grain of life) as a staple food. It is good for celiac patients and other individuals who have gluten intolerance can consume rice which is gluten-free. It contains starch such as amylopectin and amylose which is source of carbohydrates. It is good for infants due to the presence of lysine amino acid (Baniwal et al., 2021).

Many gluten-free products are available in market that made from gluten free flours such as corn, rice, and other flours but their cooking quality is not good. Pseudocereals are becoming famous due to their gluten-free property and high nutritional profile (Stamatovska et al., 2018). Pseudocereals like buckwheat and its flour also contain more phenolic compounds and have higher antioxidant values than rice and its flours (Kaur et al., 2023).

Barley belongs to family of grass like rice and wheat. It is cultivated globally in temperate or hot weather. Its origin is Eurasia, where it was produced as early as

10,000 years ago. It is also a substitute of wheat due to free from gluten protein and used in the preparation of bread. Beverages (alcoholic or nonalcoholic drinks) are the major product prepared by utilizing barley (Baniwal et al., 2021).

Pseudocereals are largely used in the food products of cereals because of the rich source of antioxidants, minerals, vitamins, dietary fiber, and essential amino acids (Stamatovska et al., 2018). These are richer in high-quality proteins, dietary fibers, unsaturated fats, and bioactive compounds (e. g., polyphenols and phytosterols) as compared to cereal grains (Nandan et al., 2024).

Maize, a major crop, was first cultivated in Central America and at the last of 15th century was introduced in Europe and other countries for food, feed, and industrial purposes. Italian Polenta, Eastern European specialties, bread and Western European specialties are the food products of maize (Revilla et al., 2022).

Pseudocereals contain high total protein content as compared to maize and rice. These have nutritional characteristics that make them better to cereals, such as the presence of essential amino acids, a high PER, and a high biological value (Kaur et al., 2023).

Pseudocereals (e. g., quinoa, amaranth, and buckwheat) have higher quantity of arginine, leucine, and lysine, which can fulfill the need of amino acids and perform physiological function in the body. Protein, dietary fiber, and carbohydrates are present in good amount in it. Nutritional profile of bread gets improved with the use of pseudocereals. Rich quantity of phytochemicals is present in buckwheat. According to the results of sensory evaluation, pseudocereals flour improving the attributes of bread such as smell, color, and taste if it is added to wheat flour. Buckwheat bread gives best results related to smell and color. Pseudocereal fortified flour especially with buckwheat may enhance the sensory attributes of bread, antioxidants, and consumer acceptance.

Quinoa, buckwheat, and amaranth are incorporated in gluten-free products as these have no gluten protein. Gluten-free bread with pseudocereals is rich in minerals, fat, protein, and fiber. Buckwheat, amaranth, and quinoa are also applied in production of noodle and pasta. Amaranth flour is used in the formulation of sugar snap cookies as a great source of protein. Wheat flour fortified with amaranth is increased the quality of cookies. In gluten-free cookies, both raw and germinated amaranth flour is incorporated. As compared to wheat and raw amaranth flour, biscuits made from germinated amaranth flour have higher number of dietary fibers and antioxidants. Wheat flour can also be fortified with quinoa of even 75% in which baking process of pancake and salt biscuit. Nutritional profile of cookies can also enhance with the use of rice flour and quinoa. Quinoa flour of 45% can be added to cookies which increase the protein content from 9.5% to 19.69% without affecting their quality.

Food industry has more focused on the production of new healthy and functional food products. Due to the great nutritional value of pseudocereals, their use in the preparation of healthy and novel food products is increasing. If staple foods are combined with pseudocereals (quinoa, buckwheat, chia, and amaranth), then pseudocereals will improve the health status of individuals due to its great nutritional composition. Nutritious products that free from gluten are highly demanded in the market that focused on the use of pseudocereals in it. Therefore, crackers, pasta, bread, and biscuits free from gluten are developed with pseudocereals (Stamatovska et al., 2018).

5.3 INDUSTRIAL PRODUCTS OF PSEUDOCEREALS

Gluten, a protein, performs technological and organoleptic functions in food. Many individuals are exposed to adverse reactions such as gluten allergy due to the increased use of gluten protein in foods (Cabanillas, 2020). IgE-mediated wheat allergy and celiac disease developed due to the consumption of gluten products. A general population have range of 0.2%–2% gluten allergies. Because of the amylase-trypsin inhibitors, non-celiac gluten sensitivity is 1.1%–6.5% prevalent with the use of gluten (Sutar & Békés, 2021). Celiac disease is an autoimmune disease in which intestine get inflamed with the damage of tissue (Valvano et al., 2020). Symptoms of gluten sensitive enteropathy, gluten sensitivity and immediate or acute allergic reaction to wheat are similar except of specific IgE or IgA anti-TG2 autoantibodies are not found in non-celiac gluten sensitivity. Gluten intake is not directly linked to celiac sprue and immediate or acute allergic reactions to wheat but these can be developed with the intake of cereal components and wheat protein. So, these gluten allergies can be managed with the use of diet that completely free from gluten (Wangtueai et al., 2020). Wheat is needed to be completely avoid in wheat allergy (Cabanillas, 2020).

Gluten protein (wheat protein) is not present in pseudocereals, for example, amaranth, chia, quinoa, tartary buckwheat, and common buckwheat. Pseudocereals have great nutritional profile that offer many health properties. According to many epidemiological studies, gluten-free products made from pseudocereals helps to heal the damaged intestine tissue. Gluten-free products of pseudocereals such like pasta, noodles, crackers, cookies, bread, and muffins have low glycemic index (GI) and predicted glycemic load as compared to refined (white) flour (Punia et al., 2022).

5.3.1 Gluten-Free Bread

Quality of nutrients and sensory attributes of gluten-free breads are not satisfactory. These gluten-free breads have large starch content due to use of rice flour and less nutrients such as protein, micronutrients and dietary fiber as in Table 5.1. According to many experimental studies, it is observed that individuals who have gluten intolerance and wheat allergy demanded nutritious, healthy, good flavor, and good texture gluten-free products. Pseudocereal foods such as quinoa, buckwheat, and amaranth are evaluated as alternative gluten-free foods which are high in biologically active components and nutrients such as dietary fiber, amino acids, and protein. These pseudocereal flours help in the management of diabetes, cancer, and cholesterol. Results showed that nutritional and physical quality of bread is getting improved. Quality of gluten-free food products can further enhance by using pseudocereal flours with other raw ingredients (Aguiar et al., 2021).

Protein content in pseudocereals is 9%–12%, and carbohydrate content in quinoa is 67%–74% than amaranth. Rheological and sensory attributes of bread get improved with the use of up to 30% buckwheat. Bread supplemented with buckwheat have more content of antioxidants which further increase antioxidants in the human body. Gluten-free bread made with amaranth incremented the wholesomeness due to the rich quantity of mineral, dietary fiber, and protein. Biscuits dough and goods are also produced with the use of two starch sources such as buckwheat and quinoa (Kaur

TABLE 5.1
Substitution of Less-Nutrient Cereals with Pseudocereals

Food Product	Cereal Mostly Used	Replaced with	Reference
Bread	Rice	Quinoa, buckwheat, and amaranth	Aguiar et al. (2021)
Pasta	Corn, rice, and other gluten-free-flours	Buckwheat, quinoa, and amaranth	Szűcs (2023)
Noodles	Rice flour and corn starch	Chia	Levent (2017)
Cookies and crackers	Wheat	Buckwheat	Xu et al. (2020)
Muffins	Wheat	Buckwheat and quinoa	Zieliński et al. (2017)
Breakfast cereal products	Wheat	Quinoa	Nirmala (2021)
Beverages	Maize, rice, and wheat	Chia, buckwheat, quinoa, and amaranth	Ignat et al. (2020)

et al., 2023). Cheese bread, which is gluten-free, has high demand in the international market. It has low content of mineral and dietary fiber due to the major ingredients such as starch and refined flour. Nutrient composition of gluten-free cheese bread like triple the amount of iron, 18 times the quantity of fiber, and 17% protein content is enhanced by supplementing amaranth flour of 10% in it (Lemos et al., 2012).

Pseudocereal flour like chia seeds are also used to increment the nutrients of food products. Bread made with 10% chia seeds has less glycemic load and GI due to soluble fiber such as chia mucilage. Linoleic acid is also present in large amount in it. Essential amino acid such as lysine that deficient in cereals but it is present in chia seeds. Soluble fiber reduces the GI by forming hydrogen bonding with water and delaying the gastric emptying as compared to insoluble fiber (Miranda-Ramos & Haros, 2020; Miranda-Ramos et al., 2020). Large quantity of phytochemicals, dietary fiber, and short chains of amylopectin that are present in pseudocereals help in reducing the GI by inhibiting the activity of amylase through impeding the starch and enzyme interaction (Xu et al., 2019; Banu & Aprodu 2020; Bączek et al., 2020; Wang et al., 2021).

5.3.2 Pasta

Pasta is becoming popular due to its taste, easy to prepare and long shelf life. It is now made with gluten-free grains due to increase demand of gluten-free products from wheat sensitive, gluten intolerant, celiac disease patients and other people who want to omit gluten-containing products for purpose of health. Many products that made from no gluten-containing flours such as corn, rice, and other flours are procurable in superstores but their cooking quality is not good as in Table 5.1. These products have low level of minerals and biologically active components as compared to wheat that intended to replace due to gluten content. It is a major challenge to prepare gluten-free pasta as gluten is the main component of it.

Pseudocereals are becoming famous due to their no gluten containing property and improved nutritional profile. Therefore, gluten-free pasta is mostly prepared from pseudocereals that contains no gluten. Ingredients in gluten-free pasta includes pseudocereals and other cereals (e. g., rye and barley), which are a good source of antioxidants, minerals, flavonoids, phenolic acids, and dietary fiber (Szűcs, 2023). Protein content of pasta is influenced its quality which plays role in physical and sensory properties. High quantity of protein and dietary fiber is present in buckwheat flour. Quality of pasta is also increased with the addition of 5% inulin (Stoican et al., 2021).

5.3.3 Noodles

Ingredients that mostly used in the preparation of gluten-free noodles are corn, starch, and rice flour but these have less nutritional value as in Table 5.1 (Levent, 2017). Pseudocereals are also used to improve the quality of noodles. Amaranth is less commonly used in making noodles due to the subpar heating tolerance, less cooking time, and final product not get properly stiffed. Noodles made with quinoa has more food loss and less flavor although it has good agglutination. Buckwheat helps in enhancing the texture of noodles and decreasing food loss during cooking as compared to quinoa. Benefits such as improved firmness, texture, cooking quality, and stability of noodles can be achieved through using all types of pseudocereals (e. g., amaranth, buckwheat, and quinoa) which reduce the weakness of each other and enhance nutritional benefits (Kaur et al., 2023).

5.3.4 Cookies and Crackers

Cookies generally have wheat flour, high saturated fat, and high sugar content as in Table 5.1. Different regions have different definition of cookies. The word biscuit is used in place of cookies in the United Kingdom and Asia while a lightly baked, dense, cake, and small leavened bread is called biscuit in America. Mixture of pseudocereal flour, legumes, and cereals such as millet, maize, rice, and sorghum are used previously in making cookie. High-sugar and fat-snap cookies are prepared from a very low gluten network as it depends on the gelatinization of starch while some gluten is required to produce hard texture of semisweet cookie. Therefore, soft wheat flour which has weak gluten strength and low gluten network is used as a major ingredient in sugar-nap cookies.

Cookie spread get reduced with the high gluten content. For these reasons, gluten-free flours are used to get desired spread and texture of cookies. In pseudocereals, buckwheat which has high nutritious value such as essential minerals, balance number of amino acids, dietary fiber, starch, and vitamin is added into maize and rice flour for preparation of cookies. Ability of fracture and hardness of cookies get reduced with the use of buckwheat in cookies. Buckwheat increases the overall acceptability and eccentricity of cookies. Buckwheat is also incorporated in the preparation of gluten-free crackers. In contrast to wheat crackers, buckwheat crackers have more antioxidant activity, total tocopherols, phenolic acid, and capacity to scavenge free radicals (Xu et al., 2020).

GI and starch digestibility of bakery products are affected by baking method, processing conditions, intensity, mixing time, and proofing time. Low GI foods help in reducing the chances of heart and blood vessel diseases, obesity, diabetes mellitus, and some kind of cancers. Nutritional value of gluten-free foods is improved with the substitutes of maize, starches, and rice. Gluten-free biscuits are prepared as an ideal product for gluten sensitive individuals (Di Cairano et al., 2021). GI also gets reduced by the presence of non-starchy polysaccharides, galactomannan, and dietary fibers. Malted flour added food products have less GI (Kaur et al., 2021).

In comparison of other cereal grains, buckwheat has high quantity of antioxidants, polyphenols, essential amino acids, and micronutrients. Quercetin and rutin are dominant polyphenols in buckwheat. Both common and tartary buckwheat have the same nutritional value, but tartary buckwheat has a bitter taste. Common buckwheat is mostly used in the preparation of cookies. Germination of tartary buckwheat flour is further enhanced the palatability and antioxidant and phytochemical properties. Malted buckwheat flour when added in cookies has low glucose-raising potential because of the high quantity of resistant starch and fiber content as compared to unmalted buckwheat flour (Molinari et al., 2018).

Polyphenols that are found in both malted and unmalted reduce the digestibility due to formation of complex structure by interacting with starch. Starch hydrolysis rate is further reduced by the interaction of polyphenols with digestive enzymes such as α-glucosidase and α -amylase. Digestion and susceptibility of starch's enzyme get reduced with amylase and protease inhibitors in buckwheat (Molinari et al., 2018). Enzymatic-hydrolysis of starch and interaction between digestive enzymes and starch get resisted when granules of starch are attached with the starch-protein and amylose–lipid complex interactions starch that formed due to high protein, fat, and lipid content in pseudocereals (Giuberti & Gallo 2018; Zhang et al., 2021).

5.3.5 Muffins

Muffins are cookies which have soft texture, sweet taste, and dense calories. In the past, eggs, milk, wheat flour, oil or fat, and sugar were used in the preparation of muffins. These traditional muffins made from wheat flour are not suitable for celiac disease patients and individuals who have wheat allergy or gluten intolerance due to the presence of gluten protein in muffins. Gluten intolerance and celiac disease patients cannot absorb several nutrients, including vitamins which require fat for absorption (vitamin D, K, E, and A), vitamin B9 (folic acid), and minerals. Nutritional value of products that contains no-gluten can be enhanced with the incorporation of nutrient-rich meal flours, diversification of meals, and supplementation of minerals (WronkoWska, 2016).

Gluten-free flours like quinoa, buckwheat, chickpea, millet, and rice can be used to make gluten-free muffins (Stoican et al., 2021). Buckwheat flour which is gluten-free provides more nutritional value when added in the preparation of muffins as compared to commercial gluten-free maize flour as in Table 5.1. Micronutrients such as manganese, zinc, magnesium, and potassium are present in higher quantity in suspension of common buckwheat flour (WronkoWska, 2016). GI is 14%–22% reduced when 20%–40% buckwheat flour is used in muffins because of the inherent flavonoids in buckwheat, which block the α-glucosidase and α-amylase during metabolism. Germinated Tartary buckwheat flour has more flavonoid content which further reduced the GI of muffins by 22%–38% (Peng et al., 2019; Bhinder et al., 2022).

5.3.6 Breakfast Cereals and Granolas

Pseudocereals like quinoa are also added to breakfast cereals as these are gluten-free and good nutritional profile as compared to wheat as in Table 5.1 (Nirmala, 2021). Extrusion cooking method is used to produce pseudocereals based granolas and breakfast cereals because of their high fat content. Starch such as rice flour must be added along with amaranth flour to improve their nutritional and product qualities. Products based on pseudocereals have low GI, which slowly absorb and digest the carbohydrates. Granolas made from amaranth or quinoa flour have good nutritional and physiochemical properties, and good acceptability in sensory evaluations. These granolas have increased shelf life, high nutritional profile, and low water activity. Sensory properties of amaranth flour were improved with the use of oligofructose and inulin (Schoenlechner, 2017). Baby breakfast cereals or hot breakfast cereals can be prepared through boiling quinoa in water (Jancurová et al., 2009). Buckwheat flour (nutraceutical food due to flavonoid content such as quercetin-3-rutinoside) is also used in the preparation of breakfast cereals, bread, and pasta (Raina & Gupta, 2015).

5.3.7 Beverage Industry

Beverages made from fermented cereals are formed to enhance flavor and nutritional value for the purpose of health (Ignat et al., 2020). Many beverage companies are making gluten-free cereal beverages due to their increasing demand. Two categories of gluten-free beverages are alcoholic gluten-free beverages and nonalcoholic gluten-free beverages. Beer is the most common alcoholic gluten-free beverages, as it is considered the third alcoholic drink that is popular in the world after tea and water. Gluten-free beer is produced by some studies in comparison to regular beer to cater gluten sensitive individuals. Many biologically active components such as prebiotics, antioxidants, probiotics, micronutrients, and fatty acids are found in gluten-free cereal-based beverages (Kerpes et al., 2017). Wheat, rice, and maize are mostly used in nonalcoholic beverages but some basic components like essential amino acids are deficient in them as in Table 5.1. Pseudocereals can improved the nutritional profile of gluten-free beverages (Ignat et al., 2020). Gluten-free beverages have many health benefits such as malted quinoa-based beverages have antidiabetic and antihypertensive properties (Queiroz et al., 2018).

5.3.8 Healthy Snacks

Worldwide prevalence of obesity is about 13% in which 40% women and 39% men were overweight in 2016 according to the WHO. In worldwide, increased prevalence of obesity is found in 5–19 years of children and adolescents and it is tripled from 4% to 18% between 1975 and 2016. Healthy eating patterns helps to lose weight in obese patients. Risk of overweight get reduced with direct relationship of energy consumption and perform physical activity to burn calories. Therefore, healthy snacks are required for children and adolescents such as crackers which are low-calorie food and nutritionally adequate and have good satiety. These types of snacks have increased shelf life, easy-to-eat and inexpensive, easily available, and can consume at any time of the day (Martín-Esparza et al., 2023). In pseudocereals, superfoods such as amaranth and quinoa are used in combination with vegetables such as zucchini, carrots, and bell pepper to produce healthy gluten-free snacks (Magro et al., 2024).

5.3.9 Animal Fodder

Animal fodder can be made from hulled and unhulled pseudocereals because of the high quantity of protein. Good results can be achieved when proper amount of buckwheat is added to animal feed. Ground and proper mixed grains to prepare animal fodder. Natural form of pseudocereal grains can be given to birds (Leiber, 2016). Animal fodder can be prepared from dry quinoa which has high amount of minerals (Zulkadir & İdikut, 2021). Rabbits and ruminants feed can be prepared from amaranth. It also helps in lowering cholesterol (Peiretti, 2018).

5.3.10 By Products of Pseudocereals

Active and intelligent films made from gelatin and polyvinyl alcohol (PVA) are produced through extract of amaranth leaf which are good source of betalains. It enhanced the practical and aesthetic qualities of films. It also used in smart packaging (Rodríguez et al., 2020). It is getting increased to replace synthetic dyes are replaced with natural pigments and amaranth is used to produce yellow and green dyes. Red pigment of amaranth plant is also used to color medicines and foods (Ramanarayanan et al., 2017). Individuals who have bone fractures, mineral deficiencies, lactating women, and growing children can consume buckwheat honey for health benefits (Belton & Taylor, 2002).

5.3.10.1 Active and Intelligent Films from Amaranth Leaf Extract (ALE) and Smart Food Packaging

Active/intelligent packaging composite food film was produced using PVA and gelatin with amaranth leaf extract (ALE) that incorporated for the purpose of fish and chicken freshness preservation. The assessment demonstrated that ALE was an excellent source of crude phenolic compounds and possessed high antioxidant betalains. Due to the presence of betalains, ALE turned color with basic pH. On the basis of the results, the impact of ALE addition on physical, mechanical, and functional properties of the film was corroborated. Coating prepared from ALE in film enhanced protection against ultraviolet light, diminished solubility in water and permeability to water vapor, and enhanced mechanical properties. The addition of ALE endowed the film with antioxidant, antibacterial, and smart characteristics. In case of preservation of chilled fish/chicken, the films were found to be active since they were able to reduce microbial activity and inhibit oxidative rancidity. This was evidenced by the film color changes from red to yellow on spoilage and associated with the increase in pH and microbial load of the spoiled samples. Hence, ALE GREEN packaging technology was adopted including PVA – gelatin films that were both active and intelligent with a likelihood of being used to package flesh food (Kanatt, 2020).

Smart packaging is trending now as it is a unique technology that is capable to control the quality and safety of food during its shelf-life time. This technology employs the office of indicators and sensors which are used in the packaging and, which are meant to track changes in the physiological alterations of the foodstuffs (as affected by microbial and chemical degradations). These indicators as a rule give knowledge, for instance, the measure of correspondence of product packed, through a color change that is easily recognizable either by the food distributor or the consumer. However, most of the existing indicators are made of nonrenewable and nonbiodegradable synthetic source. Since there is an urgent need to enhance food packaging's sustainability, the type of sensors must also meet this need (Rodrigues et al., 2021).

5.3.10.2 Natural Dyes from Amaranth

Anthocyanin, colored water-soluble natural pigments, has been widely used as photosensitizers in Dye-Sensitized Solar Cells (DSSCs). Another family of pigments is betalains with red-violet betacyanin and yellow betaxanthin, located in Caryophyllales order plants; amaranth, red beet, bougainvillea, and cacti. Another class of yellow and yellow-orange pigments called carotenoids has also been used as photosensitizers in DSSCs. Literature review has pointed out studies that have been done in order to compare the yield from anthocyanins, betalains, and chlorophyll derived from plants. Particularly, red amaranth was chosen as a candidate for extraction of two natural dyes: chlorophyll and betalain which can be used as low-cost sensitizers for wide band gap semiconductors in DSSCs according to the absorption spectra (Ramanarayanan et al., 2017).

5.3.10.3 Buckwheat Honey

Buckwheat honey is honey collected from the buckwheat flowers and has a dark brown almost black color, strong flowery smell, and taste rather like black treacle. Among honeys, this type is a good source of antioxidants that may help to decrease the level of oxidative stress brought by reactive oxygen species. The consumption of buckwheat honey, as part of the human diet, provides the body with extra antioxidant which helps potent the body's own antioxidant mechanism. Additionally, there is evidence that buckwheat honey has the ability to lower cholesterol level in the blood which can lead to a better state of heart health and even decreased high blood pressure. Furthermore, and most interestingly, buckwheat honey proved to be more effective in treating coughs associated with colds than certain over the counter cough syrups. All types of honey possess the same basic chemical composition and nutritional value, but buckwheat honey contains more macronutrients, trace elements, and vitamin C. According to some other sources, buckwheat honey is different from other types of honey in that it exhibits the very highest antibacterial and antioxidant properties (Dżugan et al., 2020).

5.3.11 Novel Source of Biologically Active Peptides

Pseudocereals like amaranth, quinoa, buckwheat, and chia are called functional foods due to their role in maintaining health. It has good quantity of protein and amino acids that are essential for our body is an important component of functional foods. Proteins in pseudocereals are also a source of biologically active peptides that perform physiological functions through release in chemical or enzymatic hydrolysis or gastrointestinal digestion. Properties such as antihypertensive, antioxidant, antidiabetic, antimicrobial, lowering cholesterol, and relieving pain are associated with biologically active peptides. These bioactive properties are due to the hydrophobicity, ionizable and polar groups, sequence of amino acids and their length and also depending on the specificity of the enzyme and substrate (Morales et al., 2021).

5.4 FORMULATION OF NOVEL FOOD PRODUCTS FROM PSEUDOCEREALS

Pseudocereals–seeds that are not true cereals but are used similarly– are gaining popularity in food products innovation for their gluten-free, high-protein, and nutrient-dense qualities. Amaranth, quinoa, and buckwheat are leading examples due to their rich nutrient profiles and functional benefits.

5.4.1 Gluten-Free Pasta

Paucar-Menacho et al. (2023) evaluated the nutritional quality of gluten-free pasta made with sprouted pseudocereal flours and cushuro. It has potential to sustain pasta production, promote health benefits and increase the antioxidant contents, bioactive components (phenolic acid, flavonoids, and polyphenols), bioavailability of minerals, and digestion of carbohydrates as in Table 5.2. Kiwicha (*Amaranthus caudatus*) and Quinoa (*Chenopodium quinoa* Wild.) were incorporated in this pasta. These were dried at 60°C for 24 hours in a climatic chamber. These were grinded in a MDNT-60XL and sieved it. Water (60g) and eggs (60g) were added in composite flour of 180g. These were mixed in bowl for 10 minutes and 100 rpm. Plastic bags were used to rest these doughs for 30 minutes. Pastaia 2 extruder (Italvisa Maquinas LTDA, and Tatui, Brazil) was used to extrude and cut these doughs. Natural air convection at 55°C was used to dry the pasta. At the end, vacuum-sealed plastic bags were used to store this dried pasta as in Figure 5.2. In comparison of control wheat pasta, pasta made with sprouted kiwicha (15%), sprouted quinoa (13%), and cushuro (8%) has high amount of soluble dietary fiber, fat, and protein content.

5.4.2 Gluten-Free Bread

Aguiar et al. (2021) prepared the gluten-free bread with pseudocereals flour and potato starch to study the effect of these flours on bread as in Figure 5.3. Gluten-free bread of 30 g formulation was prepared to analyze. Different proportions of pseudocereal flour such as 82% quinoa flour, 85% buckwheat flour and 60% amaranth flour were used with rice flour. Pseudocereal flour of 50% (quinoa, buckwheat,

TABLE 5.2
Effect of Pseudocereals on the Properties of Value-Added Food Products

Value-Added Food Products from Pseudocereals	Effect of Pseudocereals on Properties of Food Products	Reference
Gluten-free pasta	It increased the antioxidant contents, bioactive components (phenolic acids, flavonoids, and polyphenols), bioavailability of minerals, and digestion of carbohydrates	Paucar-Menacho et al. (2023)
Gluten-free bread	It increased the consistency of dough, texture, softness, volume, and acceptability of bread	Aguiar et al. (2021)
Gluten-free crackers	It will enhance the nutritional value of these crackers like soluble and insoluble fibers, magnesium, potassium, fat, and protein contents.	Turk Aslan and Isik (2022)
Tortillas	It increased the quantity of minerals (iron, zinc, magnesium, potassium, and calcium) essential amino acids, phytochemicals, and unsaturated fatty acids	Gebreil et al. (2020)
Jam	Zinc, iron, and calcium content increased by the incorporation of chia seed.	Saleem et al. (2024)

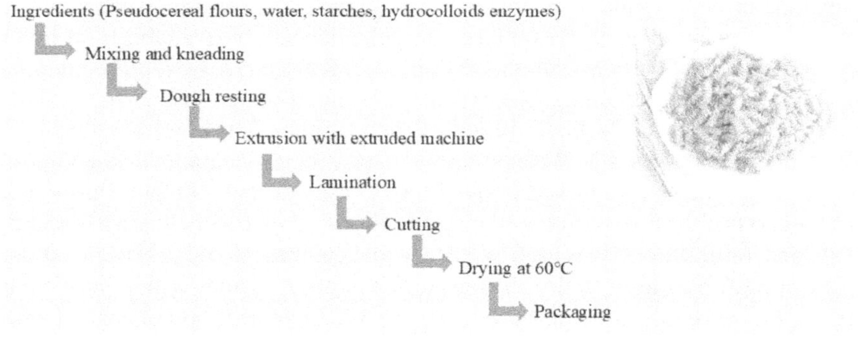

FIGURE 5.2 Flow diagram of gluten-free pasta.

and amaranth flour) and 50% rice flour has high consumer acceptability of overall liking 8. It increased the consistency of dough, texture, softness, volume, and acceptability of bread as in Table 5.2. Results showed that buckwheat flour has high dough consistency as compared to rice, quinoa, and amaranth flour which improves the quality of bread. Gluten-free bread with quinoa and buckwheat has moderate acceptability of 6.7 and 6.4 liking on a scale of 10 cm. Because of the intrinsic flavor and order of single amaranth flour bread, it has overall liking of 5.4 which has low consumer acceptability.

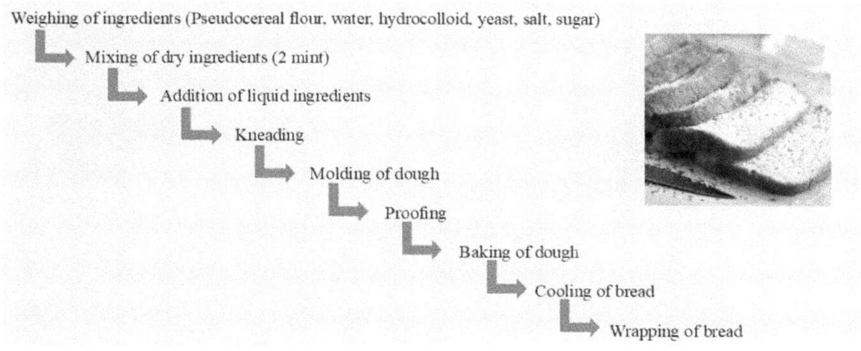

FIGURE 5.3 Flow diagram of gluten-free bread.

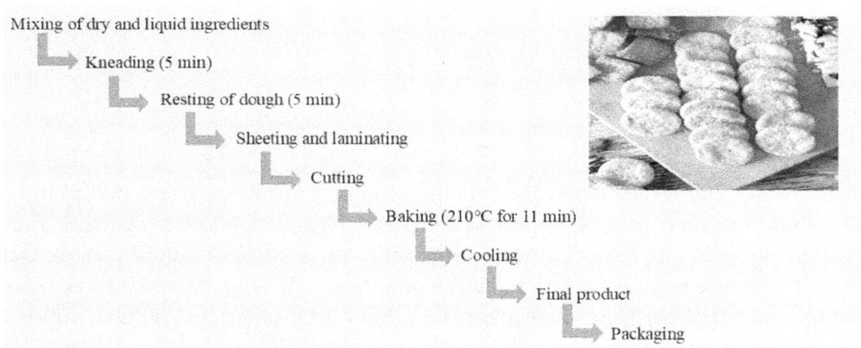

FIGURE 5.4 Flow diagram of jam.

5.4.3 Gluten-Free Crackers

Gluten-free crackers are prepared through substituting potato flour with pseudocereal flour such as quinoa, buckwheat, and amaranth flour to analyze their sensory, physical, and chemical properties. It will enhance the nutritional value of these crackers like soluble and insoluble fibers, magnesium, potassium, fat, and protein contents as in Table 5.2. Their pseudocereal-based gluten-free crackers are compared with control crackers made with corn or rice flour. Grinder was used to grind the pseudocereals (buckwheat, amaranth, and quinoa). These ground cereals were sieved through sieving machine. Raw materials without water were mixed for 1 minute. After that, water was added in dough and 5 minutes kneaded the dough. It was rested at room temperature for 5 minutes. It was thinned about 2 mm after resting period. Proper shape was made through cutting this dough. Turbo-flow ovens were used to bake the shaped sheets at 210°C for 11 minutes. Final product was packed in plastic bags as in Figure 5.4 (Turk Aslan & Isik, 2022).

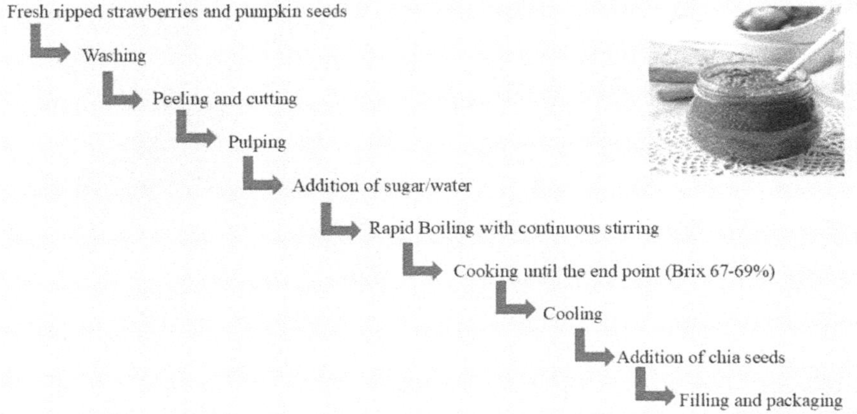

FIGURE 5.5 Flow diagram of gluten-free crackers.

5.4.4 Jam

Saleem et al. (2024) developed the functional jam of strawberry and pumpkin fortified with chia seed to assess their sensory and physiochemical properties. Fresh strawberries were cut into two parts after properly washed. Pumpkin seeds were chopped into pieces after washing and peeling. Sugar (0.4 kg), water (0.125 kg), and glucose syrup (0.2 kg) were incorporated in every 1 kg of pieces. Soluble solids of about 68% concentration or favorite consistency were obtained through boiling the mixture. It cools down at room temperature; chia seeds were added in different concentrations and stored in sterilized glass bottles, as in Figure 5.5. Zinc, iron, and calcium content increased by the incorporation of chia seeds as in Table 5.2. Sensory properties and nutritional value of strawberry and pumpkin jam improved with the addition of as high as 6%.

5.4.5 Tortillas

Tortillas are a best source of energy and good for celiac disease patients, but these lack two essential amino acids (e. g., tryptophan and lysine). To improve its nutritional status, amaranth flour is added in tortillas. Amaranth flour is added to tortillas at the levels of 25%, 50%, 75% and 100%. Water is added to the dough for mixing and then left the dough for 10 minutes. After that, divided it into pieces of 50 g. These are shaped into roller shape and baked in oven at 250°C for 5 minutes. Then cooled at room temperature as in Figure 5.6. It improved the nutritional value of tortillas, including minerals (iron, zinc, magnesium, potassium, and calcium), essential amino acids, phytochemicals, and unsaturated fatty acids, as in Table 5.2 (Gebreil et al., 2020).

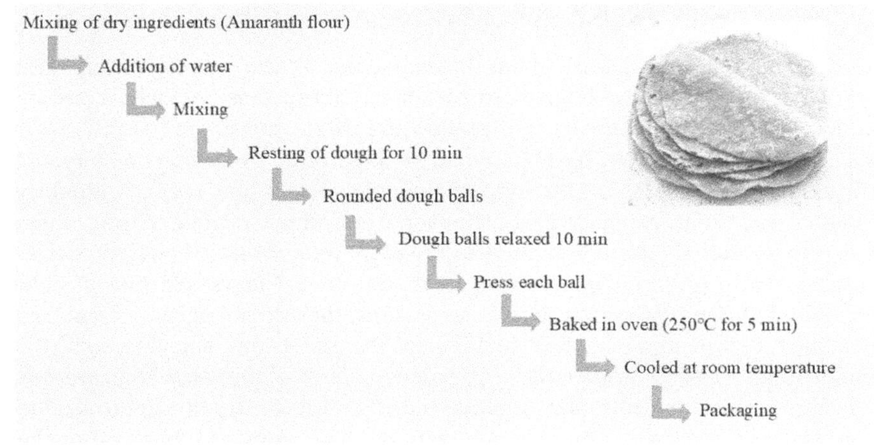

FIGURE 5.6 Flow diagram of gluten-free tortillas.

5.5 HEALTH BENEFITS OF PSEUDOCEREAL FOOD PRODUCTS

Majority population of the world eat gluten-containing staple food such as bread and pasta. Individuals who have gluten intolerance, wheat allergy, or celiac disease prefer to eat food products that contain no gluten because of the adverse reactions to gluten. Nutritional value of gluten-containing products is far better than that of gluten-free products present in the market (Larretxi et al., 2019). Therefore, nutrition-dense gluten-free products are made from cereals such as pseudocereals, sorghum, and maize that are free of gluten. Gluten-free products like pasta, breakfast cereals, granolas, cookies, crackers, beverages, and bread made from gluten-free flours especially pseudocereals have many health benefits.

Nutritional value of bread, such as bioactive ingredients (flavanols and flavan-3-ols), was incremented with the incorporation of buckwheat flour, which scavenged the free radicals due to activity of antioxidant. Another study determined that antioxidant status of gluten-free bread is improved with the addition of red and black varieties of quinoa (Verardo et al., 2018). Cookies prepared with the use of gluten-free quinoa grains have higher content of essential amino acids, fiber, and antioxidants (Martínez-Villaluenga et al., 2020).

Kahlon and Chiu (2015) studied that phytonutrients, dietary fiber, minerals, and protein contents are present in rich amount in pasta that is made with pseudocereals such as amaranth, buckwheat, and quinoa without using eggs. High content of phenolic acid, antioxidants, and low GI are properties of tartary malt cookies made with germinated tartary buckwheat as compared to tartary flour and control rice cookies (Molinari et al., 2018). Xu et al. (2020) observed that crackers prepared with wheat flour have less antioxidant activity as compared to crackers formed with the incorporation of wholegrain buckwheat flour.

5.6 CONSUMER TRUST, PERCEPTION, AND RESPONSE

The demand for high-quality foods is increasing as the population increases worldwide. More attention is given to pseudocereals such as buckwheat and quinoa and cereals. Some cereals like pseudocereals are considered suspicious by some people, although they have good nutritional profile. Consumers now focused on low-carb diets due to health concerns such as weight loss and also preferred less processed foods. Negative attitude of some consumers is created toward cereals due to suspicious attitude related to bread as they think it is overprocessed. Consumers also preferred gluten-free foods due to the increased rate of celiac disease and gluten-intolerance. For these reasons, they do not accept cereals with gluten such as wheat and have shifted toward the gluten-free ancient cereals like pseudocereals. They consider these pseudocereals as superfoods that are good for health and not causing any negative harmful effect. Organic food products are considered healthy, acceptable, and safe by consumers as compared to their substitutes (Fischer, 2023).

Healthy dietary patterns and good lifestyle are the priorities of consumers nowadays. Consumers demand high-quality gluten-free products due to health reasons such as gluten intolerance and celiac sprue. Gluten-free diet helps in improving the health of celiac disease patients. Less-fiber and gluten-free foods are followed by the individuals who have wheat allergies, gluten intolerance, herpetiformis, gluten ataxia, and dermatitis (Thakur et al., 2021). These individuals look for products that are free from gluten, but these products are very costly, low availability, and poor sensory attributes (Alvarez-Jubete et al., 2010b).

Although many gluten-free food products are available in market but these products are of poor quality due to lack of good nutritional profile (Mir et al., 2018). In comparison of products contains gluten like bakery products, gluten-free products have low quantity of fibers, proteins, and vitamin B (Thakur et al., 2021). Therefore, products with no gluten content are required to enhance their properties. Pseudocereals are used in the preparation of products without gluten protein. Hydrocolloids, emulsifiers, and enzymes are also used to increment the nutritional status of gluten-free products (Thakur et al., 2021).

Nutritional value such as polyphenols, fiber, antioxidants, iron, vitamin E, protein, and calcium content of gluten-free bread is enhanced by adding of pseudocereal flour. Gluten-free products of pseudocereals have good physical and sensory attributes (Capriles & Arêas, 2014). Food manufacturers, consumers, and researchers are paying attention to value-added pseudocereals food products. Novel food products are produced through pseudocereals due to their property of no gluten content and great nutritional profile such as high in micronutrients, protein, essential amino acids, carbohydrates, and phytochemicals. The nutritional and antioxidant effects of corn starch-based and rice-based flour are incremented by adding buckwheat flour (Alvarez-Jubete et al., 2010b). Polyphenols of pseudocereals prevent from many health issues (e. g., cancer, inflammation, oxidative stress, heart diseases, and diabetes) (Mir et al., 2018).

5.7 MARKET STRATEGY

A few gluten-free products made from pseudocereals (buckwheat, amaranth, quinoa, and chia seeds) are present in market. Value-added products of pseudocereals are limited in the market, although incorporation of pseudocereals in formulation of gluten-free products is increasing. For instance, a small number of bread, pasta, and quinoa manufacturing companies are available in countries like the US. Commercial products of buckwheat and amaranth based gluten-free products are also available in the market. Gluten intolerant and wheat-allergy individuals can get benefit by enhancing the nutritious gluten-free food products produced with pseudocereals (buckwheat, quinoa, and amaranth) in the market (Alvarez-Jubete et al., 2010a).

Pseudocereals have less allergenic effect and good nutritional value as compared to rice, maize, and wheat. Due to their functional health claims like high protein, high fiber, whole kernel grain, and less allergen response. Although pseudocereals have lot of benefits but they are less consumed worldwide due to their less availability and high prices. Production of pseudocereals must be sustained in the world to encourage the health benefits and embryonic market (Nandan et al., 2024).

Goals of food security can be fulfilled with the use of pseudocereals. Increased production of pseudocereals is important to fulfill the criteria of food security. Gluten-free products from pseudocereals, such as whole-grain gluten-free flour, could be beneficial for gluten-intolerance and celiac disease patients (Robin et al., 2015).

Effective food marketing should be created among children, adolescents, and adults through advertising campaigns on social media like Facebook, YouTube, Instagram, TikTok, television, movies, print media, celebrity and athlete endorsements, and interactive games. Digital marketing such as e-mail, social networking sites, online games, messages, and websites (Tsochantaridou et al., 2023). Target audience of pseudocereals are individuals having celiac disease, gluten intolerance, wheat allergy, overweight, obese, cancer, hypercholesterolemia, diabetes, and heart problems (Kaur et al., 2023).

5.7.1 Strategies to Improve the Acceptance of Novel Food Products of Pseudocereals

- Usage of value-added gluten-free products of pseudocereals can be increased in developing countries by climate change and understanding the molecular composition of the stored protein of them. More efforts are required to increase the consumption of pseudocereals among people as only a small number of populations use amaranth and quinoa. These elements must be considered to enhance the novel gluten-free pseudocereal products in wholesale market and superstores (Kaur et al., 2023).
- Crops with properties of climate resistance and increased nutritional value can be achieved with various techniques (e. g., genetic modification).
- Varieties of crop can also improve with the research on postharvest, nutritional (trait-specific germplasm), agro-morphological (biotic stress, abiotic, and seed shattering), and yield-related trait technologies.

- Common platform should be provided to make easily accessible these opportunities to the agriculture worker (Rodríguez et al., 2020).
- Nutritional superiority of pseudocereal must be popularized among general population through conducting awareness seminars and knowledge enrichment programs as a right food for right nutrition to switch on the desired genes.
- As compared to other cereals, market value of novel pseudocereal products should be enhanced with documented and extension activities.
- Capacity building, social acceptability, community seed bank, synergy, increased income, and livelihood help in ensuring the sustainability of activities from field to table.
- Hidden hunger of people can be alleviated with exploring the hidden potential of pseudocereals through using international funds (Langyan et al., 2024).
- Processing techniques of pseudocereals can also improve their acceptance.

5.7.2 Processing Techniques to Improve the Acceptance of Pseudocereals

Many processing techniques (e. g., microwave, cooking, extrusion, soaking, fermentation, irradiation, and dehulling) are used to remove the protein-interfering compounds and improve the nutritional value of pseudocereals. Different properties like destroying microorganisms, enhanced physical and functional properties, sterilization, and improving texture and flavor achieved during processing through heat treatments. Results that appeared in nutrients such as dietary fiber, fat, protein, and carbohydrate and anti-nutrients such as saponins, oxalates, and phytates during processing techniques are shown in Table 5.3. Digestibility and quality of protein can improve with these processing techniques. Some side effects such as denaturation of protein and changed bioavailability of essential amino acids are also reported.

Many research studies supported that nutritional value and digestibility of protein affected with the techniques of food processing in pseudocereals. Bioprocessing (biorefinery, fermentation, and enzyme-assisted processing) and conventional grain processing (baking, roasting, cooking, milling, and drying) are commonly used in food processing of pseudocereals. Some food processing techniques are mentioned below (Langyan et al., 2024).

5.7.2.1 Cooking

Nutritional value, bioavailability, and digestibility are affected by the most common and important cooking process (Kalpanadevi & Mohan, 2013). Protein digestibility is low in uncooked protein due to heat-resistant compounds. Digestibility is further affected by the interaction of non-protein components with the protein components, reduced resistance of protein from the enzymatic activity and denatured protein due to the process of cooking. The process of cooking improves the digestibility, destroying the protease inhibitors, and leaches out the unfavorable compounds. It also reduces the tannins and phytic acid contents, and eliminates trypsin inhibitors (Fawale et al., 2017). High pressure and temperature were used in extrusion cooking, which has very low cooking time and high shear forces. Because of the short cooking

TABLE 5.3
Effect of Processing Techniques on Pseudocereals

Pseudocereals	Nutrients/ Anti-nutrients	Effect of the Processing Techniques	References
Amaranth		**Nutrients**	
	Carbohydrates	Carbohydrates content reduced after germination and soaking	Thakur et al. (2021)
	Protein	Protein content increased by the techniques of germination and soaking	Sindhu et al. (2019); Thakur et al. (2021)
	Fat	Fat content reduced by using the process of germination and soaking	Cornejo et al. (2019); Sindhu et al. (2019)
	Phenols	Phenols have increased bioavailability by boiling and roasting techniques	Repo-Carrasco-Valencia et al. (2010)
	Crude fiber	Fiber have increased bioavailability by boiling and roasting techniques	Repo-Carrasco-Valencia et al. (2010)
		Anti-nutrients	
	Chymotrypsin, trypsin, and amylase inhibitors	Process of cooking eliminates the trypsin inhibitors	Fawale et al. (2017)
	Tannins	Process of cooking reduces the tannins contents	Fawale et al. (2017)
	Phytate	Process of cooking reduces the phytate contents	Fawale et al. (2017)
Quinoa		**Nutrients**	
	Carbohydrates	Soaking and germination significantly reduced the content	Thakur et al. (2021)
	Protein	Fermented quinoa sourdough markedly increased the amino acids	Rizzello et al. (2016)
	Fat	Soaking and germination significantly reduced the content	Sindhu et al. (2019); Chauhan et al. (2015); Xu et al. (2017)
	Phenols	Fermented quinoa sourdough markedly increased the polyphenols	Rizzello et al. (2016)

(Continued)

TABLE 5.3 (*Continued*)
Effect of Processing Techniques on Pseudocereals

Pseudocereals	Nutrients/ Anti-nutrients	Effect of the Processing Techniques	References
	Minerals	Iron, phosphorus, and manganese retained 100% in quinoa by steaming technique	Ramos Diaz et al. (2017)
		Anti-nutrients	
	Saponins	Wet (washing in the alkaline water) and dry techniques reduced the saponins	Ruiz et al. (2017); Filho et al. (2017)
	Phytate	Fermentation, germination, soaking, and cooking reduced the phytate content	Demir and Bilgiçli (2020); Thakur et al. (2021)
	Trypsin inhibitors	Heat treatments, cooking, and boiling reduced the trypsin inhibitors	Borges et al. (2016)
Buckwheat		**Nutrients**	
	Carbohydrates	Soaking and germination decreased the contents	Thakur et al. (2021)
	Protein	Protein content increased by fermented buckwheat flour	WronkoWska (2016)
	Fat	Fat content increased by fermented buckwheat flour	WronkoWska (2016)
		Anti-nutrients	
	Alpha-amylase and trypsin activity inhibitors	Sprouting, heat, cooking, and boiling treatments reduced the trypsin inhibitors	Handoyo et al. (2006)
	Tannins	Tannins contents reduced by the techniques of soaking and germination	Thakur et al. (2021)

time, nutrients are retained but protein is denatured due to pressure and high heat. Due to lower amylose and high lipid content, pseudocereals required a high shear force in comparison to processing of wheat and rice. Maize and rice flour which has low-fat content need to be blended with high-fat containing quinoa and amaranth flour. In extrusion cooking, buckwheat which has low fat content can be directly used (Ramos Diaz et al., 2017). Biochemical and physical properties of pseudocereals are also affected by the extrusion cooking (Kuktaite et al., 2022).

Pseudocereals are also used in the two most common cooking techniques like boiling and steam. Both these cooking techniques were observed to determine the mineral content of pseudocereals. Iron, phosphorus, and manganese retained 100% in quinoa by the steaming technique. Both boiling and steaming techniques have almost same effect on mineral contents in pseudocereals (Mota et al., 2016). Polyphenols, fiber, and mineral contents of quinoa and amaranth seeds have increased bioavailability by boiling and roasting techniques (Repo-Carrasco-Valencia et al., 2010). In comparison of raw and popped grains, amaranth has increased gruel viscosity (Muyonga et al., 2014).

5.7.2.2 Microwave and Irradiation

Frequency of range between 300 megahertz and 300 gigahertz is used in microwave energy. Molecular motion of the product continuously produces energy through non-ionizing radiation (Divekar et al., 2017). Functional properties e. g., oil and water holding capacity, emulsifying, and foaming, are improved with the use of microwaves during processing of pseudocereals. It enhanced the quality of protein, reduced the concentration of bioactive compounds, and inactivated protease inhibitors (Vagadia et al., 2018). Food under controlled environment and specific time interval is exposed to irradiation which is a safe method. It improved the quality of protein through removing unfavorable compounds and preventing various microbial diseases. But it degrades specific amino acids through negatively affecting the digestibility of protein.

5.7.2.3 Germination

A process in which embryonic axis is elongated by the absorption of water in dry/dormant seeds is called germination. This germinating process alleviates the anti-nutritional factors through activity of enzymes or leaching method in which hydrolytic enzymes enhanced the nutritional contents, changing the structural properties and biochemical changes. Increased solubility and improved digestibility of protein, and reduced polyphenols and phytic acid are recorded in germinated seeds (Albarracín et al., 2015).

Growth, development, and nutritional characteristics of germinated plant get improved by soaking in water before the process of germination. Glutamic acid decarboxylase, GABA, phenylalanine ammonia lyase activity, and rutin increased during the test of slightly acidic electrolyzed water in buckwheat (Hao et al., 2016).

5.7.2.4 Drying

Various drying techniques (e. g., freeze-, hot air-, spray-, vacuum drying, and sun) are used to extract water molecules from food. Shelf life of food is enhanced through using the spray-drying technique. It decreases the heat-labile substances of the protein and changes the functional properties and digestibility of protein. High temperature is mainly used in the spray-drying technique. Material in the form of powder is produced through hot air, which evaporate the moisture in liquid foods. In the freeze-drying technique, protein-protein interactions and aggregations increase through sublimation. It enhances the emulsifying and several functional attributes such as foaming, hydration capacity, and improves the solubility and digestibility of protein (Tontul et al., 2018).

5.7.2.5 Fermentation

Fermentation is a method in which nutritional value of crops is enhanced through utilizing various microorganisms such as bacteria and yeast. Flavor, aroma, and texture are also improved in it. It removed unwanted compounds and increased the functionality and availability of protein (Fawale et al., 2017). Aroma and palatability of pseudocereals improved through the lactic acid fermentation. Antidiabetic activity is showed in tartary buckwheat which is fermented with *L. plantarum* and *L. paracasei* (Feng et al., 2018). In an animal-based experimental study, lactic acid fermentation of pasta made with quinoa flour further enhanced the micronutrients in mouse (Carrizo et al., 2020). Polyphenols and antioxidant contents are also improved through lactic acid fermentation (Bustos et al., 2017).

Sensory and rheological properties are improved by sourdough fermentation along with reducing the saponins content (Bolívar-Monsalve et al., 2018). Fermented quinoa sourdough markedly improved the quality of bread and increased the antioxidant activity, soluble fibers, amino acids, and polyphenols (Rizzello et al., 2016). Large quantity of micro and macronutrients was showed in muffins made by fermented buckwheat flour (WronkoWska, 2016).

In an experimental study, amaranth, canihua, and quinoa grains, and flours were fermented with *Lactobacillus plantarum* 299v®. Degradation is increased more in flours than grains by this fermentation process. Endogenous phytase activity, pH, and lactic acid production increased with the incorporation of *Lactobacillus plantarum* 299v®. It also increases the bioavailability of minerals (Castro-Alba et al., 2019).

5.8 CONCLUSION

Pseudocereals have high biological value essential amino acids and high PER as compared to other cereals (e. g., maize, rice, and wheat). Their demand in market is increasing because of their use in the diet of celiac patients as they contain no gluten protein. These have been used in the formation of biscuits, bread, noodles, tortillas, baked goods, and pasta. Wheat bread is a staple diet in many countries. Although, it is a best source of carbohydrates, micronutrients, copper, and phosphorus, but deficient in threonine, tryptophan, and lysine. The lipid content in quinoa and amaranth is 2–3 times higher than wheat. Pseudocereals have been found to have a higher overall protein content than maize and rice.

REFERENCES

Aguiar, E. V., Santos, F. G., Centeno, A. C. L., & Capriles, V. D. (2021). Influence of pseudocereals on gluten-free bread quality: A study integrating dough rheology, bread physical properties and acceptability. *Food Research International*, 150, 110762.

Albarracín, M., De Greef, D., Gonzalez, R. J., & Drago, S. R. (2015). Germination and extrusion as combined processes for reducing phytates and increasing phenolics content and antioxidant capacity of Oryza sativa L. whole grain flours. *International Journal of Food Sciences and Nutrition*, 66 (8), 904–911.

Alvarez-Jubete, L., Arendt, E. K., & Gallagher, E. (2010a). Nutritive value of pseudocereals and their increasing use as functional gluten-free ingredients. *Trends in Food Science & Technology*, 21 (2), 106–113.

Alvarez-Jubete, L., Wijngaard, H., Arendt, E. K., & Gallagher, E. (2010b). Polyphenol composition and in vitro antioxidant activity of amaranth, quinoa buckwheat and wheat as affected by sprouting and baking. *Food Chemistry*, 119 (2), 770–778.

Bączek, N., Jarmułowicz, A., Wronkowska, M., & Haros, C. M. (2020). Assessment of the glycaemic index, content of bioactive compounds, and their in vitro bioaccessibility in oat-buckwheat breads. *Food Chemistry*, 330, 127199.

Baniwal, P., Mehra, R., Kumar, N., Sharma, S., & Kumar, S. (2021). Cereals: Functional constituents and its health benefits. *The Pharma Innovation*, 10 (2), 343–349.

Banu, I., & Aprodu, I. (2020). Assessing the performance of different grains in gluten-free bread applications. *Applied Sciences*, 10 (24), 8772.

Belton, P. S., & Taylor, J. R. (2002). *Pseudocereals and less common cereals: Grain properties and utilization potential*. Springer Science & Business Media.

Bhinder, S., Singh, N., & Kaur, A. (2022). Impact of germination on nutraceutical, functional and gluten free muffin making properties of Tartary buckwheat (Fagopyrum tataricum). *Food Hydrocolloids*, 124, 107268.

Borges, J., Bonomo, R., De Paula, C. D., Oliveira, L., & Cesário, M. (2016). Physicochemical and nutritional characteristics and uses of Quinoa (Chenopodium quinoa Willd.). Food and Agriculture Organization of the United Nations.

Bolívar-Monsalve, J., Ceballos-González, C., Ramírez-Toro, C., & Bolívar, G. A. (2018). Reduction in saponin content and production of gluten-free cream soup base using quinoa fermented with Lactobacillus plantarum. *Journal of Food Processing and Preservation*, 42 (2), e13495.

Bustos, A. Y., Gerez, C. L., Mohtar, L. G. M., Zanini, V. I. P., Nazareno, M. A., Taranto, M. P., & Iturriaga, L. B. (2017). Lactic acid fermentation improved textural behaviour, phenolic compounds and antioxidant activity of chia (Salvia hispanica L.) dough. *Food Technology and Biotechnology*, 55 (3), 381.

Cabanillas, B. (2020). Gluten-related disorders: Celiac disease, wheat allergy, and nonceliac gluten sensitivity. *Critical Reviews in Food Science and Nutrition*, 60 (15), 2606–2621.

Capriles, V. D., & Arêas, J. A. G. (2014). Novel approaches in gluten-free breadmaking: interface between food science, nutrition, and health. *Comprehensive Reviews in Food Science and Food Safety*, 13 (5), 871–890.

Carrizo, S. L., de LeBlanc, A. d. M., LeBlanc, J. G., & Rollán, G. C. (2020). Quinoa pasta fermented with lactic acid bacteria prevents nutritional deficiencies in mice. *Food Research International*, 127, 108735.

Castro-Alba, V., Lazarte, C. E., Perez-Rea, D., Carlsson, N. G., Almgren, A., Bergenståhl, B., & Granfeldt, Y. (2019). Fermentation of pseudocereals quinoa, canihua, and amaranth to improve mineral accessibility through degradation of phytate. *Journal of the Science of Food and Agriculture*, 99 (11), 5239–5248.

Chauhan, A., Saxena, D., & Singh, S. (2015). Total dietary fibre and antioxidant activity of gluten free cookies made from raw and germinated amaranth (Amaranthus spp.) flour. *LWT-Food Science and Technology*, 63 (2), 939–945.

Cornejo, F., Novillo, G., Villacrés, E., & Rosell, C. M. (2019). Evaluation of the physicochemical and nutritional changes in two amaranth species (Amaranthus quitensis and Amaranthus caudatus) after germination. *Food Research International*, 121, 933–939.

Demir, B., & Bilgiçli, N. (2020). Changes in chemical and anti-nutritional properties of pasta enriched with raw and germinated quinoa (Chenopodium quinoa Willd.) flours. *Journal of Food Science and Technology*, 57 (10), 3884–3892.

Di Cairano, M., Condelli, N., Caruso, M. C., Cela, N., Tolve, R., & Galgano, F. (2021). Use of underexploited flours for the reduction of glycaemic index of gluten-free biscuits: Physicochemical and sensory characterization. *Food and Bioprocess Technology*, 14 (8), 1490–1502.

Divekar, M. T., Karunakaran, C., Lahlali, R., Kumar, S., Chelladurai, V., Liu, X., Borondics, F., Shanmugasundaram, S., & Jayas, D. S. (2017). Effect of microwave treatment on the cooking and macronutrient qualities of pulses. *International Journal of Food Properties*, 20 (2), 409–422.

Dżugan, M., Grabek-Lejko, D., Swacha, S., Tomczyk, M., Bednarska, S., & Kapusta, I. (2020). Physicochemical quality parameters, antibacterial properties and cellular antioxidant activity of Polish buckwheat honey. *Food Bioscience*, 34, 100538.

Fawale, O. S., Gbadamosi, S. O., Ige, M. M., & Kadiri, O. (2017). Effects of cooking and fermentation on the chemical composition, functional, and antinutritional properties of kariya (Hildergardia barteri) seeds. *Food Science & Nutrition*, 5 (6), 1106–1115.

Feng, L., Xie, Y., Peng, C., Liu, Y., & Wang, H. (2018). A novel antidiabetic food produced via solid-state fermentation of Tartary buckwheat by L. plantarum TK9 and L. paracasei TK1501. *Food Technology and Biotechnology*, 56 (3), 373.

Filho, A. M. M., Pirozi, M. R., Borges, J. T. D. S., Pinheiro Sant'Ana, H. M., Chaves, J. B. P., & Coimbra, J. S. D. R. (2017). Quinoa: Nutritional, functional, and antinutritional aspects. *Critical Reviews in Food Science and Nutrition*, 57 (8), 1618–1630.

Fischer, A. R. (2023). Consumer attitude toward nonconventional breeding. *Developing Sustainable and Health Promoting Cereals and Pseudocereals* Chapter 19, 451–466, ISBN 9780323905664, https://doi. org/10.1016/B978-0-323-90566-4.00014-X.M

Gebreil, S. Y., Ali, M. I. K., & Mousa, E. A. M. (2020). Utilization of amaranth flour in preparation of high nutritional value bakery products. *Food and Nutrition Sciences*, 11 (5), 336–354.

Giuberti, G., & Gallo, A. (2018). Reducing the glycaemic index and increasing the slowly digestible starch content in gluten-free cereal-based foods: A review. *International Journal of Food Science & Technology*, 53 (1), 50–60.

Handoyo, T., Maeda, T., Urisu, A., Adachi, T., & Morita, N. (2006). Hypoallergenic buckwheat flour preparation by Rhizopus oligosporus and its application to soba noodle. *Food Research International*, 39 (5), 598–605.

Hao, J., Wu, T., Li, H., Wang, W., & Liu, H. (2016). Dual effects of slightly acidic electrolyzed water (SAEW) treatment on the accumulation of γ-aminobutyric acid (GABA) and rutin in germinated buckwheat. *Food Chemistry*, 201, 87–93.

Ignat, M. V., Salanţă, L. C., Pop, O. L., Pop, C. R., Tofană, M., Mudura, E., Coldea, T. E., Borşa, A., & Pasqualone, A. (2020). Current functionality and potential improvements of non-alcoholic fermented cereal beverages. *Foods*, 9 (8), 1031.

Jancurová, M., Minarovicová, L., & Dandar, A. (2009). Quinoa–a review. *Czech Journal of Food Sciences*, 27 (2), 71–79.

Kahlon, T. S., & Chiu, M.-C. M. (2015). Teff, buckwheat, quinoa and amaranth: Ancient whole grain gluten-free egg-free pasta. *Food and Nutrition Sciences*, 6 (15), 1460–1467.

Kalpanadevi, V., & Mohan, V. (2013). Effect of processing on antinutrients and in vitro protein digestibility of the underutilized legume, Vigna unguiculata (L.) Walp subsp. unguiculata. *LWT-Food Science and Technology*, 51 (2), 455–461.

Kanatt, S. R. (2020). Development of active/intelligent food packaging film containing Amaranthus leaf extract for shelf life extension of chicken/fish during chilled storage. *Food Packaging and Shelf Life*, 24, 100506.

Kaur, H., Shams, R., Dash, K. K., & Dar, A. H. (2023). A comprehensive review of pseudo-cereals: Nutritional profile, phytochemicals constituents and potential health promoting benefits. *Applied Food Research*, 3 (2), 100351.

Kaur, J., Kaur, K., Singh, B., Singh, A., & Sharma, S. (2021). Insights into the latest advances in low glycemic foods, their mechanism of action and health benefits. *Journal of Food Measurement and Characterization*, 16, 533–546. https://doi. org/10.1007/s11694-021-01179-z

Kerpes, R., Fischer, S., & Becker, T. (2017). The production of gluten-free beer: Degradation of hordeins during malting and brewing and the application of modern process technology focusing on endogenous malt peptidases. *Trends in Food Science & Technology*, 67, 129–138.

Kuktaite, R., Repo-Carrasco-Valencia, R., de Mendoza, C. C., Plivelic, T. S., Hall, S., & Johansson, E. (2022). Innovatively processed quinoa (Chenopodium quinoa Willd.) food: Chemistry, structure and end-use characteristics. *Journal of the Science of Food and Agriculture*, 102 (12), 5065–5076.

Kurćubić, V. S., Stajić, S. B., Jakovljević, V., Živković, V., Stanišić, N., Mašković, P. Z., Matejić, V., & Kurćubić, L. V. (2024). Contemporary speculations and insightful thoughts on buckwheat—A functional pseudocereal as a smart biologically active supplement. *Foods*, 13 (16), 2491.

Langyan, S., Khan, F. N., & Kumar, A. (2024). Advancement in nutritional value, processing methods, and potential applications of Pseudocereals in dietary food: A review. *Food and Bioprocess Technology*, 17 (3), 571–590.

Larretxi, I., Txurruka, I., Navarro, V., Lasa, A., Bustamante, M. Á., Fernández-Gil, M. d. P., Simón, E., & Miranda, J. (2019). Micronutrient analysis of gluten-free products: Their low content is not involved in gluten-free diet imbalance in a cohort of celiac children and adolescent. *Foods*, 8 (8), 321.

Leiber, F. (2016). Buckwheat in the nutrition of livestock and poultry. In Meiliang Zhou, Ivan Kreft, Sun Hee Woo, Nikhil Chrungoo, Gunilla Wieslander (eds), *Molecular Breeding and Nutritional Aspects of Buckwheat*, Chapter 18, . 229–238) https://doi. org/10.1016/B978-0-12-803692-1.00018-3..

Lemos, A. d. R., Capriles, V. D., Pintoe Silva, M. E. M., & Arêas, J. A. G. (2012). Effect of incorporation of amaranth on the physical properties and nutritional value of cheese bread. *Food Science and Technology*, 32, 427–431.

Levent, H. (2017). Effect of partial substitution of gluten-free flour mixtures with chia (Salvia hispanica L.) flour on quality of gluten-free noodles. *Journal of Food Science and Technology*, 54, 1971–1978.

Liu, J., Yu, L. L., & Wu, Y. (2020). Bioactive components and health beneficial properties of whole wheat foods. *Journal of Agricultural and Food Chemistry*, 68 (46), 12904–12915.

Magro, P. C., Maestro-Gaitán, I., Blázquez, M. R., Prieto, J. M., Iñiguez, F. M. S., Sobrado, V. C., & Gómez, M. J. R. (2024). Determination of nutritional signatures of vegetable snacks formulated with quinoa, amaranth, or wheat flour. *Food Chemistry*, 433, 137370.

Martín-Esparza, M. E., Raigón, M. D., García-Martínez, M. D., & Albors, A. (2023). Toward the development of potentially healthy low-energy-density snacks for children based on pseudocereal and pulse flours. *Foods*, 12 (15), 2873.

Martínez-Villaluenga, C., Peñas, E., & Hernández-Ledesma, B. (2020). Pseudocereal grains: Nutritional value, health benefits and current applications for the development of gluten-free foods. *Food and Chemical Toxicology*, 137, 111178.

Mir, N. A., Riar, C. S., & Singh, S. (2018). Nutritional constituents of pseudo cereals and their potential use in food systems: A review. *Trends in Food Science & Technology*, 75, 170–180.

Miranda-Ramos, K., Millán-Linares, M. C., & Haros, C. M. (2020). Effect of chia as breadmaking ingredient on nutritional quality, mineral availability, and glycemic index of bread. *Foods*, 9 (5), 663.

Miranda-Ramos, K. C., & Haros, C. M. (2020). Combined effect of chia, quinoa and amaranth incorporation on the physico-chemical, nutritional and functional quality of fresh bread. *Foods*, 9 (12), 1859.

Molinari, R., Costantini, L., Timperio, A. M., Lelli, V., Bonafaccia, F., Bonafaccia, G., & Merendino, N. (2018). Tartary buckwheat malt as ingredient of gluten-free cookies. *Journal of Cereal Science*, 80, 37–43.

Morales, D., Miguel, M., & Garcés-Rimón, M. (2021). Pseudocereals: A novel source of biologically active peptides. *Critical Reviews in Food Science and Nutrition*, 61 (9), 1537–1544.

Mota, C., Nascimento, A. C., Santos, M., Delgado, I., Coelho, I., Rego, A., Matos, A. S., Torres, D., & Castanheira, I. (2016). The effect of cooking methods on the mineral content of quinoa (Chenopodium quinoa), amaranth (Amaranthus sp.) and buckwheat (Fagopyrum esculentum). *Journal of Food Composition and Analysis*, 49, 57–64.

Muyonga, J. H., Andabati, B., & Ssepuuya, G. (2014). Effect of heat processing on selected grain amaranth physicochemical properties. *Food Science & Nutrition*, 2 (1), 9–16.

Nandan, A., Koirala, P., Tripathi, A. D., Vikranta, U., Shah, K., Gupta, A. J., Agarwal, A., & Nirmal, N. (2024). Nutritional and functional perspectives of pseudocereals. *Food Chemistry*, 448, 139072.

Nirmala, K. B. (2021). Development and evaluation of gluten free breakfast cereal products (*Doctoral dissertation*, DRPCAU, PUSA).

Paucar-Menacho, L. M., Vásquez Guzmán, J. C., Simpalo-Lopez, W. D., Castillo-Martínez, W. E., & Martínez-Villaluenga, C. (2023). Enhancing nutritional profile of pasta: The impact of sprouted pseudocereals and cushuro on digestibility and health potential. Foods, 12 (24), 4395.

Peiretti, P. (2018). Amaranth in animal nutrition: A review. *Livestock Research for Rural Development*, 30 (5), 1–20.

Peng, L. X., Wei, L. J., Yi, Q., Chen, G. H., Yao, Z. D., Yan, Z. Y., & Zhao, G. (2019). In vitro potential of flavonoids from tartary buckwheat on antioxidants activity and starch digestibility. *International Journal of Food Science & Technology*, 54 (6), 2209–2218.

Punia Bangar, S., Sharma, N., Singh, A., Phimolsiripol, Y., & Brennan, C. S. (2022). Glycaemic response of pseudocereal-based gluten-free food products: A review. *International Journal of Food Science & Technology*, 57 (8), 4936–4944.

Queiroz, V. A. V., daSilva Aguiar, A., deMenezes, C. B., deCarvalho, C. W. P., Paiva, C. L., Fonseca, P. C., & da Conceição, R. R. P. (2018). A low calorie and nutritive sorghum powdered drink mix: Influence of tannin on the sensorial and functional properties. *Journal of Cereal Science*, 79, 43–49.

Raina, A. P. and Gupta, V. (2015). Evaluation of buckwheat (fagopyrum species) germplasm for rutin content in seeds. *Indian Journal of Plant Physiology*, 20, 167–171.

Ramanarayanan, R., Nijisha, P., Niveditha, C., & Sindhu, S. (2017). Natural dyes from red amaranth leaves as light-harvesting pigments for dye-sensitized solar cells. *Materials Research Bulletin*, 90, 156–161.

Ramos Diaz, J. M., Sundarrajan, L., Kariluoto, S., Lampi, A. M., Tenitz, S., & Jouppila, K. (2017). Effect of extrusion cooking on physical properties and chemical composition of corn-based snacks containing amaranth and quinoa: Application of partial least squares regression. *Journal of Food Process Engineering*, 40 (1), e12320.

Repo-Carrasco-Valencia, R. A., Encina, C. R., Binaghi, M. J., Greco, C. B., & Ronayne de Ferrer, P. A. (2010). Effects of roasting and boiling of quinoa, kiwicha and kañiwa on composition and availability of minerals in vitro. *Journal of the Science of Food and Agriculture*, 90 (12), 2068–2073.

Revilla, P., Alves, M. L., Andelković, V., Balconi, C., Dinis, I., Mendes-Moreira, P., Redaelli, R., Ruiz de Galarreta, J. I., Vaz Patto, M. C., & Žilić, S. (2022). Traditional foods from maize (Zea mays L.) in Europe. *Frontiers in Nutrition*, 8, 683399.

Rizzello, C. G., Lorusso, A., Montemurro, M., & Gobbetti, M. (2016). Use of sourdough made with quinoa (Chenopodium quinoa) flour and autochthonous selected lactic acid bacteria for enhancing the nutritional, textural and sensory features of white bread. *Food Microbiology*, 56, 1–13.

Robin, F., Théoduloz, C., & Srichuwong, S. (2015). Properties of extruded whole grain cereals and pseudocereals flours. *International Journal of Food Science & Technology*, 50 (10), 2152–2159.

Rodrigues, C., Souza, V. G. L., Coelhoso, I., & Fernando, A. L. (2021). Bio-based sensors for smart food packaging—Current applications and future trends. *Sensors*, 21 (6), 2148.

Rodríguez, J. P., Rahman, H., Thushar, S., & Singh, R. K. (2020). Healthy and resilient cereals and pseudo-cereals for marginal agriculture: Molecular advances for improving nutrient bioavailability. *Frontiers in Genetics*, 11, 49.

Ruiz, K. B., Khakimov, B., Engelsen, S. B., Bak, S., Biondi, S., & Jacobsen, S.-E. (2017). Quinoa seed coats as an expanding and sustainable source of bioactive compounds: An investigation of genotypic diversity in saponin profiles. *Industrial Crops and Products*, 104, 156–163.

Saleem, M. F., Ahmed, S. A., Galali, Y., Sebo, N. H., Yildrim, A., & Najmdaddin, B. S. (2024). Physiochemical and sensory properties of pumpkin and strawberry jams fortified with chia seed (Salvia hispanica L). *Cihan University-Erbil Scientific Journal*, 8 (1), 29–35.

Schmidt, D., Verruma-Bernardi, M. R., Forti, V. A., & Borges, M. T. M. R. (2023). Quinoa and amaranth as functional foods: A review. *Food Reviews International*, 39 (4), 2277–2296.

Schoenlechner, R. (2017). Pseudocereals in gluten-free products. In Claudia Monika Haros, Regine Schonlechner (eds), *Pseudocereals: Chemistry and Technology*, Chapter 9, 193–216). https://doi. org/10.1002/9781118938256. ch9.

Shahbaz, M., Raza, N., Islam, M., Imran, M., Ahmad, I., Meyyazhagan, A., Pushparaj, K., Balasubramanian, B., Park, S., & Rengasamy, K. R. (2023). The nutraceutical properties and health benefits of pseudocereals: A comprehensive treatise. *Critical Reviews in Food Science and Nutrition*, 63 (29), 10217–10229.

Sindhu, R., Beniwal, S. K., & Devi, A. (2019). Effect of grain processing on nutritional and physico-chemical, functional and pasting properties of amaranth and quinoa flours. *Indian Journal of Traditional Knowledge (IJTK)*, *18* (3), 500–507.

Stamatovska, V., Nakov, G., Uzunoska, Z., Kalevska, T., & Menkinoska, M. (2018). Potential use of some pseudocereals in the food industry. *ARTTE*, 6 (1), 54–61.

Stoican, E.-C., Moşoiu, C.-E., & Israel-Roming, F. (2021). Gluten-free products and possibilities of new formulation for improving textural and nutritional characteristics-review. *Scientific Bulletin Series F. Biotechnologies*, 25 (1), 2285–1364.

Szűcs, K. (2023). The existence of gluten-free and functional pasta in Hungary. *Analecta Technica Szegedinensia*, 17 (3), 13–18.

Tontul, İ., Kasimoglu, Z., Asik, S., Atbakan, T., & Topuz, A. (2018). Functional properties of chickpea protein isolates dried by refractance window drying. *International Journal of Biological Macromolecules*, 109, 1253–1259.

Thakur, P., Kumar, K., Ahmed, N., Chauhan, D., Rizvi, Q. U. E. H., Jan, S., Singh, T. P., & Dhaliwal, H. S. (2021). Effect of soaking and germination treatments on nutritional, anti-nutritional, and bioactive properties of amaranth (amaranthus hypochondriacus L.), quinoa (chenopodium quinoa L.), and buckwheat (fagopyrum esculentum L.). *Current Research in Food Science*, 4, 917–925.

Tsochantaridou, A., Sergentanis, T. N., Grammatikopoulou, M. G., Merakou, K., Vassilakou, T., & Kornarou, E. (2023). Food advertisement and dietary choices in adolescents: An overview of recent studies. *Children*, 10 (3), 442.

Turk Aslan, S., & Isik, F. (2022). Effects of pseudocereal flours addition on chemical and physical properties of gluten-free crackers. *Food Science and Technology*, 42, e52521.

Vagadia, B. H., Vanga, S. K., Singh, A., Gariepy, Y., & Raghavan, V. (2018). Comparison of conventional and microwave treatment on soymilk for inactivation of trypsin inhibitors and in vitro protein digestibility. *Foods*, 7 (1), 6.

Valvano, M., Longo, S., Stefanelli, G., Frieri, G., Viscido, A., & Latella, G. (2020). Celiac disease, gluten-free diet, and metabolic and liver disorders. *Nutrients*, 12 (4), 940.

Verardo, V., Glicerina, V., Cocci, E., Frenich, A. G., Romani, S., & Caboni, M. F. (2018). Determination of free and bound phenolic compounds and their antioxidant activity in buckwheat bread loaf, crust and crumb. *LWT*, 87, 217–224.

Wang, X., Lao, X., Bao, Y., Guan, X., & Li, C. (2021). Effect of whole quinoa flour substitution on the texture and in vitro starch digestibility of wheat bread. *Food Hydrocolloids*, 119, 106840.

Wangtueai, S., Phimolsiripol, Y., Vichasilp, C., Regenstein, J. M., & Schöenlechner, R. (2020). Optimization of gluten-free functional noodles formulation enriched with fish gelatin hydrolysates. *LWT*, 133, 109977.

WronkoWska, H. Z. M. (2016). Gluten-free muffins based on fermented and unfermented buckwheat flour–content of selected elements. *Journal of Food and Nutrition Research* (ISSN 1336-8672), 55 (2), 108–113.

Xu, J., Zhang, Y., Wang, W., & Li, Y. (2020). Advanced properties of gluten-free cookies, cakes, and crackers: A review. *Trends in Food Science & Technology*, 103, 200–213.

Xu, J. M., Fan, W., Jin, J. F., Lou, H. Q., Chen, W. W., Yang, J. L., & Zheng, S. J. (2017). Transcriptome analysis of Al-induced genes in buckwheat (Fagopyrum esculentum Moench) root apex: New insight into Al toxicity and resistance mechanisms in an Al accumulating species. *Frontiers in Plant Science*, 8, 1141.

Xu, X., Luo, Z., Yang, Q., Xiao, Z., & Lu, X. (2019). Effect of quinoa flour on baking performance, antioxidant properties and digestibility of wheat bread. *Food Chemistry*, 294, 87–95.

Yadav, L. (2022). Pseudocereals: A novel path towards healthy eating. In Viduranga Y. Waisundara (ed), *Pseudocereals*. IntechOpen.

Zhang, S., Chen, S., Geng, S., Liu, C., Ma, H., & Liu, B. (2021). Effects of tartary buckwheat bran flour on dough properties and quality of steamed bread. *Foods*, 10 (9), 2052.

Zieliński, H., Ciesarová, Z., Kukurová, K., Zielinska, D., Szawara-Nowak, D., Starowicz, M., & Wronkowska, M. (2017). Effect of fermented and unfermented buckwheat flour on functional properties of gluten-free muffins. *Journal of food science and technology*, *54* (6), 1425–1432.

Zulkadir, G., & İdikut, L. (2021). The impact of various sowing applications on the nutritional value of quinoa dry herb. *Journal of Food Processing and Preservation*, 45 (12), e15730.

6 Extraction and Characterization of Pseudocereal Bioactive Compounds with their Potential Health Benefits and Applications in Functional Foods

Sabeera Muzzaffar and Munazah Sidiq

6.1 INTRODUCTION

Pseudocereals are dicotyledons distinguished from cereals based on their structure and function. These are abundant sources of proteins and essential amino acids but lack gluten, which plays an important role in agro-industries. These act as an alternative to gluten-free flour for individuals with celiac diseases. In addition, from an environmental point of view, pseudocereals enhance the variety of natural resources. They are eco- friendly, functional, economical, and budget-friendly based on their resistance to adverse conditions, excellent nutritional profile, and agro-economic. Pseudocereals have high biological value protein, roughage, and minerals. In addition, they also contain biologically active substances that are advantageous to human health (Kaur et al., 2023). Moreover, the nutritional composition of pseudocereals exceeds those of cereals, including the availability of amino acids and high biological value. They also have health-healing properties including lower chance of diabetes mellitus, cancer, heart-related conditions, hypertension, and oxidative stress brought on by the existence of phenolic substances (Mir et al., 2018). In comparison with rice and its flour, pseudocereal like buckwheat and its flour possess better phenolic composition and antioxidant activity. The oxidative load caused due to oxidation chain reactions and free radical inhibition may be increased or decreased by the consumption of pseudocereals. Furthermore, strong correlation exists between oxidative stress and chronic illness. Numerous research studies have concluded that the antioxidant activity of pseudocereal, in particular quinoa, might explain why pseudocereals might be included in our diet. As a result, numerous research carried out

DOI: 10.1201/9781003584858-6

in recent years has outlined fresh domestic and commercial uses for pseudocereals, which enable the maintenance of nutritional properties in these innovative goods and broaden their appeal to various customers. Traditionally, the main constituents of gluten-free food products have been rice and maize. Pseudocereals are becoming more popular as a substitute for gluten-free recipes. During the last decade, research on the cauterization of numerous gluten-free foods, including pseudocereals, has been expanded, primarily food such as savory and bakery food items, including drinks, pasta, and snacks. Pseudocereals possess acquired prominence and interest in the pharmaceutical and nutraceutical sectors because of their superior nutritional profile to cereals.

6.2 BIOACTIVE COMPOSITION OF PSEUDOCEREAL

Pseudocereals like buckwheat, amaranth, and quinoa are rich in bioactive compounds including phenols, flavonoids, fatty acids, and vitamins which are proven to have positive impacts on human health, such as preventing and reducing a number of degenerative disorders.

Buckwheat, a pseudocereal belonging to the Polygonaceae family, is commonly cultivated in mountainous areas, particularly in Russia and China. The Global buckwheat production is approximately 3.8 million tons, with Russia leading at 1.5 million tons, followed by China at 0.9 million tons (FAO STAT data, 2019). The crop is also cultivated in France (8.3%), USA (5.7%), Poland (5.4%), Brazil (3.5%), and Japan (1.0%). The seeds are mostly used for making of breakfast cereals like groats, flour for bakery products, and also other nutrient-rich products including bread, tea, honey, and sprouts (Małgorzata et al., 2018). Buckwheat, a functional pseudocereal enriched with nutraceutical components can alleviate health-related issues including malnutrition and celiac disease. Buckwheat-based food products have gained much interest in the market due to their appealing sensory and techno-functional qualities providing health benefits and also being suitable for those suffering from gluten sensitivity. The inclusion of buckwheat in product formulation can help address various illnesses and create nutrient-rich and gluten-free products (Begemann et al., 2021). Biologically active compounds possessing health-beneficial properties for managing chronic diseases like bioactive peptides, proteins, dietary fiber, vitamins, flavonoids, d-fagomine, fagopyrins, fagopyritols, and phenolics are present in buckwheat (Zhou, et al., 2015). In addition, its seeds are a rich source of well-balanced amino acids, including albumins, globulins, prolamins, and glutelins (Jin et al., 2022). These macromolecules are gluten-free and therefore appropriate for individuals with celiac disease. The biological value of such proteins is outstanding but the existence of anti-nutritional factors like tannins and proteases limits their edibility (Mattila et al., 2018). Among polyphenolic compounds, flavonoids and phenolic acids enhance buckwheat's nutraceutical potential. Flavonoids like rutin, orientin. isoorientin, vitexin, isovitexin, and quercitin are found in buckwheat. Among all pseudocereal, rutin is the only bioactive compound in buckwheat with high anti-inflammatory, antioxidant, and anticancer activity (Zhu, 2016). Fagopyrin, a light-sensitive polyphenolic compound is toxic due to its photosensitive nature but is effective against diabetes and also has laxative, antibiotic, and antiviral properties. Phenolic compounds include vanillic, caffeic, gallic, chlorogenic, *p*-coumaric, and *p*-hydroxybenzoic acids (Xiong et al., 2022). Li et al. (2016) reported that these polyphenolic compounds are effective against liver cells.

Quinoa (*Chenopodium quinoa*) is an annual herbaceous dicotyledon of the Amaranthaceae family. The Andean countries of South America were the first to cultivate the crop. The nutritional value, resistance to drought, and its ability to thrive in high-salinity environment led to the crop cultivation. From 8 countries in 1980 to 40 in 2010 and more than 100 in 2021, the number of countries cultivating quinoa has increased rapidly. Production is dominated by South American nations, namely Bolivia, Ecuador, and Peru, who collectively provide over 80% of global production (Bazile et al., 2016). The primary edible component of quinoa are quinoa grains, which are rich in proteins, essential amino acids, vitamins, and minerals. The grains are also gluten-free. Attributing to its nutritional qualities and health advantages, quinoa is regarded as a novel nutritious food sometimes also being identified as a "*super-food*." The UN General Assembly proclaimed 2013 as the "International Year of Quinoa" in recognition of the significance of the plant. The consumption of bioactive-rich food helps in alleviating oxidative stress and metabolic disorders like cancer, heart disease, and other cardiovascular diseases. The carotenoids, polyphenols, bioactive proteins and peptides, saponins, phytoecdysteroids, and anti-nutrient components including tannins and phytic acid are some of the bioactives found in quinoa.

Triterpenes having hydroxyl groups and a 27–30-carbon ring structure are called phytosterols. Due to their structural resemblance to human and animal cholesterol, phytosterols compete with them for intestinal absorption and reduce the formation of atherogenic lipoproteins in the liver and intestines, which lowers serum cholesterol levels. Phytosterols have been shown to have cholesterol-lowering qualities as well as antioxidant, anti-inflammatory, and anticancer benefits. The overall phytosterol content of quinoa has been observed to range from 38.8 to 82.5 mg/100 g, with the main constituents being β-sitosterol, campesterol, brassicasterol, and stigmasterol which is higher than in barley, corn, and millet (Islam et al., 2017).

Phytoecdysteroids are polyhydroxylated steroids that play a role in plant defense against insects. Apart from their protective effects, these have also been reported to strengthen dermal thickness, boost keratinocyte differentiation, and accelerate wound healing. Quinoa, whose phytoecdysteroid content ranges from 13.8 to 57.0 mg/100 g dry weight, is one of the most abundant food sources of these compounds. Twenty-hydroxyecdysone (20HE), which makes up 62%–90% of all quinoa phytoecdysteroids, is the most prevalent of the 13 phytoecdysteroids that have been found in the grain (Kumpun et al., 2011). Makisterone A, 24-epi-makisterone A, 24(28)-dehydromakisterone A and polypodine B, 24,25 dehydroinokosterone, 25,27-dehydroinokosterone, and 5b-hydroxy-24(28)-dehydromakisterone A are further minor phytoecdysteroids found in quinoa.

Secondary metabolites called saponins are mostly found in seeds, leaves, roots, fruits, and stems and are widely spread across the kingdom of plants. They are made to shield plants from insects, birds, and dangerous microbes (Singh and Kaur, 2018). Sapogenin (triterpenoid aglycone) in combination with sugar chains makes up saponin which is being classified as mono-, di- and tridesmosidic based on the number of sugar chains. The outer seed of quinoa is rich in saponins, and the saponins with bitter characteristics range from 140 to 2,300 mg/100 g dry weight, whereas those with sweet traits are found in the range of 20 to 40 mg/100 g dry weight (Guclu-Ustundag and Mazza, 2007). The complex combination of triterpene glycosides that makes up quinoa saponins includes oleanolic acid, hederagenin, phytolaccagenic acid, serjanic acid, and 3b, 23, 30-trihydroxyolean-12-en-28-oic acid. Saponins are the antinutritional factors present in the seed coat of quinoa interfering with the palatability and digestibility thereby making the removal of saponin as an important step before consumption. They are eliminated either by the dry method, which involves toasting the grains and then rubbing them to remove the outer layers, or by the wet approach, which involves washing and rubbing in cold water (Risi and Galwey, 1991). A number of biological activities including antiviral, antimicrobial, antioxidant, anti-inflammatory, cytotoxic, analgesic, hypocholesterolemic, hypoglycemic, antithrombotic, diuretic, modulatory of the absorption of minerals and vitamins, neuroprotective, and immune stimulatory have been found linked with saponins irrespective of their indigestible nature. The ability of quinoa saponins to prevent the growth of *Candida albicans* (Woldemichael and Wink, 2001) and *Botrytis cinerea* (Stuardo and Martin, 2008) was first documented in research. Additionally, 3T3-L1 adipocyte development was suppressed and cell viability was decreased by quinoa saponins (Yao et al., 2015). Based on these findings, it has been reported that quinoa saponins are a natural bioactive component that may be useful in modulating adipose tissue mass because they have the ability to decrease adipogenesis. While there is less information regarding the biological activities of quinoa saponins in animals, it has been shown that these substances can increase humoral and cellular immune responses in mice that have had subcutaneous ovalbumin immunization (Verza et al., 2012).

6.2.1 Bioactive Proteins and Peptides

Quinoa proteins have been shown to have biological effects in addition to their nutritional qualities. A protein-rich fraction of quinoa has been found to check the increase in cholesterol levels in the liver and plasma of mice by restricting the reabsorption of bile acid in small intestine and also by regulating cholesterol catabolism and production (Aluko and Monu, 2003). Moreover, quinoa proteins are now known to be a source of bioactive peptides, despite the paucity of research on the subject. As per reports provided, Aluko and Monu were the first researchers to study the antioxidant and inhibitory properties of Angiotensin-converting enzyme of quinoa hydrolysates produced using the enzyme Alcalase. In addition of this it was also discovered that quinoa protein hydrolyzates Papain had dipeptidyl peptidase IV (DPP-IV) inhibitory activity and antioxidant activity (Rizzello et al, 2017). Moreover, a finding of (Vilcacundo et al, 2017) reported the antidiabetic qualities of quinoa protein despite its ability to identify the peptide sequence causing the benefits. Peptides from various sequenced quinoa proteins have been found in fermented doughs (Vilcacundo et al, 2017) and gastrointestinal digests in recent years (Li et al., 2018) and are hence becoming a food constituent for both prevention and management of chronic diseases associated with free radical damage, high blood pressure, and diabetes due to their multipurpose activities including antioxidant, antidiabetic, and chemo preventive activities shown for some of the peptides. To validate their effects in vivo, more research needs to be done on the bioavailability and mode of action of these peptides.

6.3 EXTRACTION OF BIOACTIVE COMPOUNDS

Extraction of bioactive compounds from plant sources can be done using conventional extraction techniques. Most of these methods are based on the extracting ability of solvent and the application of heat and/or mixing. The most commonly used methods for extraction of biologically active compounds from plant sources include hydrodistillation, maceration, and Soxhlet extraction (Blicharski and Oniszczuk, 2017). Both conventional and novel methods can be used for the production of bioactives from plants. Maceration, percolation, infusion, decoction, Soxhlet extraction, and hot reflux extraction (Lang and Wai, 2001), but these methods have a drawback of time consumption as these methods take a long time and also have various draws backs therefore, creating rapid, safe, and eco-friendly technologies for analyzing and isolating bioactive compounds is essential. Isoflavones have been extracted using novel methods like ultrasound-assisted extraction (UAE), microwave-assisted extraction, supercritical fluid extraction, and pressurized liquid extraction with green solvents like ionic liquids and supercritical carbon dioxide. These techniques employ organic solvents, are time efficient, and also result in higher yields and better quality (Shang et al., 2019). Natural deep eutectic solvents, or DESs, have drawn much interest lately as effective environment-friendly solvents for extracting bioactive substances from natural resources. These technologies have been chosen based on their benefits, drawbacks, modes of operation, and types of equipment that can be extracted. The primary goals of the extraction process are to (1) extract the desired bioactive compounds from complex plant samples; (2) improve analytical methods' selectivity; (3) boost the sensitivity of bioassay by increasing the concentration of the targeted compounds; (4) transform the bioactive compounds into a form better suited for separation and detection; and (5) provide a robust, repeatable procedure that is unaffected by changes in the sample matrix (Azmir et al., 2013). Nonetheless, the fundamental workings of every solvent extraction procedure are as follows: (1) the solvent enters the solid matrix; (2) the target component dissolves in the solvents; (3) the compounds spread out of the solid matrix; and (4) the extracted compounds are gathered. According to Garavand et al. (2019), the main variables affecting extraction efficiency are the solid/solvent ratio, temperature during the extraction process, sample particle size, type of solvents employed, and extraction duration. For the solvent to permeate the sample, the particle size must be tiny, and a high temperature will result in a larger yield. However, an excessively high temperature may result in the loss of volatile chemicals during the extraction process. Since the extraction process reaches an equilibrium condition after a certain time, too much length cannot affect the extraction. A moderate ratio of substance to solvent is ideal; a too-high ratio could prolong the extraction process (Zwingelstein et al., 2020). While lab-scale extractions only need a modest number of raw ingredients and solvents, industrial scaling up of the extraction necessitates careful consideration of both economy and productivity. During the large-scale extraction process, several elements are considered, including energy consumption, kinetics, economics, instrumentation, and process type (continuous or batch) (Belwal et al., 2020).

6.3.1 Conventional Extraction

Maceration, Soxhlet extraction, percolation, infusion, digestion, decoction, and heat reflux extraction are some of the conventional techniques used to extract bioactive chemicals from plant materials (Mehmood et al., 2019). Because they need more mild or polar solvents to extract directly from biomass, these extraction methods are known as solvent extraction methods. Ethanol, methanol, hexane, n-butane, petroleum ether, and water are among the several solvent types employed in these procedures. In addition to the traditional extraction methods mentioned above, hydrodistillation, steam distillation, steam water distillation, and steam diffusion are some procedures that use water as a solvent. The amount of solvent utilized, the kinds of equipment used, and the substances extracted set different methods apart. Because Soxhlet extraction uses less solvent than maceration, it is more appropriate, although it can degrade some amount of natural compounds because of high temperature.

The type of solvent and its properties are the primary determinants of the conventional extraction processes (Gu et al., 2019; Zengin et al., 2020). Lipophilic substances and certain pigments, such as carotenoids and chlorophyll, are extracted by lower-polar solvents like petroleum ether, chloroform, etc. Nonetheless, bioactive chemicals are typically extracted using highly polar solvents (Ares et al., 2018). Furthermore, because of their acidified state, they are ideal for the extraction of flavonoids and anthocyanins (Okolie et al., 2019).

Fruits that are high in nutrients and good for human health include oranges, lemons, and bananas. Minerals and phenolic chemicals found in the peels of these fruits have antibacterial qualities and may be utilized medicinally (Parashar et al., 2014). To extract bioactive components from fruit peels, Saleem and Saeed (2020) used traditional extraction methodologies with methanol, ethyl acetate, ethanol, and distilled water. They asserted that the separated compounds possessed antibacterial properties. The microbiological inhibition concentration of the bacterium (*Klebsiella pneumoniae*) was 130 µg/ml. Water was considered a solvent at the time for the extraction of yellow lemon peels. The solvents used for the extraction were ethanol, methanol, distilled water, and ethyl acetate.

6.3.2 Enzyme Assisted Extraction

Bioactive compounds that are tightly bound inside the cell wall and challenging to extract using conventional techniques are usually extracted with the aid of enzymes. The enzymes in this method hydrolyze the cell wall, allowing the bioactive compounds in the cell plasma to escape the cell. (Nadar et al., 2018; Panja, 2018). Polygalacturonase, xylanase, pectin esterase, cellulase, hemicellulase, amylase, b-galactosidase, protease, 1,4-glycosidase, tannase, papainase, and tyrosinase are a few of the most often utilized enzymes for this purpose (Barbosa et al., 2020; Liu et al., 2016). This enzymatic hydrolysis of the cell wall enhances the release of bioactive compounds, thereby increasing the yield. The process is carried out at a low temperature ranging from 15°C to 45°C making it a useful method for extracting heat-sensitive bioactive compounds. Temperature greater than 60°C can induce structural changes, leading to molecular vibrations potentially altering the structural conformation of proteins and breaking the intercross links of peptide chains. Although enzyme-assisted extraction increases the efficiency of extraction, the enzyme–substrate model used may influence the effectiveness of the reaction and therefore customized protocols are required to be optimized (Nayak et al., 2025).

6.3.3 Ultrasound-Assisted Extraction

UAE uses ultrasound energy and solvents to extract bioactive compounds (Oliveira et al., 2016). These are mechanical waves with a frequency range of >20 kHz greater than the audible hearing frequency of humans (20 Hz–20 kHz). These waves passing through any media (solid, liquid, and gas) comprise a sequence of compression and rarefaction cycles displacing the original position of molecules. The negative pressure generated during high-intensity sound waves exceeds the attractive force holding the molecules together, resulting in the formation of cavitation bubbles. The bubbles merge and subsequently collapse during the compression phase, resulting in hot spots and localized conditions increasing the temperature to 5,000 K and pressure to 1,000 atm. The basic mechanism involved in UAE is acoustic cavitation. The collapsing bubbles and the sound waves together induce fragmentation, pore formation, increased absorption, and swelling index in the cellular matrix of plants. Ultrasound leads to erosion which is localized damage to plant tissues. The eroded part promotes the contact with the solvent thereby increasing the extraction yield. Sonoporation (presence of pores) causes the release of bioactive compounds from the cell membrane. Furthermore, the generation and collapse of cavitation bubbles encourage turbulence within the fluid resulting in the breakdown of the cell wall and contributing to the release of bioactive compounds. Moreover, the water absorption of the pomace is also accelerated by ultrasound resulting in the accessibility of water as a solvent for bioactive compounds to be extracted. The increased yield of bioactive compounds via UAE is attributed to the combined effect of all the above mechanisms (Kumar et al., 2021).

Various aspects promoting to UAE include ultrasonic power, frequency, solvent used in ultrasonic extraction, pH of the solvent, solid-liquid ratio, temperature, and time of UAE.

6.4 IDENTIFICATION AND CHARACTERIZATION OF BIOACTIVE COMPOUNDS

Following extraction, the bioactive compounds are separated, purified, and identified for confirming the presence of a particular compound and its quantity. These compounds are then classified based on their functional activities which are assessed using various biologically active analyses. Different analytical methods are employed for separating, purifying, and identifying bioactive compounds. However, the method is selected based on its simplicity, specificity, and speed (Altemimi et al., 2017). Chromatographic techniques including column chromatography are used to separate and isolate the compound of interest from the extract. The separation and identification of compounds occur based on molecular size, boiling points, ionic strength, adsorption properties, and so on (Zhang et al., 2018). Adsorption column chromatography is one of the chromatographic methods where target molecules are separated based on their affinity toward adsorbent phases like aluminum oxide, silica gel, silver nitrate, and polyamide. Conversely, counter-current chromatography and partition chromatography rely on liquid–liquid extraction where one liquid phase acts as a stationary phase and the other as a mobile phase. Additionally, in membrane filtration and gel chromatography, the separation occurs based on size where the small-sized molecules can pass through, and the larger size is retained within the membrane. In the case of ion-exchange chromatography separation takes place based on net surface charge.

Different methods are used to analyze the phenolic content, vitamins, and fatty acids. The compound quantification is done based on the standards including gallic acid, tannic acid, oleic acid, catechin, galacturonic acid, and so on. Subsequently, various bioactivity assays including radical scavenging (DPPH, ABTS), carotene bleaching, oxygen reducing power, and antimicrobial activity are employed (Ivanovic et al., 2020).

To determine molecular bonding and chemical composition after isolation, a variety of spectroscopic techniques are used. The spectroscopic approach is based on the molecule's absorption of electromagnetic radiation, which produces a spectrum that can be utilized for identification. Nuclear magnetic resonance (NMR), Fourier transform infrared spectroscopy (FTIR), UV-visible spectroscopy, and mass spectroscopy (MS) are some of the several spectroscopic techniques (Sasidharan et al., 2011). The combination of sensitive and quick analytical techniques with the spectroscopic method allows for the quick identification of bioactive substances. FTIR, NMR, MS, and other techniques are employed in conjunction with high-performance liquid chromatography (HPLC)/GC, which is the most used method for both qualitative and quantitative isolation and extract identification (Trigo et al., 2020). Furthermore, a variety of detectors, such as the Evaporative Light Scattering Detector, Electron Capture Detector, UV Diode Ray (DAD), NMR, MS, and others, are used to detect polyphenols, alkaloids, quinones, flavonoids (chromophores), saponins, and terpenes (non-chromophores). The bioactive compound's characteristics, sensitivity, and specific information, such as structural details and quantification, all affect the choice of detectors (Ivanovic et al., 2020).

Using capillary electrophoresis, Kreft et al. (1999) isolated and identified the bioactive component rutin from buckwheat. The isolation, content, and different methods of evaluation of bioactive proteins from pseudocereals were also reported by Amanda Malik and Awanish (2022). Tandem MS, electrospray ionization, and ultra-performance liquid chromatography (UPLC-ESI-TMS). In tartary buckwheat, acids such as gallic acid, hydroxybenzoic acid, p-hydorxy benzoic acid, vanillic acid, salicylic acid, caffeic acid, coumaric acid, and protocatechuic acid were found (Liu et al., 2019). For characterization of phenolics in amaranth, high-performance liquid chromatography coupled with photodiode array detection (HPLC-DAD) was employed (Pasko et al., 2008). Both qualitative and quantitative research for characterization of phenolics in quinoa were conducted using HPLC technique (Mattila et al., 2018).

6.5 HEALTH BENEFITS OF PSEUDOCEREALS

Being a rich repository of bioactive compounds pseudocereals offer various health benefits like anticancerous, antidiabetic, and cholesterol lowering. Additionally, vitamins such as vitamin E and B complex and minerals including zinc, magnesium, and potassium are also present. As pseudocereals lack gluten and are therefore safe for patients with celiac diseases. These are also rich in soluble dietary fiber aiding in digestion and also lowering in blood cholesterol. The various health benefits of pseudocereals are mentioned below:

6.5.1 Antidiabetic Activity

A typical Western diet is characterized by high consumption of sugary and fatty foods. There is currently insufficient data to support the claim that the type and quantity of dietary carbohydrates affects the postprandial glucose and insulin responses, affecting the risk of liver and biliary cancer (Fedirko et al., 2013). The glycemic index (GI) signifies an increase in an individual's blood glucose level after food consumption, which may or may not affect the carbohydrate movement toward the bloodstream concerning insulin production. Elevation in GI results in type II diabetes wherein insulin deficiency occurs within the body due to a decrease in pancreatic function. To date, few studies support the ability of amaranth seeds and oil as an antidiabetic (Kim et al., 2006). Moreover, it has also been reported that diets with low GI reduce insulin resistance besides preventing the risk of diabetes, cardiovascular diseases, and different types of cancers. Based on cooking time the GI of quinoa lies between 35 and 53. For 150 g of quinoa when subjected to cooking, refrigeration, and reheating in an oven for 1.5 minutes then its GI was reported to be 53 suggesting that after overcooking quinoa pseudocereal retains a low GI value. Further, a study by Berti et al. (2005) reported that quinoa consumption significantly reduced triglycerides and free fatty acids. High levels of free fatty acids alter glucose regulation increasing triglycerides. Thus, consuming pseudocereals helps reduce the amount of free fatty acids and maintain better health.

6.5.2 Anticancerous

Cancer, one of the chronic diseases, is considered to be affected by the diet. According to estimated reports, normal cells can generate into clinically identified tumors within the period of 10–30 years, and food intake may either accelerate or retard the growth of cancerous cells (Thakur et al., 2021). It has also been reported that the consumption of pseudocereals like buckwheat helps reduce the risk of cancers (Shen et al., 2008). The intake of antioxidant-rich pseudocereals helps in retarding the development of cancerous cells by shielding the DNA from oxidative damage. Studies have shown that the ethanolic and methanolic extracts of tartary and common buckwheat can protect the DNA from injury caused by hydroxyl radicals (Cao et al., 2008). Various colorectal cancers have shown preventive responses to rutin and its aglycone quercetin by inhibiting cell growth by the mechanism of apoptosis (cell death), demonstrating antioxidant/anti-inflammatory properties at levels as low as 0.5 μmol/L. Flavones from golden buckwheat have been reported to inhibit gastric cancer cells (Zhang et al., 2013). Moreover, ethanolic extracts of quinoa show an inhibitory effect on rat prostate cancer cells (Gawlik-Dziki et al., 2013). In addition, proteins and peptides of buckwheat pseudocereal have been reported to have chemopreventive effects on various cancer cell lines. These effects are associated with the downregulation of mRNA expression contributing to the development of colon cancer cells (Buzzi et al., 2009) and also fragmentation of DNA, cytosolic chromosome expression, and disruption of mitochondrial membrane potential (de Mejia and Dia, 2010).

6.5.3 Hypocholesterolemic

Due to the hypocholesterolemic activity, pseudocereals are becoming the subject of interest for research. Consumption of pseudocereals and their products has been reported to reduce blood glucose levels by 14.5%, triglycerides by 4.85%, total cholesterol by 4%–13%, and low-density lipoprotein (LDL) by 9.05% (Wieslander et al., 2011). The tendency to expel bile acid is connected with the buckwheat proteins lipid-lowering properties. It has been researched that including buckwheat protein in the diet significantly reduces the cholesterol level in the serum, gall bladder, and liver of hamsters (Tomotake et al., 2002). Various human studies have examined the hypolipidemic activity of food products developed from quinoa-like cereal bars. It has been reported that regular intake of cereal bars by 22 persons in the age group of 18–45 years for 30 days shows a pronounced decline in triglycerides, total cholesterol, and LDL and also reduces body weight and blood pressure (Farinazzi-Machado et al., 2012). As per the study of Berger et al. (2005), hamsters fed with a diet consisting of amaranth oil (5%) and *Amaranthus* grain (20%) showed a decrease in non-HDL cholesterol. In addition, a study by Shin et al. (2004) reported that squalene concentration of amaranth enhances the steroid excretion from the body through feces, thereby contributing to the hypocholesterolemic effect. Furthermore, a 50% reduction in LDL and total cholesterol in rabbits with hypercholesterolemic fed with extruded amaranth and defatted flour has been reported (Plate and Arêas, 2002).

6.5.4 Anti-inflammatory

Anti-inflammation is the ability of a substance to decrease the inflammation in the body usually caused by injury or irritation hallmarked by redness, swelling, or pain. Anti-inflammatory drugs or agents prevent inflammation in the body by altering the production of chemicals that cause inflammation. Pseudocereals are one of the anti-inflammatory agents that prevent the body from inflammation. The polyphenolic composition of quinoa has shown the downregulation of interleukin 1 (IL-1), IL-8, and tumor necrosis factor cytokines in cultured colonic epithelial cells of experimental rats reduces inflammation and supports gastrointestinal health (Noratto et al., 2015). The saponins present in quinoa seed have also been documented to be linked with the inhibition of inflammatory cytokines depicting that quinoa seeds could be used as functional food for the prevention of inflammation (Yao et al., 2014). Betacyanins and selenium of amaranth seed sprout are reported to have anti-inflammatory properties by significantly lowering the pro-inflammatory markers in IL-6 in raw macrophages and also prevents movement of nuclear kappa factor B (NF-ķB) toward nucleus (Tyszka-Czochara et al., 2016).

6.6 APPLICATIONS OF PSEUDOCEREALS IN FUNCTIONAL FOODS

From ancient times, pseudocereals have been used in the formulation of extruded and popped snacks, beverages, soups, and tortillas but in recent years their use has been widely spread in baked goods, plant-based meat, fermented, and non-fermented dairy beverages. Pseudocereals have now been employed as a constituent in the development of bread so as to enhance its nutritional profile. A study by Chlopicka et al. (2012) documented the influence of the addition of 15% and 30% of pseudocereal flour including buckwheat, quinoa, and amaranth on the antioxidant capacity of bread. Addition of 30% flour was found significantly effective in enhancing the properties of bread. The incorporation of buckwheat flour to wheat flour was found to be more effective than other pseudocereal flour in increasing the antioxidant potential of bread formulated from pseudocereal flour. The sensory parameters including color were also improved in bread fortified with buckwheat flour.

The usage of pseudocereal in gluten free products has also been studied. A study by Alvarez-Jubete et al. (2010) documented the usage of pseudocereal in development of gluten free bread. The gluten free bread exhibited notably higher level of fats, protein, and minerals compared to normal bread. Similarly, pseudocereal flour has also been used as a formulating agent in pasta and noodles. Due to rich polyphenolic composition buckwheat flour can find its use in formulating cookies, crackers, and biscuits. Gluten free cookies with varying ratios of rice and buckwheat flour were developed by Torbica et al. (2012). Furthermore, a study by Starowicz et al. (2017) developed a rye-buckwheat ginger cake fortified with rutin for enhancement of nutritional profile. Additionally, Atef et al. (2012) studied the mixing of quinoa meal with wheat flour at 75% without altering the baking properties and those with 100% were found acceptable.

6.7 CONCLUSION

The extraction, identification, and characterization of bioactive compounds from pseudocereals exhibits promising possibilities for functional foods. The bioactive compounds with numerous health benefits substantially contributes to better wellness and nutrition. The regular intake of pseudocereals into diet helps in prevention and management of different health issues thereby providing a natural and sustainable approach for better health. With progress in research, pseudocereal-based functional foods have a lot of potential for the advancement of public health and nutrition science.

REFERENCES

Altemimi, A., Lakhssassi, N., Baharlouei, A., Watson, D. G., Lightfoot, D. A. (2017 Sep 22). Phytochemicals: Extraction, isolation, and identification of bioactive compounds from plant extracts. *Plants (Basel)*, 6(4), 42.

Aluko, R.E., & Monu, E. (2003). Functional and bioactive properties of quinoa seed protein hydrolysates. *Food Chemistry and Toxicology*, 68, 1254–1258.

Alvarez-Jubete, L., Arendt, E. K. & Gallagher, E. (2010). Nutritive value of pseudocereals and their increasing use as functional gluten – free ingredients. *Trends in Food Science & Technology*, 21, 106–113.

Amanda, M. M., & Awanish, S. (2022). Pseudocereals proteins- A comprehensive review on its isolation, composition and quality evaluation techniques. *Food Chemistry Advances*, 1, 277–283.

Ana M. Ares, Silvia Valverde, José L. Bernal, María J. Nozal, & José Bernal (2018). Extraction and determination of bioactive compounds from bee pollen. *Journal of Pharmaceutical and Biomedical Analysis*, 147, 110–124.

Atef, A., Abou-Zaid, M. A., & El-Faham Wafaa Y. S. (2012). Use of Quinoa Meal to Produce Bakery Products to Celiac and Autism Stuffs. *International Journal of Science and Research*, 3(9), 1344–1354.

Azmir, J., Zaidul, I. S. M., Rahman, M., Sharif, K., Mohamed, A., Sahena, F., Jahurul, M., Ghafoor, K., Norulaini, N., & Omar, A. (2013) Techniques for extraction of bioactive compounds from plant materials: A review. *Journal of Food Engineering*, 117(4), 426–436.

Barbosa, J. R., dos Santos, F. M. M., da Silva, M. L. H., & de Carvalho, J. R. N. (2020). Polysaccharides of mushroom *Pleurotus* spp.: New extraction techniques, biological activities and development of new technologies. *Carbohydrate Polymers*, 229, 115550.

Bazile, D., Jacobsen, S.-E., & Verniau A. (2016). The global expansion of quinoa: Trends and limits. *Frontiers in Plant Science*, 7, 622.

Begemann, J., Ostovar, S., & Schwake-Anduschus, C. (2021). Facing tropane alkaloid contamination in millet–analytical and processing aspects. *Quality Assurance and Safety of Crops & Foods*, 13(2), 79–86.

Belwal, T., Chemat, F., Venskutonis, P. R., Cravotto, G., Jaishwal, D. K., Bhatt, I. D., Devkota, H. P., & Luo, Z. (2020). Recent advances in scaling-up of non-conventional extraction techniques: Learning from successes and failures. *Trends Analytical Chemistry*, 127, 115895.

Berger, A., Gremaud, G., Baumgartner, M., Rein, D., Monnard, I., Kratky, E., Geiger, W., Berti, C., Riso, P., Brusamolino, A., & Porrini, M. (2005). Effect on appetite control of minor cereal and pseudocereal products. *British Journal of Nutrition*, 94(5), 850–858.

Berti, C., Riso, P., Brusamolino, A., & Porrini, M. (2005). Effect on appetite control of minor cereal and pseudocereal products. *British Journal of Nutrition*, 94(5), 850–858.

Blicharski, T., & Oniszczuk, A. (2017). Extraction methods for the isolation of isoflavonoids from plant material. *Open Chemistry*, 15, 34–45.

Buzzi, N., Colicheo, A., Boland, R. & de Boland, A. R. (2009). MAP kinases in proliferating human colon cancer Caco-2 cells. *Molecular and Cell Biochemistry*, 328, 201–208.

Cao, W., Chen, W., Suo, Z., & Yao, Y. (2008). Protective effects of ethanolic extracts of buckwheat groats on DNA damage caused by hydroxyl radicals. *Food Research International*, 41, 924–929.

Chlopicka, J., Pasko, P., Gorinstein, S., Jedryas, A., & Zagrodzki, P. (2012). Total phenolic and total flavonoid content, antioxidant activity and sensory evaluation of pseudocereal breads. *LWT - Food Science and Technology*, 46, 548–555.

Oliveira, C.F.d., Giordani, D., Lutckemier, R., Gurak, P. D., Olivera, F.C., Marczak, L. D. F. (2016). Extraction of pectin from passion fruit peel assisted by ultrasound, LWT - Food Science and Technology.

de Mejia, E. G., & Dia, V. P. (2010) The role of nutraceutical proteins and peptides in apoptosis, angiogenesis, and metastasis of cancer cells. *Cancer Metastasis Reviews*, 29, 511–528.

FAO STAT data. (2019). Online database. https://faostat.fao.org. 2019

Farinazzi-Machado, F. M. V., Barbalho, S. M., Oshiiwa, M., Goulart, R., & Pessan, J. S. (2012). Use of cereal bars with quinoa (Chenopodium quinoa W.) to reduce risk factors related to cardiovascular diseases. *Food Science and Technology*, 32(2), 239–244.

Fedirko, V., Lukanova, A., Bamia, C., Trichopolou, A., Trepo, E., Nöthlings, U., Schlesinger, S., Aleksandrova, K., Boffetta, P., Tjønneland, A., & Johnsen, N. F. (2013). Glycemic index, glycemic load, dietary carbohydrate, and dietary fiber intake and risk of liver and biliary tract cancers in Western Europeans. *Annals of Oncology*, 24, 543–553.

Garavand, F., Rahaee, S., Vahedikia, N., & Jafari, S. M. (2019). Different techniques for extraction and micro/nanoencapsulation of saffron bioactive ingredients. *Trends Food Science and Technology*, 89, 26–44.

Gawlik-Dziki, U., Świeca, M., Sułkowski, M., Dziki, D., Baraniak, B., & Czyż, J. (2013). Antioxidant and anticancer activities of Chenopodium quinoa leaves extracts – in vitro study. *Food and Chemical Toxicology*, 57, 154–160.

Gu, L. B., Zhang, G. J., Du, L., Du, J., Qi, K., Zhu, X. L., Zhang, X. Y., & Jiang, Z. H. (2019). Comparative study on the extraction of *Xanthoceras sorbifolia* Bunge (yellow horn) seed oil using subcritical n-butane, supercritical CO_2, and the Soxhlet method. *LWT*, 111, 548–554.

Guclu-Ustundag, O., & Mazza, G. (2007). Saponins: Properties, applications and processing. *Critical Reviews in Food Science Nutrition*, 47, 231–258.

Ivanović, M., Islamčević Razboršek, M., Kolar, M. (2020). Innovative extraction techniques using deep eutectic solvents and analytical methods for the isolation and characterization of natural bioactive compounds from plant material. *Plants*, 9, 1428.

Islam, M. A., Jeong, B. G., Jung, J., Shin, E. C., Choi, S. G., & Chun, J. (2017). Phytosterol determination and method validation for selected nuts and seeds. *Food Analytical Methods*, 10, 3225–3234.

Jin, J., Ohanenye, I. C., & Udenigwe, C. C. (2022). Buckwheat proteins: Functionality, safety, bioactivity, and prospects as alternative plant-based proteins in the food industry. *Critical Reviews in Food Science and Nutrition*, 62(7), 1752–1764.

Kaur, H., Shams, R., Dash, K. K., & Dar, A. H. (2023). A comprehensive review of pseudo-cereals: Nutritional profile, phytochemicals constituents and potential health promoting benefits. *Applied Food Research*, 3, 100351.

Kim, H. K., Kim, M. J., & Cho, H. Y., Kim, E. K., & Shin, D. H. (2006). Antioxidative and anti-diabetic effects of amaranth (Amaranthus esculantus) in streptozotocin-induced diabetic rats. *Cell Biochemistry and Function*, 24, 195–199.

Kreft, S., Knapp, M., & Kreft, I. (1999). Extraction of rutin from buckwheat seeds and determination by capillary electrophoresis. *Journal of Agricultural and Food Chemistry*, 47(11), 4649–4652.

Kumar, K., Srivastav, S., Sharanagat, V. S. (2021). Ultrasound assisted extraction (UAE) of bioactive compounds from fruit and vegetable processing by-products: A review. *Ultrasonics Sonochemistry* 70, 105325,

Kumpun, S., Maria, A., Crouzet, S., Evrard-Todeschi, N., Girault, J. P., & Lafont, R. (2011). Ecdysteroids from Chenopodium quinoa Willd an ancient Andean crop of high nutritional value. *Food Chemistry*, 125, 1226–1234.

Lang, Q., & Wai, C. M. (2001). Supercritical fluid extraction in herbal and natural product studies—A practical review. *Talanta*, 53, 771–782.

Li, F., Zhang, X., Zheng, S., Lu, K., Zhao, G., & Ming, J. (2016). The composition, antioxidant, and antiproliferative capacities of phenolic compounds extracted from Tartary buckwheat bran (*Fagopyrum tartaricum* (L.) Gareth). *Journal of Functional Foods*, 22, 145–155.

Li, X., Da, S., Li, C., Xue, F., & Zang, T. (2018). Effects of high-intensity ultrasound pretreatment with different levels of power output on the antioxidant properties of alcalase hydrolyzates from Quinoa (*Chenopodium quinoa* Willd.) protein isolate. *Cereal Chemistry*, 95, 518–526.

Liu, J. J., Gasmalla, M. A. A., Li, P., & Yang, R. (2016). Enzyme-assisted extraction processing from oilseeds: Principle, processing and application. *Innovative Food Science and Emerging Technology*, 35, 184–193.

Liu, M., Sun, W., Ma, Z., Zheng, T., Huang, L., Wu, Q. & Chen, H. (2019). Genome-wide investigation of the AP2/ERF gene family in tartary buckwheat (Fagopyum Tataricum). *BMC Plant Biology*, 19(1), 1–19.

Małgorzata, S., Georgios, K., & Henryk, Z. (2018). Sensory analysis and aroma compounds of buckwheat containing products: A review. *Critical Reviews in Food Science and Nutrition*, 58(11), 1767–1779.

Mattila, P. H., Pihlava, J. M., Hellström, J., Nurmi, M., Eurola, M., Mäkinen, S., & Pihlanto, A. (2018). Contents of phytochemicals and anti-nutritional factors in commercial protein-rich plant products. *Food Quality and Safety*, 2(4), 213–219.

Mehmood, A., Ishaq, M., Zhao, L., Yaqoob, S., Safdar, B., Nadeem, M., Munir, M., & Wang, C. (2019). Impact of ultrasound and conventional extraction techniques on bioactive compounds and biological activities of blue butterfly pea flower (*Clitoria ternatea* L.). *Ultrasonics Sonochemistry*, 51, 12–19.

Mir, N. A., Riar, C. S., & Singh, S. (2018). Nutritional constituents of pseudo cereals and their potential use in food systems: A review. *Trends in Food Science and Technology*, 75, 175–180.

Nadar, S. S., Rao, P., & Rathod, V. K. (2018). Enzyme assisted extraction of biomolecules as an approach to novel extraction technology: A review. *International Food Research Journal*, 108, 309–330.

Nayak, M., & Patra, D. K. (2025). Dose-dependent inhibition of photosynthesis and redox alterations in cymbopogon nardus exposed to cadmium and chromium: evidence through the activity of RUBISCO. *International Journal of Phytoremediation*, 27(9), 1223–1238.

Noratto, G., Carrion-Rabanal, R., Medina, G., & Mencia, A. (2015). Quinoa protective effects against obesity-induced intestinal inflammation. *FASEB Journal*, 29, 602–609.

Okolie, C. L., Akanbi, T. O., Mason, B., Udenigwe, C. C., & Aryee, A. N. (2019). Influence of conventional and recent extraction technologies on physicochemical properties of bioactive macromolecules from natural sources: A review. *International Food Research Journal*, 116, 827–839.

Panja, P. (2018). Green extraction methods of food polyphenols from vegetable materials. *Current Opinion in Food Science*, 23, 173–182.

Parashar, S., Sharma, H., & Garg, M. (2014). Antimicrobial and antioxidant activities of fruits and vegetable peels: A review. *Journal of Pharmacogny and Phytochemicals*, 3(1), 160–164.

Pasko, P., Sajewicz, M., Gorinstein, S., & Zachwieja. Z. (2008). Analysis of selected phenolic acids and flavonoids in Amaranthus cruentus and Chenopodium quinoa seeds and sprouts by HPLC. *Acta Chromatographica*, 20(4), 661–672.

Plate, A. Y., & Arêas, J. A. (2002). Cholesterol-lowering effect of extruded amaranth (Amaranthus caudatus L.) in hypercholesterolemic rabbits. *Food Chemistry*, 76(1), 1–6.

Priyanka, T., Krishan, K., & Harcharan, S. D. (2021). Nutritional facts, bio-active components and processing aspects of pseudocereals: A comprehensive review. *Food Bioscience*. 42(24), 101170.

Risi, J., & Galwey, N. W. (1991). Genotype × environment interaction in the Andean grain crop quinoa (*Chenopodium quinoa*) in temperate environments. *Plant Breed*, 107, 141–147.

Rizzello, C. G., Lorusso, A., Russo, V., Pinto, D., Marzani, B., & Gobbetti, M. (2017). Improving the antioxidant properties of quinoa flour through fermentation with selected autochthonous lactic acid bacteria. *International Journal of Food Microbiology*, 241, 252–261.

Saleem, M., & Saeed, M. T. (2020). Potential application of waste fruit peels (orange, yellow lemon and banana) as wide range natural antimicrobial agent. *Journal of King Saud University Science*, 32(1), 805–810.

Sasidharan, S., Chen, Y., Saravanan, D., Sundram, K. M., Yoga Latha, L. (2011). Extraction, isolation and characterization of bioactive compounds from plants' extracts. *African Journal of Traditional, Complementary and Alternative Medicines*, 8(1), 1–10. Epub 2010 Oct 2. PMID: 22238476; PMCID: PMC3218439.

Shang, X., Dou, Y., Zhang, Y., Tan, J. N., Liu, X., & Zhang, Z. (2019). Tailor-made natural deep eutectic solvents for green extraction of isoflavones from chickpea (*Cicer arietinum* L.) sprouts. *Industrial Crops Products*, 140, 111724.

Shen, M., Chapman, R. S., He, X., Liu, L. Z., Lai, H., Chen, W., & Lan, Q. (2008). Dietary factors, food contamination and lung cancer risk in Xuanwei, China. *Lung Cancer*, 61(3), 275–282.
Shin, D., Heo, H., Lee, Y., & Kim, H. (2004). Amaranth squalene reduces serum and liver lipid levels in rats fed a cholesterol diet. *British Journal of Biomedical Science*, 61(1), 11–14.
Singh, B., & Kaur, A. (2018). Control of insect pests in crop plants and stored food grains using plant saponins: A review. *Journal of Food Science and Technology*, 87, 93–101.
Starowicz, M., Ciesarová, Z. & Henryk Zieliński, H. (2017). Analysis of the antioxidative properties and Maillard reaction products in ginger cakes enriched with rutin. *Madridge Journal of Food Technology*, 2(1), 44–52.
Stuardo, M., & San Martin, R. (2008). Antifungal properties of quinoa (*Chenopodium quinoa* Willd) alkali treated saponins against *Botrytis cinerea*. *Industrial Crops and Products*, 27, 296–302.
Tomotake, H., Shimaoka, I., Kayashita, J., Nakajoh, M., & Kato, N. (2002). Physicochemical and functional properties of buckwheat protein product. *Journal of Agricultural and Food Chemistry*, 50(7), 2125–2129.
Torbica, A., Hadnađev, M., & Hadnađev, D. T. (2012). Rice and buckwheat flour characterisation and its relation to cookie quality. *Food Research International*, 48(1), 277–283.
Tyszka-Czochara, M., Pasko, P., Zagrodzki, P., Gajdzik, E., Wietecha-Posluszny, R., & Gorinstein, S. (2016). Selenium supplementation of amaranth sprouts influences betacyanin content and improves anti-inflammatory properties via NFκB in murine RAW 264.7 macrophages. *Biological Trace Element Research*, 169(2), 320–330.
Verza, S. G., Silveira, F., Cibulski, S., Kaiser, S., Ferreira, F., Gosmann, G., Roehe, P. M., & Ortega, G. G. (2012). Immunoadjuvant activity, toxicity assays, and determination by UPLC/Q-TOF-MS of triterpenic saponins from *Chenopodium quinoa* seeds. *Journal of Agricultural and Food Chemistry*, 60, 3113–3118.
Vilcacundo, R., Martinez-Villaluenga, C., & Hernandez-Ledesma, B. (2017). Release of dipeptidyl peptidase IV, α-amylase and α-glucosidase inhibitory peptides from quinoa (*Chenopodium quinoa* Willd.) during *in vitro* simulated gastrointestinal digestion. *Journal of Functional Foods*, 35, 531–539.
Wieslander, G., Fabjan, N., Vogrincic, M., Kreft, I., Janson, C., Spetz-Nyström, U., Vombergar, B., Tagesson, C., Leanderson, P., & Norbäck, D. (2011). Eating buckwheat cookies is associated with the reduction in serum levels of myeloperoxidase and cholesterol: A double blind crossover study in day-care centre staffs. *The Tohoku Journal of Experimental Medicine*, 225(2), 123–130.
Woldemichael, G. M., & Wink, M. (2001). Identification and biological activities of triterpenoid saponins from *Chenopodium quinoa*. *Journal of Agricultural and Food Chemistry*, 49, 2327–2332.
Xiong, Y., Zhang, P., Warner, R. D., Shen, S., & Fang, Z. (2022). Cereal grain-based functional beverages: From cereal grain bioactive phytochemicals to beverage processing technologies, health benefits and product features. *Critical Reviews in Food Science and Nutrition*, 62(9), 2404–2431.
Yao, Y., Yang, X., Shi, Z., & Ren, G. (2014). Anti-inflammatory activity of saponins from Quinoa (*Chenopodium quinoa* Willd.) seeds in lipopolysaccharide-stimulated RAW 264.7 macrophages cells. *Journal of Food Science*, 79(5), 1018–1023.
Yao, Y., Zhu, Y., Gao, Y., Shi, Z., Hua, Y., & Ren, G. (2015). Suppressive effects of saponin-enriched extracts from quinoa on 3T3-L1 adipocyte differentiation. *Food Function*, 6, 3282–3290.
Zengin, G., Cvetanović, A., Gašić, U., Tešić, Ž., Stupar, A., Bulut, G., Sinan, K. I., Uysal, S., Picot-Allain, M. C. N., & Mahomoodally, M. F. (2020). A comparative exploration of the phytochemical profiles and bio-pharmaceutical potential of Helichrysum stoechas subsp. barrelieri extracts obtained via five extraction techniques. *Process Biochemistry*, 91, 113–125.

Zhang, H. X., Qu, J., Wang, Z. X., & Hu, D. W. (2013). Inhibitory effect of red clover flavone from golden buckwheat on migration ability of human gastric cancer cell line SGC7901 and its mechanism. *Journal of Jilin University (Medical Edition)*, 39, 78–81.

Zhang, Q. W., Lin, L. G. & Ye, W. C. (2018). Techniques for extraction and isolation of natural products: a comprehensive review. Chinese Med*icine*, 13, 20.

Zhou, X., Hao, T., Zhou, Y., Tang, W., Xiao, Y., Meng, X., & Fang, X. (2015). Relationships between antioxidant compounds and antioxidant activities of Tartary buckwheat during germination. *Journal of Food Science and Technology*, 52(4), 2458–2463.

Zhu, F. (2016). Chemical composition and health effects of Tartary buckwheat. *Food Chemistry*, 203, 231–245.

Zwingelstein, M., Draye, M., Besombes, J.L., Piot, C., & Chatel, G. (2020). Viticultural wood waste as a source of polyphenols of interest: Opportunities and perspectives through conventional and emerging extraction methods. *Journal of Waste Management*, 102,782–794.

7 Traditional and Novel Techniques for Extraction of Pseudocereal Starch and Proteins and Their Effect on Functional Properties

*Umar Garba, Balarabe B. Ismail,
Zahrau Bamalli Nouruddeen,
Alkasim Kabir Yunusa, Aminu Abdullahi,
Hafizu Ibrahim Kademi, Aanchal Sharma,
and Gulzar Ahmad Nayik*

7.1 INTRODUCTION

Globally, consumer awareness has increased significantly on the dietary impact on human health and well-being. Therefore, the need for foods with additional health benefits beyond basic nutrition has surged. It is therefore, not surprising that the term "functional foods" has continued to generate an increased popularity in different parts of the world. The term was first coined in the year 1980s in Japan to describe a specific class of whole, fortified, enriched, or enhanced foods that contain ingredients, such as vitamins, minerals, bioactive compounds, antioxidants, and probiotics, with the capacity to impact specific body functions in addition to being nutritious (Qureshi et al., 2024). Functional foods provide a great number of physiological benefits, and prevent or reduce the risk of chronic disease, or otherwise optimize health when consumed at efficacious levels as part of a regular diet (Hasler, 2002; Vignesh et al., 2024).

As the concept of functional foods continues to evolve considerably, broad categories of foods, such as fruits, vegetables, beverages, dairy products, cereals, legumes, oilseeds, fishes, and herbs and spices, have been classified as functional foods based on their unique health-promoting properties. Cereals have been known to play a crucial role in providing the nutritional demands of the human population (Pirzadah & Malik, 2020). They are also considered the primary class of functional foods because

DOI: 10.1201/9781003584858-7

of their rich nutrient and phytochemical profile, such as vitamins, fiber, and bioactive compounds (Guo et al., 2022; Vignesh et al., 2024). Pseudocereals, in particular, are far better than common cereals in terms of nutrients and health benefits, given their richness in high-quality proteins, dietary fibers, unsaturated fats, and bioactive compounds such as polyphenols and phytosterols that are associated with numerous health benefits, including antidiabetic, hypolipidemic, antiobesity, anticancer, anti-inflammatory, antihypertensive, and hepatoprotective properties (Nandan et al., 2024). Some of the most popular pseudocereals are quinoa (*Chenopodium quinoa* Willd), amaranth (*Amaranthus* L. spp.), and buckwheat (*Fagopyrum esculentum* Moench and *Fagopyrum tataricum* (L.) Gaertn). The pseudocereals can be ground to produce flour for pasta and bakery goods. The cereal can also be suitable for patients with celiac disease (CD) because of their gluten-free nature (Ismail et al., 2024). Nevertheless, despite their rich micronutrients, large-scale production and consumption remain generally underexplored (Graziano et al., 2022; Malik & Singh, 2022; Nandan et al., 2024).

A significant number of researches have focused on macronutrients, such as proteins and starch, from unconventional sources for further utilization in various novel food applications such as protein drinks, gluten-free products, smoothies, meat analogues, and sourdough bread (Vidaurre-Ruiz et al., 2023). Pseudocereals have 10–20% proteins with greater nutritional quality than cereals and many legumes. For example, the quality of buckwheat protein exceeds 90%, with essential amino acids and sulfur-containing amino acids being the most common (Alonso-Miravalles & O'Mahony, 2018). Chia seeds are higher in glutamic acid, aspartic acid, and arginine, whereas quinoa seeds are rich in essential amino acids, particularly methionine, lysine, and threonine, which are lacking in most cereal grains. On the other hand, pseudocereals such as buckwheat seeds, quinoa, and amaranth have a starch content ranging from 60% to 75% of the total dry weight. However, the amylose content of starch in buckwheat (18.3%–47% of total starch) is higher than in quinoa (11%–12% amylose) and amaranth (7.8%–34.3% amylose) (Nandan et al., 2024).

Several traditional and novel techniques have been developed for the extraction of pseudocereal protein and starch, including wet milling, alkaline, ultrasound-assisted, enzyme-assisted, organic solvent-based, and ultrasound-assisted enzymatic extraction methods (Jan et al., 2017; Wang et al., 2021; Yang et al., 2024). The selection of the method for extraction depends on the type of pseudocereal and target recovery rate. Preparation of protein-enriched fractions involves dry or wet fractionation techniques, which are the two principal approaches. Wet milling is commonly used to extract starch from pseudocereals (Kringel et al., 2020). Starch extraction from small-size grains like amaranth involves soaking followed by alkali or acid treatment, while in quinoa, alkaline steeping and enzymatic treatment methods have been used to recover starch constituents (Jan et al., 2017). In some recent studies, a combination of methods has been used, which demonstrated an improved extraction efficiency, solubility, stability and functionality of the protein and starch extracted from pseudocereals. This chapter will discuss the traditional and novel techniques for extraction of pseudocereals starch and protein and their effects on functional foods.

7.2 EXTRACTION OF PSEUDOCEREAL STARCH

For many years, pseudocereals have been processed using traditional methods, such as hand-operated wooden or stone pestles and mortars. The grains are grinded and sieved to remove coarse materials, producing flour and meal (Haros & Wronkowska, 2017). Milling, which is a key step in starch extraction, is classified into either dry or wet milling, with the primary difference being whether water is used or not. The dry milling uses minimal or no water and focuses on separating the anatomical parts of the grain, such as the endosperm, germ, and pericarp. In contrast, wet milling aims to separate the chemical components of the grain, such as starch, protein, fiber, and oil (Haros & Wronkowska, 2017). The goal of dry milling is to maximize flour yield, while wet milling seeks to isolate each component in its purest form (Haros & Wronkowska, 2017).

Dry Milling

Dry milling involves a series of repeated steps of grinding, sifting, and separation to extract maximum yield of flour, primarily from the endosperm (Haros & Wronkowska, 2017). The process starts with cleaning of the grains to remove foreign materials. Modern equipment used for grain's cleaning explores some parameters such as size, color, shape, solubility, specific weight, and response to magnetic force for this purpose (Haros & Wronkowska, 2017). Water may be added to the grains in a process called conditioning before grinding to toughen the pericarp and germ, preventing them from disintegrating during milling. However, harsh dry-milling conditions can damage starch granules, affecting the properties and yield of the extracted starch (Kringel et al., 2020).

Wet Milling

Wet milling is a more complex process involving physical, biochemical, and chemical operations to separate the grain's components (Wronkowska & Haros, 2014). Wet milling is initially designed for industrial starch production, but now aims to achieve optimal efficiency and maximum separation of each grain fraction (Haros & Wronkowska, 2017). It usually begins with soaking the grains in water solutions containing alkali or acid at a specific temperature, followed by mechanical separation of the various grain fractions based on the physical properties. While starch is the primary product of wet milling, the process also yields protein- and fiber-rich fractions, which are valuable for many technological and food industry applications (Haros & Wronkowska, 2017).

Wet milling process for pseudocereals is adapted for corn and the process involves biochemical, chemical, and mechanical operations to separate the grain into its main components (Haros & Wronkowska, 2017). The first step involves cleaning the grains to remove extraneous materials, followed by steeping the cleaned grains in an aqueous solution of SO_2 or alkali at a predetermined temperature, which varies depending on the pseudocereal and the method used (Haros & Wronkowska, 2017). Steeping is a critical step, as it softens the grains and allows water and additives to diffuse into them, thereby facilitating subsequent separation (Resio et al., 2006). Alkaline steeping was found to be more effective than water steeping for extraction of quinoa starch (Jan et al., 2017).

During steeping, some solid components of the grain dissolve, forming a nutritive fraction that is concentrated and dehydrated (Haros & Wronkowska, 2017). In some cases, the grains absorb high amounts of water, thereby becoming soft enough to disintegrate by hand. The first milling step produces a suspension, which is milled again to pulverize the endosperm particles and leave the fibrous material intact. The suspension is filtered through screens of decreasing mesh size, allowing starch and protein fractions to pass through. These fractions are then separated by centrifugation, with the lighter protein fraction concentrated and dehydrated. The starch forms a sediment, which is concentrated and washed (Haros & Wronkowska, 2017).

Wet milling methods have been developed on laboratory and pilot scales to evaluate starch yield and quality, as well as the effects of operative variables and potential modifications to traditional methods. These methods allow for the use of small sample sizes to quantify the yield and quality of milling products (Haros & Wronkowska, 2017).

7.2.1 Traditional Techniques for the Extraction of Pseudocereal Starch

Pseudocereals such as buckwheat, quinoa, and amaranth, despite not belonging to the grass family, are cultivated for their starchy seeds, and their importance as high energy sources (Jan et al., 2017). Buckwheat belongs to the *Fagopyrum* genus (Polygonaceae family), grain quinoa to the *Chenopodium* genus (Amaranthaceae family/Chenopodioideae subfamily), and amaranth to the *Amaranthus* genus (Amaranthaceae family) (Haros & Wronkowska, 2017). Starch yields in pseudocereals vary due to some factors such as steeping solution, temperature, time, cultivar, location, and agricultural practices. Table 7.1 provides starch yields for selected pseudocereals as a percentage of the material's weight.

TABLE 7.1
Yield of Starch in Some Selected Pseudocereals

S/N	Pseudocereals	Steeping Method	Sample	Starch Yield (%)	References
1.	Amaranth	Acid	Seeds	45.0 ± 1.35	Resio et al. (2006)
		Alkali	Whole grain	17.56 ± 0.13	Villarreal et al. (2013)
			Whole grain flour	45.29 ± 0.21	
2.	Buckwheat	Acid	Seeds with hull	48.7 ± 1.8	Wronkowska and Haros (2014)
			Dehulled	58.8 ± 3.5	
		Acid	Wholegrain flour (Aktaş cv.)	54.1 ± 3.90	Arslan et al. (2022)
			Wholegrain flour (*Güneş* cv.)	50.3 ± 0.28	
			Groat flour (*Aktaş* cv.)	81.1 ± 0.64	
			Groat flour (*Güneş* cv.)	76.0 ± 0.95	

(Continued)

TABLE 7.1 (*Continued*)
Yield of Starch in Some Selected Pseudocereals

S/N	Pseudocereals	Steeping Method	Sample	Starch Yield (%)	References
		Alkali (0.10%)	Defatted flour	28.9 ± 0.5	Jindal and Saxena (2015)
		Alkali (0.15%)		34.8 ± 0.6	
		Alkali (0.20%)		45.3 ± 0.8	
		Alkali (0.25%)		40.6 ± 0.4	
		Alkali (0.30%)		34.4 ± 0.5	
3.	Quinoa	Water	Flour	32.97 ± 0.23	Jan et al. (2017)
			Seeds	28.15 ± 0.16	
		Alkali (0.20%)	Seeds	37.78 ± 0.28	
			Flour	39.15 ± 0.35	
		Alkali (0.25%)	Seeds	43.77 ± 0.37	
			Flour	48.52 ± 0.47	
		Alkali (0.30%)	Seeds	40.89 ± 0.31	
			Flour	43.53 ± 0.38	
		Acid		64.3 ± 0.70	Ballester-Sánchez et al. (2019)

7.2.1.1 Amaranth

Amaranth is a pseudocereal with small and highly nutritious seeds. It includes cultivars such as *Amaranthus hypochondriacus* Annapurna, *Amaranthus hypochondriacus* Durga, and *Amaranthus paniculatus* (Kumawat2022). Extraction of starch from amaranth seed typically involves soaking in water or alkaline solution (Figure 7.1). Different procedures have been proposed, varying in temperature and chemical concentrations (Kringel et al., 2020). Alkaline soaking using sodium metabisulfite or sodium hydroxide effectively separates protein from starch. However, high alkali concentrations can damage starch granules, thereby reducing the yield. Although alkaline methods (AMs) offer operational advantages like lower reagent costs and shorter processing times, enzymes can be introduced to lower the alkali concentrations, (Kringe et al., 2020).

Starch content of amaranth ranges from 55 to 60 g/100 g (Villarreal et al., 2013). Kumawat. (2022) reported a maximum starch yield of 28.5% using 0.30% alkali, while Villarreal et al. (2013) reported the yields of 17.56% and 24.47% for whole grain and whole grain flour, respectively. Higher yields in flour were attributed to pre-grinding of the grain before alkaline treatment.

7.2.1.2 Buckwheat (*Fagopyrum esculentum* Moench)

Buckwheat, a pseudocereal from the Polygonaceae family, is a rich source of starch with good functional properties (Jindal & Saxena, 2015). The two main species are common buckwheat (*F. esculentum*) and Tartary buckwheat (*F. tataricum*) (He et al.,

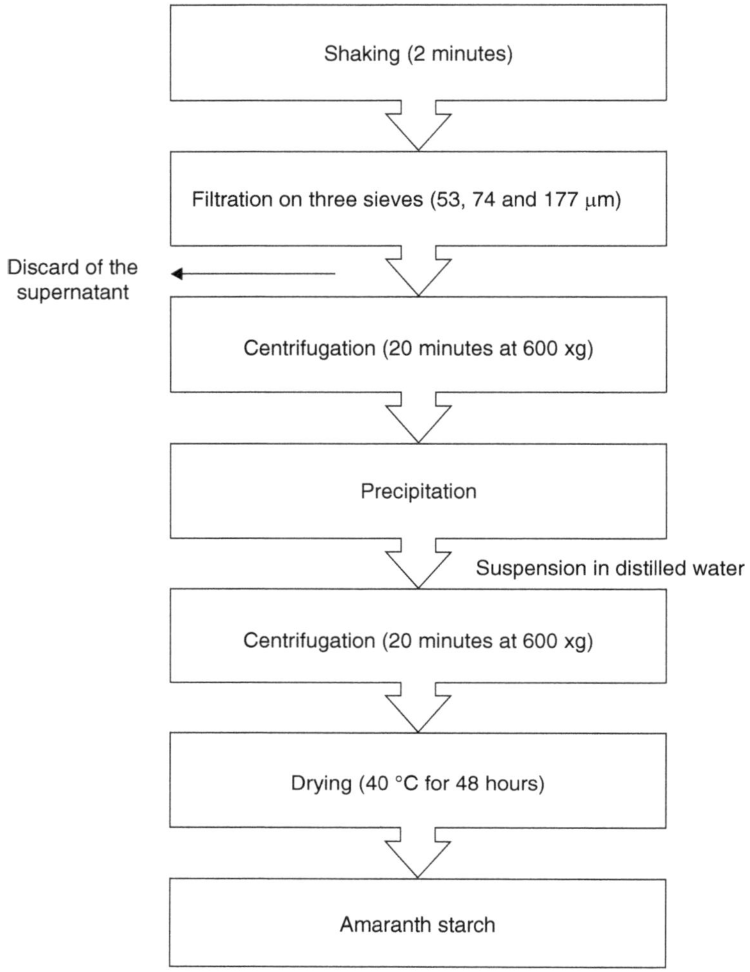

FIGURE 7.1 Isolation of amaranth starch (Resio et al., 2006).

2023). Starch yields in buckwheat range from 28.9% to 45.3%, with higher yields achieved at increased alkali concentrations (Jindal & Saxena, 2015). Wronkowska and Haros (2014) reported yields of 48.7% and 58.8% for hulled and dehulled buckwheat seeds, respectively. Arslan et al. (2022) found starch content ranging from 50.3% to 81.1% in two buckwheat cultivars.

7.2.1.3 Quinoa

Quinoa is a pseudocereal, native to South America, which is mostly consumed as flour or whole grain (Jan et al., 2017). Its starch constitutes range between 52.2% and 69.2% of dry matter (Kringel et al., 2020). The extraction methods for quinoa starch are presented in Figure 7.2. The starch yields vary between 32.97% and 48.45% for quinoa flour and 28.15% to 43.77% for seeds (Jan et al., 2017). Increasing alkali

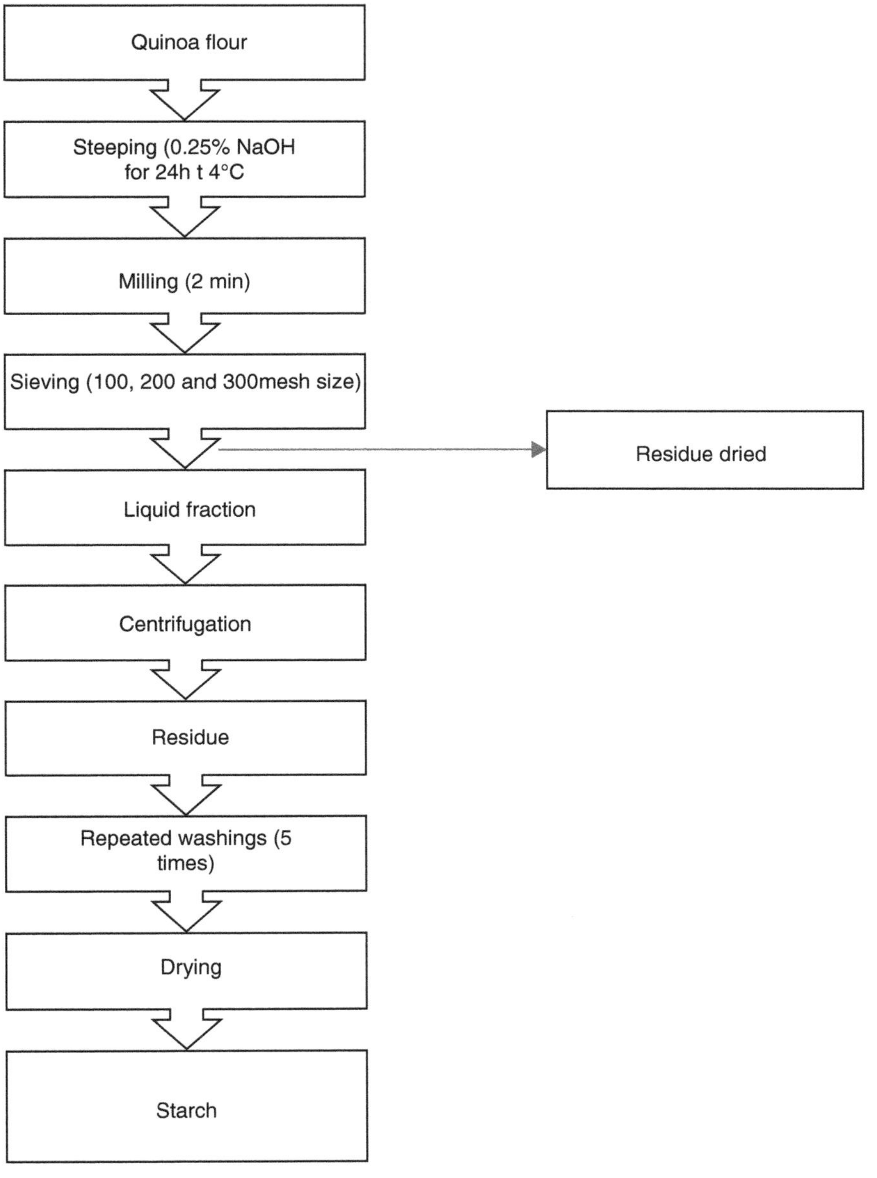

FIGURE 7.2 Isolation of starch from quinoa seeds (Jan et al., 2017).

concentration to 0.25% improves the yield, but further increase reduces the yield due to swelling of starch granule or gelatinization, and increasing slurry viscosity (Jan et al., 2017). Ballester-Sánchez et al. (2019) reported a higher yield of 62.2% under specific steeping conditions of 1.25 hours in SO_2 solution at 30°C.

The quinoa's small starch granules and endogenous substances could increase slurry viscosity and complicate its wet milling (Haros & Wronkowska, 2017; Kringel et al., 2020). Wet milling using stone mills or rollers produces whole quinoa flour, but separating the fractions remains challenging (Ballester-Sánchez et al., 2019). Milling wet quinoa grains in a knife mill has been proposed as an alternative, although it still faces separation issues (Kringel et al., 2020).

In conclusion, traditional starch extraction methods for pseudocereals, such as dry and wet milling, are effective but require careful optimization to maximize the yield and quality. Advances in processing techniques, including the use of enzymes and modified milling methods, continue to improve the efficiency of starch extraction from pseudocereals.

7.2.2 Novel Techniques for Extracting Starch from Pseudocereals

The increasing interest in pseudocereals such as quinoa (*Chenopodium quinoa*), buckwheat (*Fagopyrum esculentum*), and amaranth (*Amaranthus* spp.) is attributed to their unique starch properties, which have high amylose content, small granule size, and low digestibility, making them valuable in food and non-food applications (Ciudad-Mulero et al., 2019; Thakur et al., 2021a,b). However, extracting starch from these grains poses unique challenges, primarily because of their complex cellular structure and the strong association between starch granules and proteins (Li & Zhu, 2017, 2018). Traditional methods often do not provide high yields and purity, which necessitate the exploration of other innovative techniques (Guzel et al., 2023; Schönlechner & Bender, 2023). Examples of the various alternative extraction methods with high yield and purity are presented in Table 7.2.

7.2.2.1 Pre-extraction Considerations

Before adopting any extraction method, several critical factors must be considered to optimize starch recovery (Contreras-Jiménez et al., 2019; Ruales et al., 1993). These pretreatment steps can significantly enhance the efficiency of subsequent extraction processes. The raw material particle size can significantly impact extraction efficiency, with the optimal particle size ranging between 250 and 500 μm. Additionally, the moisture content of seed 12%–14% and conditioning temperature of 35°C–45°C play a crucial role in preparing the material for extraction (Alvarez-Jubete et al., 2009).

7.2.2.2 Enzymatic Techniques

Recently, many researchers focused on the use of enzymatic hydrolysis due to its highly specific and eco-friendly approach to starch extraction (Inouchi et al., 1999; Linsberger-Martin et al., 2012). For example, starch from *Amaranthus caudatus* was extracted by dispersing flour in water and treating it with protease to hydrolyze proteins, followed by α-amylase to break down other polysaccharides. The resulting suspension was centrifuged, and the starch pellet was washed and dried. This method achieved a starch recovery of 72%, with protein contamination levels reduced to less than 0.5%, thus preserving the granules' structural integrity as confirmed by scanning electron microscopy (Capriles et al., 2008).

A study on *Amaranthus cruentus* compared two starch extraction methods: the AM and an alkaline method with protease digestion (AMP). Both methods used

TABLE 7.2
Comparison of Starch Extraction Methods from Various Sources

Source Material	Extraction Method	Operating Conditions	Yield	Purity (%)	Notable Effects	References
Quinoa	Enzymatic	Protease treatment (pH 7.0, 45°C, 2 hours)	72.0%	99.5	High purity starch	Junejo et al. (2022)
Quinoa	UAE	25 kHz, 40°C, 20 minutes	82.0%	98.0	Enhanced extraction efficiency	He at al. (2022)
Amaranth	Enzymatic (AMP)	NaOH (0.05 g/100 mL) + protease	116.7%	–	Increased yield vs whole grain	Villarreal et al. (2013)
Amaranth	UAE	60 kHz, 10–30 minutes	–	–	SDS↑(37.20%), RS↑(20.98%), Enhanced functionality	Yadav et al. (2024)
Amaranth	SCF	50°C, 3,630 psi	70.2	–	Protein↑(16.17%), WAC↑(126.33%)	Delgado-García et al. (2022)
Amaranth	Enzymatic	Alcalase (pH 8.0, 50°C, 3 hours)	75.5%	97.8	High purity, preserved structure	Shen et al. (2024)
Buckwheat	MAE	400 W, 6 minutes	82.0%	96.5	Rapid extraction	Zhou (2015)
Quinoa	Hybrid UAE-MAE	25 kHz + 300 W, 40°C	85.0%	98.5	Highest combined yield and purity	Li et al. (2022)
Amaranth	Modified centrifugation	5,000 g, 15°C, 30 minutes	76.0%	96.0	Good yield-purity balance	Bet et al. (2019)

NaOH solutions, with the AM at 0.25 g/100 mL and AMP at 0.05 g/100 mL, followed by wet milling, enzymatic hydrolysis, and multi-stage centrifugation. Milling increased starch yield, recovery, and purity compared to whole grain extraction, with whole grain flour yielding 116.7% and 123.6% more than whole grain. However, protein, fiber, and ash contents decreased by about 44.4%, 34.8%, and 30.4%. The extracted starch exhibited an A-type diffraction pattern, a crystallinity of ~39%, and a gelatinization temperature around 74°C, highlighting the benefits of combining enzymatic hydrolysis with alkaline soaking for starch extraction from pseudocereals (Middlewood & Carson, 2012; Reddy et al., 2024; Villarreal et al., 2013).

7.2.2.3 Ultrasound/Microwave Processing

Ultrasound-assisted extraction (UAE) has emerged as an innovative technique that leverages acoustic cavitation to disrupt the cellular matrix, facilitating the release of starch granules (Zhu & Li, 2019). In a study on quinoa *(Chenopodium quinoa)*, flour was suspended in distilled water and subjected to ultrasound treatment at 25 kHz for 20 minutes at 40°C. This method yielded 82% starch, a significant improvement over conventional water-based extraction methods, which typically recover about 68%

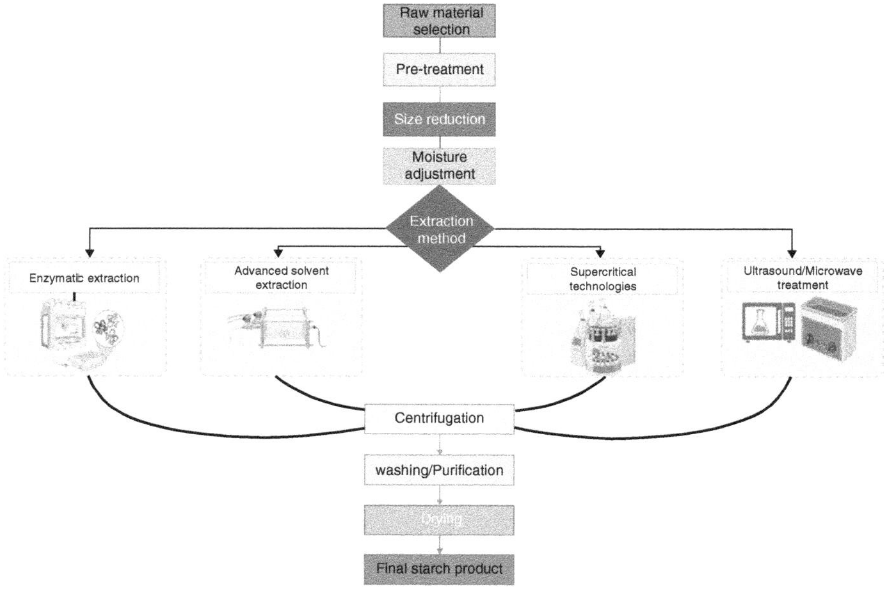

FIGURE 7.3 Workflow of starch extraction techniques for pseudocereals.

starch (Estivi et al., 2022). The flowchart for the typical extraction of pseudocereal starch using the various methods explained above is presented in Figure 7.3.

Recent studies on extraction of amaranth starch have demonstrated that ultrasonic treatment (60 kHz for 10–30 minutes) significantly affects both the structural and functional properties of the extracted starch. Microscopic analysis of the sample revealed that sonicated starch exhibited granule agglomeration and surface fissures, leading to enhanced functional properties like increased solubility, amylose content, swelling capacity and improved water and oil absorption capabilities. Notably, the treatment resulted in significant changes to starch digestibility profiles, with a decrease in rapidly digestible starch content and concurrent increases in slowly digestible starch (SDS) from 32.81% to 37.20% and resistant starch from 9.87% to 20.98%. While the crystalline pattern remained largely unchanged, the modification in digestibility profiles suggests potential applications in functional food development (Yadav et al., 2024).

7.2.2.4 Supercritical Technologies

Supercritical fluid extraction (SFE) has gained attention as an advanced technique for high-purity starch recovery (Perez-Rea & Antezana-Gomez, 2024; Villalobos et al., 2024). Recent studies with amaranth have demonstrated optimal extraction conditions at 50°C and 3630 psi, achieving an extraction yield of 70.2 mg oil/g flour. Under these conditions, the defatted flour exhibited enhanced protein content (16.17 g/100 g flour) and improved techno-functional properties, including increased water absorption (126.33 g H_2O/100 g flour) and oil absorption (119.09 g oil/100g flour) capacities. Notably, the extracted oils showed high polyunsaturated fatty acid content, while the defatted flour maintained desirable functional properties (Delgado-García et al., 2022; Vatansever et al., 2021). These findings suggest that SFE offers a viable alternative to traditional organic solvent extraction methods, providing both high-quality starch and valuable oil by-products (Modupalli et al., 2024).

7.2.2.5 Advanced Centrifugation and Solvent Extraction

Centrifugation-based methods are one of the most traditional and scalable approaches to starch extraction (Li et al., 2019; Steffolani et al., 2013). This technique has been extensively applied to quinoa, where flour is suspended in a diluted sodium chloride solution to facilitate protein solubilization, followed by homogenization and centrifugation. The resulting starch pellet is washed repeatedly with water and ethanol to remove residual proteins and lipids (Ando et al., 2002).

7.3 EXTRACTION OF PSEUDOCEREAL PROTEINS

Protein extraction techniques are important for the isolation and purification of proteins from plants and animals while maintaining their structural and functional integrity (Malik & Singh, 2022). These methods are broadly categorized into physical, chemical, and enzymatic approaches. Physical methods, such as milling, sieving, and membrane filtration, are energy-efficient and preserve native protein characteristics; however, the yield is low. Chemical techniques, involving solvents, salts, and pH adjustments, offer higher yields but may compromise protein functionality. Enzymatic methods on the other hand use specific enzymes to hydrolyze polysaccharides or disrupt protein matrices, provide high precision and purity but are cost intensive (Malik & Singh, 2022).

The yield of protein extraction can be enhanced by increasing protein solubility or breaking down the protein-containing matrix. However, structural and compositional changes, isomerization, bonding, and degradation can affect the techno-functional and physiological properties of proteins (Deleu et al., 2019). Advanced technologies, such as UAE and reverse micelle systems, have emerged to improve efficiency and reduce processing times, making them valuable for applications requiring high purity and functionality.

7.3.1 Physical Extraction of Pseudocereal Proteins

Physical methods, also known as traditional dry fractionation techniques, are preferred for their ability to produce safe and economically stable products, albeit with slightly lower yields (Schutyser & Van der Goot, 2011). These methods are less likely to alter food properties compared to chemical or enzymatic processes (Fabian & Ju, 2011). Dry fractionation, a key physical technique, is highly energy efficient and retains the native characteristics of proteins. It involves two main steps: (1) separating starch particles from protein-rich bodies through milling and classification and (2) isolating components through turbo-segregation or electrostatic segregation based on physicochemical differences (Laguna et al., 2018).

For example, protein-rich fractions from buckwheat flour are extracted by milling the flour into bran, dispersing it in water, and processing it in a colloid mill.

Enzymes such as α-amylase are added to form a slurry, which is then heated to deactivate the enzymes. The resulting hydrolysate is segregated, and the retentate is lyophilized to produce polyphenols rich protein powder. This method yields 1.29–1.45 folds proteins than other techniques, with polyphenol contents increasing from inner to outer fractions (Tang, 2007).

UAE is another physical method that enhances protein recovery. Buckwheat seeds are ground, sieved, and dispersed in a buffer solution. Ultrasound treatment at 15 kHz and 500 W is applied, followed by high-speed mixing to separate the supernatant. Proteins are precipitated by adjusting the pH to the isoelectric point (pI), neutralized, and dried. Studies have shown that ultrasound-assisted methods improve protein solubility, especially at pH less than 7.0 (Tang, 2007).

Research on physical methods for isolating proteins from pseudocereals such as amaranth, chia, and quinoa is limited. However, quinoa proteins can be enriched through milling fractionation and solvent extraction. Quinoa bran is suspended in water, and the pH is adjusted to 7–12. Soluble proteins are separated by centrifugation, precipitated by acid, and freeze-dried. Optimum pH ranges are crucial for maintaining protein stability and functionality (Scilingo et al., 2002).

7.3.2 Chemical Methods of Protein Extraction

Chemical methods are widely used for their efficiency and cost-effectiveness. They involve alkaline or acidic conditions to solubilize and precipitate proteins. For instance, quinoa proteins are extracted by immersing defatted flour in alkaline water, followed by centrifugation and acid precipitation. High pH values yield proteins with 90%–93% purity, while lower pH values produce 82%–88% purity. Temperature and pH are factors that significantly influence the efficiency of protein extraction (Ruiz et al., 2016).

Amaranth proteins are isolated by suspending defatted flour in water at specific pH, followed by centrifugation, precipitation, and freeze-drying. These isolates exhibit excellent functional properties, such as foaming and emulsification (Fidantsi & Doxastakis, 2001). However, chemical methods can lead to the loss of beneficial protein properties and formation of anti-nutritional factors like phytic acid (Rahma et al., 2000).

Buckwheat proteins, primarily albumin and globulin, are extracted using AMs at pH 8.0–8.5. Enzymatic extraction depends on factors such as enzyme specificity and hydrolysis degree, which influence the functional properties of the resulting hydrolysates (Tang et al., 2009). Chemical methods, such as reverse micelle systems, have also been employed to enhance protein diffusion and recovery (Wu et al., 2018).

7.3.3 Enzymatic Extraction of Pseudocereal Proteins

Enzymatic methods are highly effective for obtaining pure protein isolates with selective modifications. These methods use enzymes to hydrolyze polysaccharides, enriching the solid phase with proteins. For example, amaranth flour is treated with amylases and glucoamylases to liquefy starch into glucose, followed by centrifugation and dehydration to isolate proteins (Mokrejs et al., 2011).

Enzymes such as Flavorzyme and Alcalase achieve high protein recovery rates (87.6% and 81.4%, respectively) by targeting specific bonds within protein chains (Hamada, 1999). Despite their efficiency, enzymatic methods are costly. Advances in immobilization techniques, which allow enzyme reuse, could reduce expenses and enhance the economic feasibility of these methods.

7.3.4 Traditional Techniques for Protein Extraction

Traditional methods for extracting proteins from pseudocereals such as quinoa, amaranth, and buckwheat are based on physicochemical processes that take advantage of the solubility of proteins in specific pH ranges or solvent systems. These techniques are widely used due to their simplicity, cost-effectiveness, and ability to produce protein isolates with preserved functional properties. Key traditional methods employed for protein extraction from pseudocereals are given below:

7.3.4.1 Alkaline Extraction and Isoelectric Precipitation

Alkaline extraction followed by isoelectric precipitation is one of the most commonly used traditional method for isolating proteins from pseudocereals. The process begins with grinding the pseudocereal grains into fine flour to increase the surface area for extraction. An optional defatting step may be included to remove lipids, which can interfere with protein solubility. The flour is then mixed with water or a buffer solution adjusted to an alkaline pH (typically between 8 and 10) using a strong base such as sodium hydroxide. The alkaline environment disrupts hydrogen bonds and ionic interactions, enhancing protein solubility. The mixture is centrifuged to separate the insoluble components (e.g., starch and cell wall materials) from the protein-rich supernatant. The pH of the supernatant is then adjusted to the pI of the target proteins (usually between pH 4.5 and 5.5) using an acid like hydrochloric acid. At this pH, proteins lose their net charge, aggregate, and precipitate out of the solution. The precipitated proteins are collected by centrifugation, washed to remove impurities, and dried to obtain the final protein isolate (Singh et al., 2013).

7.3.4.2 Salt Extraction and Dialysis

Salt extraction is another traditional method that exploits the solubility of pseudocereal proteins in salt solutions. In this process, the flour is mixed with a salt solution, typically containing 0.1–0.5 M sodium chloride (NaCl). The ionic strength of the solution helps solubilize proteins by shielding electrostatic interactions. The mixture is then centrifuged to separate the protein-rich supernatant from insoluble materials. To recover the proteins, the supernatant undergoes dialysis against distilled water to remove the salt. As the salt concentration decreases, the proteins lose solubility and precipitate. The precipitated proteins are collected by centrifugation or filtration and dried to obtain the final product (Singh et al., 2013).

7.3.4.3 Water Extraction (Mild Extraction)

Water extraction, also known as mild extraction, is used to isolate water-soluble proteins such as albumins from pseudocereals. In this method, the flour is mixed with water at a neutral or slightly acidic medium (pH 6–7) and stirred at room temperature or slightly elevated temperatures (e.g., 40°C) to facilitate protein dissolution. The mixture is then centrifuged to separate the insoluble components from the protein-rich supernatant. The proteins in the supernatant can be recovered by adjusting the pH to induce precipitation or by freeze-drying the supernatant directly to obtain a protein-rich powder. This method is particularly suitable for extracting proteins that are highly soluble in water and do not require harsh chemical treatments (Wang et al., 2015).

These traditional techniques remain popular due to their simplicity, scalability, and ability to produce protein isolates with minimal structural and functional alterations. However, they may require optimization to achieve higher yields and purity, depending on the specific pseudocereal and its protein composition.

7.4 FUNCTIONAL PROPERTIES OF PSEUDOCEREALS STARCH

In recent years, pseudocereals have garnered significant attention due to their unique nutritional profiles and potential applications in gluten-free food products. Pseudocereals such as buckwheat, quinoa, and amaranth offer viable alternatives to traditional cereals, making their functional properties a critical area of study. These grains are not botanically related to grasses but are valued for their seeds, which can be used similarly to true cereals. Their nutritional profiles are remarkable, with carbohydrates constituting 60%–80% of the dry weight of the seeds (Alonso-Miravalles & O'Mahony, 2018). Specifically, starch makes up 58.1%–64.2% of quinoa, 65.0%–75.0% of amaranth, and 54.5%–57.4% of buckwheat by dry weight (Huda et al., 2021). Table 7.3 provides a detailed comparison of the nutritional properties of pseudocereal starches.

Pseudocereal starches, like those of true cereals, are primarily composed of amylose and amylopectin. However, the amylose-to-amylopectin ratio in pseudocereals is lower, typically around 11% less than in cereals (Poshadri et al., 2023). This difference influences their texture and digestibility. Additionally, the high cell wall matrix content in pseudocereal endosperms makes their starch less accessible to enzymes, resulting in slower digestion (Huda et al., 2021). The carbohydrate fraction of pseudocereals also includes mono- and disaccharides such as glucose, fructose, arabinose, xylose, sucrose, and maltose (Poshadri et al., 2023). Buckwheat contains significantly lower values of simple carbohydrates (0%–8%) compared to quinoa and amaranth (3%–5%) and traditional cereals (1%–2%) (Ayub et al., 2021).

TABLE 7.3
Nutritional Properties of Pseudocereal Starch

Parameters (g/100 g)	Quinoa	Amaranth	Buckwheat
Energy (kcal/100 g)	368	371	343
Carbohydrates	64.2	65.2	71.5
Starch (% seed DW)	58.1–64.2	65–75	54.5–57.4
Dietary fiber (% DW)	7.0–26.5	2.7–17.3	17.8

Source: Huda et al. (2021); Ayub et al. (2021).

Dietary fiber is another critical component of pseudocereals, comprising 4%–12% of non-starchy polysaccharides. Buckwheat, for instance, contains 20%–30% more soluble fiber than true cereals and lacks the anti-nutrient phytic acid (Usman et al., 2022). Quinoa and amaranth also provide substantial dietary fiber, with quinoa containing 7%–26% and amaranth 8%–20%. The insoluble fiber in these grains is primarily composed of cellulose, xyloglucans, and pectic polysaccharides, while soluble fiber includes arabinose-rich pectic polysaccharides and xyloglucans (Punia Bangar et al., 2022). The glycemic indexes of pseudocereals are relatively low, with chia seeds at 28.53, buckwheat at 52.35, amaranth at 47.65, and quinoa at 61.50, making them suitable for managing blood glucose levels and reducing the risk of chronic diseases (Punia Bangar et al., 2022).

7.4.1 Swelling Power

The swelling power of pseudocereal starches is a key functional property that influences their application in food technology. Swelling power refers to the ability of starch granules to absorb water and swell, which is critical for processes like gelatinization. Martin Ramos Diaz (2012) found that the viscosity of wild rice, quinoa, and amaranth starches at high temperatures differed from that of traditional starches like wheat and barley. These pseudocereals exhibited lower swelling and viscosity due to their low amylose content. The study highlighted that the branched structure of amylopectin in these starches stabilizes water molecules more effectively than higher amylose starches, enhancing their water-binding capacity.

Rulahnia and Khatkar (2020) further investigated the swelling power of quinoa starch, noting that it was higher than that of wheat, barley, rice, amaranth, potato, and corn starches. The study attributed this to the strength and composition of the micellar network within the starch granules, with amylose content playing a significant role in inhibiting swelling. Waxy starches, which lack amylose, tend to swell more than their amylose-containing counterparts.

De Bock et al. (2021) examined the swelling behavior of wholemeal flours from buckwheat, quinoa, and amaranth, emphasizing the role of temperature in swelling capacity. Higher temperatures were found to promote greater water absorption, with the molecular structure of starch granules significantly influencing swelling behavior. These findings underscore the importance of understanding the interplay between starch composition, thermal behavior, and water absorption capacity in optimizing the functional properties of pseudocereal starches for food applications.

7.4.2 Gelatinization

Gelatinization is a critical functional property of starch, involving the disruption of the granular structure and the absorption of water when heated. This process is essential for determining the cooking properties of pseudocereal starches. Berezina et al. (2020) highlighted the importance of the carbohydrate–amylase complex in evaluating the technological properties of flour, including its saccharification level and flavor development.

Rulahnia and Khatkar (2020) studied the textural properties of quinoa starch gels and reported a hardness of 596.1 g/force. The firmness of starch gels is influenced by retrogradation, a process involving the recrystallization of amylopectin and water syneresis. Starches with longer amylopectin chains and higher amylose content tend to form harder gels (Mua & Jackson, 1997). The mechanical properties of starch gels are also affected by factors such as the rheological properties of the amylose matrix, the stiffness of gelatinized starch granules, and interactions between the gel's continuous and dispersed phases (Rulahnia & Khatkar, 2020).

Fredriksson et al. (1998) demonstrated that higher amylopectin content related with increased starch crystallinity, leading to higher onset, peak, and conclusion temperatures during gelatinization. The enthalpy of gelatinization, which measures the loss of molecular order within the granule, provides insights into the crystallinity of starch. Understanding the pasting properties of starch is crucial for predicting its behavior during heating and cooling, which are common in food processing.

7.4.3 Retrogradation

Retrogradation is the process by which gelatinized starch recrystallizes upon cooling, affecting the texture and shelf life of starchy foods. Wang et al. (2016) studied the retrogradation behavior of gelatinized starch, noting that enthalpy changes during retrogradation do not fully capture the process. At low water contents, retrograded starch exhibits higher enthalpy changes than native starch, likely due to the melting of recrystallized starch. The study also observed increased viscosity and the formation of B-type crystalline polymorphs during retrogradation, which are critical for determining the functional properties of starch in food processing.

Uncu et al. (2017) explored the thermal and rheological properties of quinoa-based breakfast purees, highlighting the influence of gelatinization and melting processes on the thermodynamic behavior of quinoa starch. Their findings suggest that quinoa starch's thermal properties make it a suitable gluten-free alternative for individuals with CD, offering both nutritional and functional benefits.

Tao et al. (2024) investigated the retrogradation behavior of damaged wheat starch under different milling treatments and water contents. The authors observed that higher water and damaged starch content facilitated recrystallinity and long-range ordered structures in starch pastes during storage. Increased water content enhanced water mobility, promoting retrogradation processes and reducing setback viscosity and amylose leaching. These findings emphasize the complex interplay between starch source, processing conditions, and water content in determining retrogradation behavior.

In conclusion, the functional properties of pseudocereal starches, including swelling power, gelatinization, and retrogradation, are influenced by their unique composition and structure. Understanding these properties is essential for optimizing their use in food technology and developing innovative, nutritious products. Advances in research continue to shed light on the complex interactions between starch components, processing conditions, and functional behavior, paving the way for broader applications of pseudocereals in the food industry.

7.5 FUNCTIONAL PROPERTIES OF PSEUDOCEREAL PROTEIN

Pseudocereals, such as amaranth, quinoa, and buckwheat, are recognized as excellent sources of high-quality protein, minerals, and bioactive compounds that promote human health (Kaur et al., 2023). These grains are nutritionally superior to traditional cereals due to their balanced essential amino acid profile, high protein efficiency ratio (PER), and high biological value (Mir et al., 2018). The protein content of pseudocereals ranges from 9% to 21%, making them a valuable resource for functional food development (Langyan et al., 2023). Their gluten-free nature and high nutritional value have made them increasingly popular in human diets, particularly for individuals with CD or gluten intolerance (Caeiro et al., 2022). Compared to cereals, pseudocereals have higher protein content, with quinoa, amaranth, and buckwheat containing 7%–16% protein. Their proteins are hypoallergenic, gluten-free, and nutritionally superior to those of millet and major cereals (Krkošková & Mrazova, 2005).

Pseudocereals also outperform maize and rice in overall protein content. Buckwheat, for instance, contains higher contents of globulins and albumins but lower levels of prolamins and glutelins compared to wheat (Krkošková & Mrazova, 2005). Amaranth proteins consist of albumins, globulins, glutelins, and some prolamins, with globulins exhibiting emulsifying properties (Dakhili et al., 2019). Protein content of quinoa constitutes 12%–23% of the grain, and is primarily composed of 11S globulins (37%) and 2S albumins (35%) and stabilized by disulfide bonds (Cotovanu et al., 2020). The majority of quinoa proteins are found in the endosperm and embryo, which account for about 30% of the seed's weight (Cotovanu et al., 2020). According to the Food and Agriculture Organization (FAO), amaranth proteins nearly meet the daily recommended protein intake due to their balanced amino acid profile (Mlakar et al., 2009). Quinoa proteins have a biological value comparable to beef and higher digestibility than grains such as rice, wheat, and maize (Ng & Wang, 2021). Buckwheat proteins, which have a biological value of 92.3%, surpass wheat (62.5%) and defatted dried milk (81.5%) (Litoiu et al., 2025).

7.5.1 Solubility

Protein solubility is a critical functional property that influences other characteristics such as emulsification, foaming, and gelation. Shevkani et al. (2014a) studied the solubility of amaranth, quinoa, and chia proteins across a pH range of 2–9. The solubility of amaranth protein isolates was minimal at pH 5 and increased at higher or lower pH values, consistent with findings by Marcone and Kakuda (1999). This behavior is attributed to the globulin fraction, which dominates the protein composition (Shevkani et al., 2014b).

Bolontrade et al. (2013) investigated the effect of ionic strength on protein solubility, finding revealed that higher ionic strength reduced solubility, particularly in acidic environments. In contrast, Mahajan and Dua (2002) observed a salting-in effect with NaCl, which increased protein solubility in amaranth meal. Abugoch et al. (2008) extracted quinoa protein isolates at pH 9 and 11, noting that solubility was higher for the pH 9 isolate. Similarly, Vázquez-Ovando et al. (2013) found that protein solubility of chia was lowest at pH 4, near its pI. Timilsena et al. (2016a) demonstrated that drying methods affect protein solubility of chia, with spray-dried isolates showing the highest solubility due to reduced denaturation.

7.5.2 Color

The color of pseudocereal flours and protein isolates is an important quality parameter. Shevkani et al. (2014a) characterized amaranth protein isolates, reporting high brightness (L* = 81.87–87.22) and slight red (a* = 0.24–1.37) and yellow (b* = 10.99–13.61) tints. Steffolani et al. (2015) observed that protein isolates of quinoa were less luminous than amaranth isolates, possibly due to pigment co-extraction during isolation. Spray-dried chia protein isolates exhibited higher luminosity (L* = 86.9) compared to freeze-dried (L* = 80.6) and vacuum-dried (L* = 71.2) samples (Timilsena et al., 2016a).

7.5.3 Gelling Properties

The ability of proteins to form gels is essential for developing textured foods. Avanza et al. (2005a) found that amaranth protein isolates formed self-supporting gels at concentrations ≥7% w/v and temperatures ≥70°C. Bejarano-Luján et al. (2010) demonstrated that extraction methods influence gel properties, with higher temperatures (80°C–90°C) producing more structured gels. Ruiz et al. (2016a) observed that quinoa protein isolates extracted at pH 8–9 formed dense gels, while those at pH 10–11 did not. Olivos-Lugo et al. (2010) reported stable gels from chia protein isolates at 20% w/v. Cold gelation, induced by pH reduction or salt addition, was studied by Mäkinen et al. (2015). Quinoa protein isolates denatured at pH 10.5 formed finer, more organized gels compared to those denatured at pH 8.5.

7.5.4 Water Holding Capacity (WHC)

Water holding capacity (WHC) is a measure of a protein's ability to retain water under external force. Shevkani et al. (2014b) reported WHC values for amaranth protein isolates, while Nasir et al. (2015) and Steffolani et al. (2015) provided data for quinoa isolates. Abugoch et al. (2008) found that quinoa protein isolates extracted at pH 9 had lower WHC (1.7 mL/g) than those extracted at pH 11 (2.6 mL/g). Timilsena et al. (2016a) observed that freeze-dried chia protein isolates had higher WHC (2.9 g water/g) compared to vacuum-dried (2.1 g/g) and spray-dried (2.3 g/g) isolates.

7.5.5 Oil Absorption Capacity (OAC)

Oil absorption capacity (OAC) reflects a protein's ability to bind fat through hydrophobic interactions. Shevkani et al. (2014b) reported OAC values of 3.6–6.4 g oil/g for amaranth protein isolates. Steffolani et al. (2015) found that quinoa protein isolates from Bolivian cultivars had higher OAC than those from Peruvian cultivars. Olivos-Lugo et al. (2010) reported an OAC of 4.04 g oil/g for chia protein isolates, with vacuum-dried powders showing the highest OAC (3.6 g/g) (Timilsena et al., 2016b).

7.5.6 Emulsifying Ability

Emulsifying stability (ES) and emulsifying activity (EA) are critical for food applications. Protein isolates from amaranth and chia exhibited high ES at pH 8, with values approaching 100% (Cordero-De-Los-Santos et al., 2005; Vázquez-Ovando et al., 2013). Mahajan and Dua (2002) found that NaCl and $NaHCO_3$ improved ES in amaranth meal protein emulsions. Shevkani et al. (2014b) reported peak emulsifying activity index (EAI) values at pH 9 for amaranth protein isolates. Quinoa protein isolates also showed high EAI values at pH 8 (Aluko & Monu, 2003).

7.5.7 Foaming Capacity

Foaming capacity (FC) and foaming stability (FS) are influenced by pH and protein solubility. Shevkani et al. (2014b) and Mahajan and Dua (2002) studied FC and FS in amaranth protein isolates, noting that NaCl and $NaHCO_3$ reduced FC but enhanced FS. Bolontrade et al. (2013) found that high ionic strength improved foam stability at pH 2. Timilsena et al. (2016a) reported that spray-dried chia protein isolates had higher FC and FS than freeze-dried and vacuum-dried isolates.

7.5.8 Amino Acid Profile

Pseudocereal proteins are rich in essential amino acids (Table 7.4), particularly lysine, methionine, and cysteine, which are often deficient in cereals. Amaranth and

TABLE 7.4
Amino Acid Profile of Pseudocereal Starch

Total Amino Acids (g/100 g)	Quinoa	Amaranth	Buckwheat
Histidine	0.407	0.389	0.309
Leucine	0.84	0.879	0.832
Isoleucine	0.504	0.582	0.498
Lysine	0.766	0.747	0.672
Phenylalanine	0.593	0.542	0.520
Threonine	0.421	0.558	0.506
Methionine	0.309	0.226	0.172
Tryptophan	0.167	0.181	0.192
Valine	0.594	0.679	0.678
Cysteine	0.203	0.191	0.229
Tyrosine	0.267	0.329	0.241
Arginine	1.09	1.06	0.982
Alanine	0.588	0.799	0.748
Aspartic acid	1.13	1.26	1.130
Glutamic acid	1.86	2.26	2.050
Glycine	0.694	1.64	1.030
Proline	0.773	0.698	0.507
Serine	0.567	1.15	0.685

quinoa proteins have high lysine content, making them suitable for infants and children (Mir et al., 2018). Quinoa proteins contain 2–40 g/100 g lysine and 0.30–9.10 g/100 g methionine (Nowak et al., 2016). Amaranth proteins have a PER of 2.2, comparable to casein (2.4) (Belton & Taylor, 2002). Buckwheat proteins are rich in lysine (0.624 g/100 g) and arginine (1.496 g/100 g) but low in glutamic acid and proline (Nowak et al., 2016). Pseudocereal proteins also contain bioactive peptides with antihypertensive, anti-inflammatory, and antioxidative properties (Iftikhar & Khan, 2019). These peptides further enhance the nutritional and functional value of pseudocereals, making them a vital component of gluten-free and health-promoting diets.

7.5.9 Protein profile Assessment with Sodium Dodecyl Sulfate-Polyacrylamide Gel Electrophoresis (SDS-PAGE)

The analysis of protein profiles using sodium dodecyl sulfate-polyacrylamide gel electrophoresis (SDS-PAGE) reveals a similar protein band of 50 kDa in pseudocereals when observed under nonreducing conditions, which aligns with the globulin and glutelin fractions (Mir et al., 2018). In quinoa samples (Figure 7.4), bands around 50 kDa represent the 11S globulin fraction, known as chenopodin. Chenopodin comprises subunits of approximately 49 and 57 kDa that assemble into a hexamer through non-covalent interactions (Mäkinen et al., 2016). When quinoa proteins were loaded

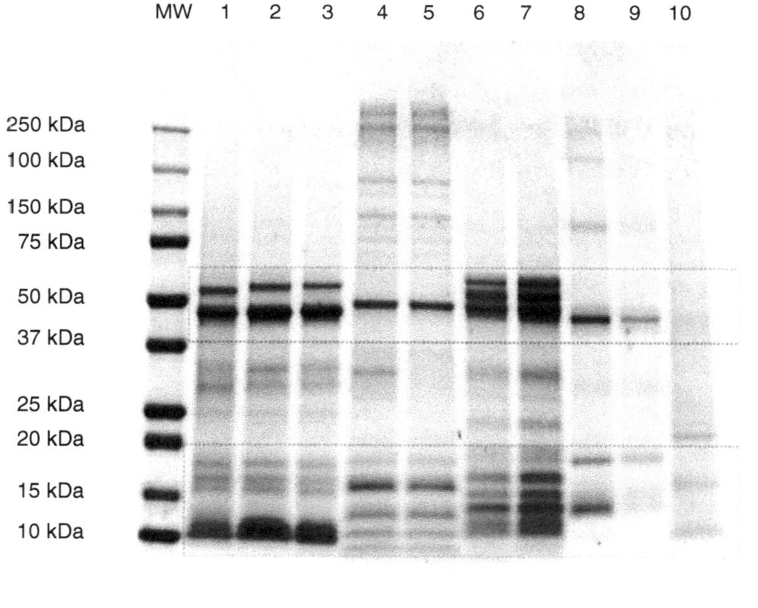

FIGURE 7.4 SDS-PAGE patterns for quinoa wholegrain flour, quinoa dehulled flour (1), quinoa protein-rich flour (2), buckwheat dehulled flour (3), buckwheat protein-rich flour (4), amaranth wholegrain flour (5), rice flour (6), rice protein concentrate (7), and maize flour (8). Each gel's first lane contains a molecular weight marker (Loreto and James, 2018).

directly with the sample loading solution, two bands were observed, one with a molecular weight (MW) less than and another greater than 50 kDa. The increased intensity of the lower MW band indicates that this subunit is more abundant in the chenopodin protein (Amagliani et al., 2016). Chenopodin (~50 kDa) did not show separation into lower MW bands when subjected to a protein extraction solution that included SDS, urea, and thiourea (i.e., under nonreducing conditions), suggesting that disulfide bonds are the main connection between the subunits (Amagliani et al., 2017).

Similar to quinoa, amaranth samples showed a 50 kDa band, indicating the presence of 11S globulin or amarantin (Janssen et al., 2017). This band may also arise from another glutelin-type protein sharing characteristics with 11S globulin (Vasco-Mendez & Paredes-López, 1994). Buckwheat samples exhibited a main band around 50 kDa, likely representing either the primary 13S legume-like globulin or the minor 8S vicilin-like globulin (Milisavljevic et al., 2004). In quinoa treated with DTT, disulfide bonds between the basic or β- (MW ~17 and 19 kDa) and acidic or α- (MW ~28 and 34 kDa) subunits were disrupted, leading to lower MW proteins (Avila Ruiz et al., 2016). Similarly, amaranth showed that reducing conditions separate the acidic or α- (34–36 kDa) and basic or β- (22–24 kDa) amarantin subunits linked by disulfide bonds (Janssen et al., 2017). Buckwheat's 13S globulin consists of a smaller basic subunit (16–29 kDa) and a larger acidic subunit (30–38 kDa) connected by a disulfide bond (Choi & Ma., 2005).

In the three gels, all of the buckwheat, amaranth, and quinoa samples displayed bands that matched low MW proteins (~10–15 kDa). The albumin fraction, which contains a lot of pseudocereals, may be related to this (Valencia-Chamorro, 2003). Amaranth showed high MW proteins (~250 kDa) in addition to globulin and albumin proteins, which under reducing conditions were resolved into lower MW bands. Abugoch et al. (2009) discovered that a considerable quantity of aggregated polypeptides with MW greater than 60 kDa were present in amaranth glutelin. The albumin-1 fraction, which was previously reported to have a MW of 34 kDa may be represented by the band on the gels at about 37 kDa for some samples that does not change under decreasing conditions.

7.6 CONCLUSION

Pseudocereals have received a growing interest by the consumers given their richness in high-quality proteins, dietary fibers, unsaturated fats, and bioactive compounds (e.g., polyphenols and phytosterols) that are associated with numerous health benefits, including antidiabetic, antiobesity, hypolipidemic, anti-inflammatory, antihypertensive, anticancer, and hepatoprotective properties. The most popular pseudocereals such as quinoa, amaranth, and buckwheat, and other pseudocereals, were reported to be particularly rich in high quality proteins and starch that can be further exploited as an ingredient in functional food products. Over the years, pseudocereals have been considered an important non-conventional source of proteins and starch for further utilization in various novel food applications. Traditional and innovative techniques for the extraction of protein and starch have been presented, with the innovative techniques presenting greater capacity in terms of extraction efficiency, purity, and functional properties. The proteins and starch extracted from pseudocereals demonstrated enhanced functional and bioactive properties which are important for developing new functional products. Evidently, huge market potential exists for functional foods formulated with pseudocereals. However, given that the cultivation of pseudocereals is limited to certain regions, it is important to meet the global needs for pseudocereals by supporting worldwide cultivation.

REFERENCES

Abugoch, L. E., Romero, N., Tapia, C. A., Silva, J. & Rivera, M. (2008). Study of some physicochemical and functional properties of quinoa (*Chenopodium quinoa* Willd) protein isolates. *Journal of Agricultural and Food Chemistry*, 56, 4745–4750.

Abugoch, L., Castro, E., Tapia, C., Añón, M. C., Gajardo, P., & Villarroel, A. (2009). Stability of quinoa flour proteins (*Chenopodium quinoa Willd.)* during storage. *International Journal of Food Science and Technology*, 44(10), 2013–2020.

Amagliani, L., O'Regan, J., Kelly, A. L., & O'Mahony, J. A. (2016). Physical and flow properties of rice protein powders. *Journal of Food Engineering*, 190, 1–9.

Alonso-Miravalles, L., & O'Mahony, J. A. (2018). Composition, Protein Profile and Rheological Properties of Pseudocereal-Based Protein-Rich Ingredients. *Foods*, 7(5), 73. https://doi.org/10.3390/foods7050073.

Aluko, R. & Monu, E. (2003). Functional and bioactive properties of quinoa seed protein hydrolysates. *Journal of Food Science*, 68, 1254–1258.

Alvarez-Jubete, L., Arendt, E. K., & Gallagher, E. (2009). Nutritive value and chemical composition of pseudocereals as gluten-free ingredients. *International Journal of Food Sciences and Nutrition*, 60(SUPPL.4), 240–257. https://doi.org/10.1080/09637480902950597

Amagliani, L., O'Regan, J., Kelly, A. L., & O'Mahony, J. A. (2017). Composition and protein profile analysis of rice protein ingredients. *Journal of Food Composition and Analysis*, 59, 18–26.

Ando, H., Chen, Y. C., Tang, H., Shimizu, M., Watanabe, K., & Mitsunaga, T. (2002). Food Components in Fractions of Quinoa Seed. *Food Science and Technology Research*, 8(1), 80–84. https://doi.org/10.3136/FSTR.8.80

Arslan, A., Haros C.M., Yalçın, E., Güneş, A. (2022). Wet milling of buckwheat cultivars and some quality properties of the fractions. *Food Science Technology International,* 28(4), 320–330.

Avanza, M., Puppo, M. & Añón, M. (2005a). Rheological characterization of amaranth protein gels. *Food Hydrocolloids*, 19, 889–898.

Avila Ruiz, G., Arts, A., Minor, M., & Schutyser, M. (2016). A hybrid dry and aqueous fractionation method to obtain protein-rich fractions from quinoa (Chenopodium quinoa Willd). *Food and Bioprocess Technology*, 9, 1502–1510.

Ayub, M., Castro-Alba, V., & Lazarte, C. E. (2021). Development of an instant-mix probiotic beverage based on fermented quinoa with reduced phytate content. *Journal of Functional Foods*, 87, 104831.

Ballester-Sánchez, J., Fernández-Espinar, M. T., & Haros, C. M. (2020). Isolation of red quinoa fibre by wet and dry milling and application as a potential functional bakery ingredient. *Food Hydrocolloids*, 101, 105513.

Bejarano-Luján, D. L., Lopes da Cunha, R. & Netto, F. M. (2010). Structural and rheological properties of amaranth protein concentrate gels obtained by different processes. *Food Hydrocolloids*, 24, 602–610.

Belton, P. S., & Taylor, J. R. (Eds.). (2002). *Pseudocereals and less common cereals: grain properties and utilization potential*. Springer Berlin, Heidelberg. doi: https://doi.org/10.1007/978-3-662-09544-7

Berezina, N. A., Nikitin, I. A., Khmeleva, E. V., Glebova, N. V., & Makarova, N. A. (2020). Features of technological characteristics of cereal and pseudocereal flour. In *BIO Web of Conferences* (Vol. 17, p. 00121). EDP Sciences.

Bet, C. D., de Oliveira, C. S., Colman, T. A. D., Bisinella, R. Z. B., Beninca, C., Lacerda, L. G., Ramos, A. P., & Schnitzler, E. (2019). Aqueous extraction of organic amaranth starch and their by-products. *Journal of Thermal Analysis and Calorimetry*, 138, 2733–2749. https://doi.org/10.1007/s10973-019-08374-7

Bolontrade, A. J., Scilingo, A. A., & Añón, M. C. (2013). Amaranth proteins foaming properties: Adsorption kinetics and foam formation—Part 1. *Colloids and Surfaces B: Biointerfaces*, 105, 319–327.

Calzetta Resio, A. N., Tolaba, M. P., & Suárez, C. (2006). Effects of steeping conditions on wet-milling attributes of amaranth. *International Journal of Food Science and Technology*, 41(2), 70–76.

Capriles, V. D., Coelho, K. D., Guerra-Matias, A. C., & Arêas, J. A. G. (2008). Effects of processing methods on amaranth starch digestibility and predicted glycemic index. *Journal of Food Science*, 73(7). https://doi.org/10.1111/J.1750-3841.2008.00869.X

Choi, S. M., & Ma, C. Y. (2005). Conformational study of globulin from common buckwheat (Fagopyrum esculentum Moench) by Fourier transform infrared spectroscopy and differential scanning calorimetry. *Journal of Agricultural and Food Chemistry*, 53(20), 8046–8053.

Ciudad-Mulero, M., Fernández-Ruiz, V., Matallana-González, M. C., & Morales, P. (2019). Dietary fiber sources and human benefits: The case study of cereal and pseudocereals. *Advances in Food and Nutrition Research*, 90, 83–134. https://doi.org/10.1016/bs.afnr.2019.02.002

Contreras-Jiménez, B., Torres-Vargas, O. L., & Rodríguez-García, M. E. (2019). Physicochemical characterization of quinoa (Chenopodium quinoa) flour and isolated starch. *Food Chemistry*, 298. https://doi.org/10.1016/j.foodchem.2019.124982

Coţovanu, I., Batariuc, A., & Mironeasa, S. (2020). Characterization of quinoa seeds milling fractions and their effect on the rheological properties of wheat flour dough. *Applied Sciences*, 10(20), 7225.

Cordero-De-Los-Santos, M., Osuna-Castro, J., Borodanenko, A. & ParedesLópez, O. (2005). Physicochemical and functional characterisation of amaranth (Amaranthus hypochondriacus) protein isolates obtained by isoelectric precipitation and micellisation. *Food Science and Technology International*, 11, 269–280.

Dakhili, S., Abdolalizadeh, L., Hosseini, S. M., Shojaee-Aliabadi, S., & Mirmoghtadaie, L. (2019). Quinoa protein: Composition, structure and functional properties. *Food Chemistry*, 299, 125161.

De Bock, P., Daelemans, L., Selis, L., Raes, K., Vermeir, P., Eeckhout, M., & Van Bockstaele, F. (2021). Comparison of the chemical and technological characteristics of wholemeal flours obtained from amaranth (Amaranthus sp.), quinoa (Chenopodium quinoa) and buckwheat (Fagopyrum sp.) seeds. *Foods*, 10(3), 651. https://doi.org/10.3390/foods10030651

Deleu, L. J., Lambrecht, M. A., Van de Vondel, J., & Delcour, J. A. (2019). The impact of alkaline conditions on storage proteins of cereals and pseudo-cereals. *Current Opinion in Food Science*, 25, 98–103.

Delgado-García, Y., Luna, S., López-Malo, A., & Morales Camacho, J. (2022). Effect of supercritical carbon dioxide on physicochemical and techno-functional properties of amaranth flour. *Chemical Engineering and Processing – Process Intensification*, 178, 109031. https://doi.org/10.1016/j.cep.2022.109031

Estivi, L., Brandolini, A., Condezo-Hoyos, L., & Hidalgo, A. (2022). Impact of low-frequency ultrasound technology on physical, chemical and technological properties of cereals and pseudocereals. In *Ultrasonics Sonochemistry* (Vol. 86). Elsevier B.V. https://doi.org/10.1016/j.ultsonch.2022.106044

Fabian, C., & Ju, Y. H. (2011). A review on rice bran protein: its properties and extraction methods. *Critical Reviews in Food Science & Nutrition,* 51(9):816–827.

Fidantsi, A., & Doxastakis, G. (2001). Emulsifying and foaming properties of amaranth seed protein isolates. *Colloids and Surfaces B: Biointerfaces*, 21(1–3), 119–124.

Fredriksson, H., Silverio, J., Andersson, R., Eliasson, A. C., & Åman, P. J. C. P. (1998). The influence of amylose and amylopectin characteristics on gelatinization and retrogradation properties of different starches. *Carbohydrate Polymers*, *35*(3–4), 119–134.

Graziano, S., Agrimonti, C., Marmiroli, N., & Gullì, M. (2022). Utilisation and limitations of pseudocereals (quinoa, amaranth, and buckwheat) in food production: A review. *Trends in Food Science & Technology*, 125, 154–165. https://doi.org/10.1016/j.tifs.2022.04.007

Guo, H., Wu, H., Sajid, A., & Li, Z. (2022). Whole grain cereals: The potential roles of functional components in human health. *Critical Reviews in Food Science and Nutrition*, 62(30), 8388–8402. https://doi.org/10.1080/10408398.2021.1928596

Guzel, M., Guzel, N., Turksoy, S., Rusu, A. V., & Trif, M. (2023). Methodologies for the development of cereals and pseudocereals for improved quality and nutritional value. In *Developing Sustainable and Health-Promoting Cereals and Pseudocereals: Conventional and Molecular Breeding* (pp. 205–231). https://doi.org/10.1016/B978-0-323-90566-4.00017-5

Haros, C. M., & Wronkowska, M. (2017). Pseudocereal dry and wet milling: processes, products and applications. *Pseudocereals: Chemistry and Technology*, 140–162. https://doi.org/10.1002/9781118938256.ch7

Hamada, J. S. (1999). Use of proteases to enhance solubilization of rice bran proteins. *Journal of Food Biochemistry*, 23(3), 307–321.

Hasler, C. M. (2002). Functional foods: Benefits, concerns and challenges: A position paper from the American Council on Science and Health. *The Journal of Nutrition*, 132(12), 3772–3781. https://doi.org/10.1093/jn/132.12.3772

He, X., Wang, B., Zhao, B., & Yang, F. (2022). Ultrasonic-Assisted Extraction of Quinoa (Chenopodium quinoa) Protein and effect of heat treatment on its in vitro digestion characteristics. *Foods* (MDPI) 11(5), 771.

He, Q., Ma, D., Li, W., Xing, L., Zhang, H., Wang, Y., ... & Du, H. (2023). High-quality Fagopyrum esculentum genome provides insights into the flavonoid accumulation among different tissues and self-incompatibility. *Journal of Integrative Plant Biology*, *65*(6), 1423–1441.

Huda, M. N., Lu, S., Jahan, T., Ding, M., Jha, R., Zhang, K., & Zhou, M. (2021). Treasure from garden: Bioactive compounds of buckwheat. *Food chemistry*, 335, 127653.

Iftikhar, M., & Khan, M. (2019). Amaranth. In Jing Wang, Baoguo Sun and Rong Tsao (Eds.), *Bioactive Factors and Processing Technology for Cereal Foods* (pp. 217–232). Springer, Singapore. https://doi.org/10.1007/978-981-13-6167-8_13

Inouchi, N., Nishi, K., Tanaka, S., Asai, M., Kawase, Y., Hata, Y., Konishi, Y., Yue, S., & Fuwa, H. (1999). Characterization of amaranth and quinoa starches. *Journal of Applied Glycoscience*, 46(3), 233–240. https://doi.org/10.5458/JAG.46.233

Ismail, B. B., Gambo, A., Garba, U., & Ayub, K. A. (2024). Chapter 1: Gluten-free pasta's consumer appeal and qualities. In A. Gull, G. A. Nayik & C. Brennan (Eds.), *Development of Gluten-Free Pasta* (pp. 1–18): Academic Press.

Jan, K. N., Panesar, P. S., Rana, J. C., & Singh, S. (2017). Structural, thermal and rheological properties of starches isolated from Indian quinoa varieties. *International Journal of Biological Macromolecules*, 102, 315–322. https://doi.org/10.1016/j.ijbiomac.2017.04.027

Jindal N., Saxena, D. C. (2015). Process standardization for isolation of starch from buckwheat (*Fagopyrum esculentum*) flour. *Journal of Agricultural Engineering and Food Technology,* 2(1), 25–28.

Junejo, S. A., Wang, J., Liu, Y., Jia, R., Zhou, Y., & Li, S. (2022). Multi-scale structures and functional properties of quinoa starch extracted by alkali, wet-milling, and enzymatic methods. *Foods*, 11, 2625.

Kaur, H., Shams, R., Dash, K. K., & Dar, A. H. (2023). A comprehensive review of pseudo-cereals: Nutritional profile, phytochemicals constituents and potential health promoting benefits. *Applied Food Research*, *3*(2), 100351.

Krkošková, B., & Mrázová, Z. (2005). Prophylactic components of buckwheat. *Food Research International*, 38(5), 561–568.

Kringel, D. H., El Halal, S. L. M., Zavareze, E. d. R., & Dias, A. R. G. (2020). Methods for the extraction of roots, tubers, pulses, pseudocereals, and other unconventional starches sources: A review. *Starch - Stärke*, 72(11–12), 1900234. https://doi.org/10.1002/star.201900234

Kumawat, K.R. (2022). Grain Amaranth : A New Millennium Crop of Pseudo-Cereals. Acessed from: https://agriarticles.com/grain-amaranth-pseudo-cereal/. Acessed on 12 Jan 2026.

Laguna, O., Barakat, A., Alhamada, H., Durand, E., Baréa, B., Fine, F., … & Lecomte, J. (2018). Production of proteins and phenolic compounds enriched fractions from rapeseed and sunflower meals by dry fractionation processes. *Industrial Crops and Products*, 118, 160–172.

Langyan, S., Yadava, P., Khan, F. N., Dar, Z. A., Singh, R., Kumar, A. (2022). Sustaining protein nutrition through plant-based foods. *Frontiers in Nutrition,* 18(8), 772573. https://doi.org/10.3389/fnut.2021.772573.

Li, G., & Zhu, F. (2017). Amylopectin molecular structure in relation to physicochemical properties of quinoa starch. *Carbohydrate Polymers*, 164, 396–402. https://doi.org/10.1016/j.carbpol.2017.02.014

Li, G., & Zhu, F. (2018). Quinoa starch: Structure, properties, and applications. *Carbohydrate Polymers*, 181, 851–861. https://doi.org/10.1016/J.CARBPOL.2017.11.067

Li, G., Xu, X., & Zhu, F. (2019). Physicochemical properties of dodecenyl succinic anhydride (DDSA) modified quinoa starch. *Food Chemistry*, 300. https://doi.org/10.1016/j.foodchem.2019.125201

Li, M., Tian, Y., & Dhital, S. (2022). Starch extraction: Principles and techniques. Edited by Luis Bello-Pérez, José Alvarez-Ramírez, Sushil Dhital. In *Starch and Starchy Food Products*, pp. 17–40. CRC Press, Boca Raton, USA.

Linsberger-Martin, G., Lukasch, B., & Berghofer, E. (2012). Effects of high hydrostatic pressure on the RS content of amaranth, quinoa and wheat starch. *Starch/Staerke*, 64(2), 157–165. https://doi.org/10.1002/STAR.201100065

Liţoiu, A. A., Păucean, A., Lung, C., Zmuncilă, A., & Chiş, M. S. (2025). An Overview of Buckwheat—A Superfood with Applicability in Human Health and Food Packaging. *Plants*, 14(14), 2200.

Loreto, A. M., & James, A. O. (2018). Composition, protein profile and rheological properties of Pseudocereal-based protein-rich ingredients. *Foods*, 7.5 (2018), 73. https://doi.org/10.3390/foods7050073

Mahajan, A. & Dua, S. (2002). Salts and pH induced changes in functional properties of amaranth (Amaranthus tricolor L.) seed meal. *Cereal Chemistry*, 79, 834–837

Mäkinen, O. E., Zannini, E. & Arendt, E. K. (2015). Modifying the cold gelation properties of quinoa protein isolate: Influence of heat-denaturation pH in the alkaline range. *Plant Foods for Human Nutrition*, 70, 1–7.

Mäkinen, O. E., Zannini, E., Koehler, P., & Arendt, E. K. (2016). Heat-denaturation and aggregation of quinoa (Chenopodium quinoa) globulins as affected by the pH value. *Food Chemistry*, 196, 17–24.

Malik, A. M., & Singh, A. (2022). Pseudocereals proteins-A comprehensive review on its isolation, composition and quality evaluation techniques. *Food Chemistry Advances*, 1, 100001.

Marcone, M. & Kakuda, Y. (1999). A comparative study of the functional properties of amaranth and soybean globulin isolates. *Food/Nahrung*, 43, 368–373.

Martin Ramos Diaz, J. (2012). Use of amaranth, quinoa and kañiwa in extruded corn snacks. Master's degree in Food ScienceAdvisor: Kirsi Jouppila.

Middlewood, P. G., & Carson, J. K. (2012). Extraction of amaranth starch from an aqueous medium using microfiltration: Membrane characterisation. *Journal of Membrane Science*, 405–406, 284–290. https://doi.org/10.1016/j.memsci.2012.03.027

Milisavljević, M. D., Timotijević, G. S., Radović, S. R., Brkljačić, J. M., Konstantinović, M. M., & Maksimović, V. R. (2004). Vicilin-like storage globulin from buckwheat (Fagopyrum esculentum Moench) seeds. *Journal of Agricultural and Food Chemistry*, 52(16), 5258–5262.

Mir, N. A., Riar, C. S., & Singh, S. (2018). Nutritional constituents of pseudo cereals and their potential use in Fanwood systems: A review. *Trends in Food Science & Technology*, 75, 170–180.

Mlakar, S. G., Turinek, M., Jakop, M., Bavec, M., & Bavec, F. (2009). Nutrition value and use of grain amaranth: potential future application in bread making, *Agriculture,* 6, 43–53.

Modupalli, N., Sunil, C. K., & Venkatachalapathy, N. (2024). Unit operations in pseudocereals and heritage grains processing. In *Unit Operations in Food Grain Processing* (pp. 425–451). https://doi.org/10.1016/B978-0-443-18965-4.00014-5

Mokrejs, P., Janacova, D., Kolomaznik, K., & Svoboda, P. (2011). Modeling isolation of Amaranth protein by enzymatic breakdown of polysaccharides. *Rasayan Journal Chemical*, 4(1), 180–188.

Mua J. P., Jackson D. S. (1997) Relationships between functional attributes and molecular structures of amylose and amylopectin fractions from corn. *Journal of Agricultural and Food Chemistry,* 45:3848–3854.

Nandan, A., Koirala, P., Tripathi, A. D., Vikranta, U., Shah, K., Gupta, A. J., & Nirmal, N. (2024). Nutritional and functional perspectives of pseudocereals. *Food Chemistry*, 448, 139072.

Nasir, M. A., Pasha, I., Butt, M. S., & Nawaz, H. (2015). Biochemical characterization of quinoa with special reference to its protein quality. *Pakistan Journal of Agricultural Sciences*, 52(3), 731–737.

Ng, C. Y., & Wang, M. (2021). The functional ingredients of quinoa (*Chenopodium quinoa*) and physiological effects of consuming quinoa: A review. *Food Frontiers*, 2(3), 329–356.

Nowak, K., Majsterek, S., Ciesielska, N., Sokołowski, R., Klimkiewicz, K., & Zukow, W. (2016). Rola nasion szałwii hiszpańskiej (chia) w żywieniu pacjentów geriatrycznych= The role of chia seeds in nutrition in geriatric patients. *Journal of Education, Health and Sport*, 6(3), 35–40.

Olivos-Lugo, B., Valdivia-López, M. & Tecante, A. (2010). Thermal and physicochemical properties and nutritional value of the protein fraction of Mexican chia seed (*Salvia hispanica L.*). *Food Science and Technology International*, 16, 89–96.

Perez-Rea, D., & Antezana-Gomez, R. (2024). The functionality of pseudocereal starches. In *Starch in Food: Structure, Function, and Applications* (pp. 377–403). https://doi.org/10.1016/B978-0-323-96102-8.00013-9

Pirzadah, T. B., & Malik, B. (2020). Pseudocereals as super foods of 21st century: Recent technological interventions. *Journal of Agriculture and Food Research*, 2, 100052. https://doi.org/10.1016/j.jafr.2020.100052

Poshadri, A., Deshpande, H. W., Machewad, G. M., Kshirsagar, R. B., Gadhe, K. S., & Kadam, S. D. (2023). Functional properties of selected composite gluten-free pseudocereals flour. *Food and Humanity*, 1, 1200–1205.

Punia Bangar, S., Sharma, N., Singh, A., Phimolsiripol, Y., & Brennan, C. S. (2022). Glycaemic response of pseudocereal-based gluten-free food products: A review. *International Journal of Food Science & Technology*, 57(8), 4936–4944.

Qureshi, I., Habib, M., Bashir, K., Jan, K., & Jan, S. (2024). Introduction to Functional Foods and Nutraceuticals. In K. Bashir, K. Jan & F. J. Ahmad (Eds.), *Functional Foods and Nutraceuticals: Chemistry, Health Benefits and the Way Forward* (pp. 1–15). Springer International Publishing.

Rahma, E. H., Dudek, S., Mothes, R., Görnitz, E., & Schwenke, K. D. (2000). Physicochemical characterisation of mung bean (Phaseolus aureus) protein isolates. *Journal of the Science of Food and Agriculture*, 80(4), 477–483.

Reddy, M. V. K., Dubey, P. K., Mishra, A. A., & Ahada Sabeel, V. (2024). Potential processing techniques for safe utilisation of pseudo cereals in the food system. *Journal of Food Composition and Analysis*, 135, 106609. https://doi.org/10.1016/j.jfca.2024.106609

Resio, A. C., Aguerre, R. J., & Suarez, C. (2006). Hydration kinetics of amaranth grain. *Journal of Food Engineering*, 72(3), 247–253.

Ruales, J., Valencia, S., & Nair, B. (1993). Effect of processing on the physico-chemical characteristics of quinoa flour (Chenopodium quinoa, Willd). *Starch - Stärke*, 45(1), 13–19. https://doi.org/10.1002/STAR.19930450105

Ruiz, G. A., Opazo-Navarrete, M., Meurs, M., Minor, M., Sala, G., van Boekel, M., ... & Janssen, A. E. (2016a). Denaturation and in vitro gastric digestion of heat-treated quinoa protein isolates obtained at various extraction pH. *Food Biophysics*, 11(2), 184–197.

Ruiz, G. A., Xiao, W., van Boekel, M., Minor, M. & Stieger, M. (2016b). Effect of extraction pH on heat-induced aggregation, gelation and microstructure of protein isolate from quinoa (Chenopodium quinoa Willd). *Food Chemistry*, 209, 203–210.

Rulahnia, K., & Khatkar, B. S. (2020). Isolation and characterization of starch from quinoa (Chenopodium Quinoa Willd.). *International Journal of Innovative Research and Advanced Studies*, 7, 1–6.

Schönlechner, R., & Bender, D. (2023). Pseudocereals. In *ICC Handbook of 21st Century Cereal Science and Technology* (pp. 191–198). https://doi.org/10.1016/B978-0-323-95295-8.00032-0

Schutyser, M. A. I., & Van der Goot, A. J. (2011). The potential of dry fractionation processes for sustainable plant protein production. *Trends in Food Science & Technology*, 22(4), 154–164.

Scilingo, A. A., Ortiz, S. E. M., Martínez, E. N., & Añón, M. C. (2002). Amaranth protein isolates modified by hydrolytic and thermal treatments. Relationship between structure and solubility. *Food Research International*, 35(9), 855–862.

Shen, W.; Yang, J.; Wang, Z. & Liu, B. (2024). Structural characterization and physicochemical properties of grain amaranth starch (recent characterization/review that compiles Alcalase/enzymatic isolation conditions). *Food Chemistry: X*, 23, 101723. https://doi.org/10.1016/j.fochx.2024.101723

Shevkani, K., Singh, N., Kaur, A. & Rana, J. C. (2014a). Physicochemical, pasting, and functional properties of amaranth seed flours: Effects of lipids removal. *Journal of Food Science*, 79, C1271–C1277.

Shevkani, K., Singh, N., Rana, J. C. & Kaur, A. (2014b). Relationship between physicochemical and functional properties of amaranth (Amaranthus hypochondriacus) protein isolates. *International Journal of Food Science & Technology*, 49, 541–550.

Shittu, T., Raji, A. & Sanni, L. (2007). Bread from composite cassava-wheat flour: I. Effect of baking time and temperature on some physical properties of bread loaf. *Food Research International*, 40, 280–290.

Singh, N., Kaur, L., & Singh, B. (2013). Extraction technologies for recovery of value-added products from food processing by-products. *Critical Reviews in Food Science and Nutrition*, 53(7), 697–718.

Steffolani, M. E., León, A. E., & Pérez, G. T. (2013). Study of the physicochemical and functional characterization of quinoa and kañiwa starches. *Starch/Staerke*, 65(11–12), 976–983. https://doi.org/10.1002/STAR.201200286

Steffolani, M. E., Villacorta, P., Morales-Soriano, E., Repo-Carrasco, R., León, A. E. & Perez, G. T. (2015). Physicochemical and functional characterization of protein isolated from different quinoa varieties (Chenopodium quinoa Willd). *Cereal Chemistry*, 93, 275–281.

Tao, H., Fang, X. H., Cai, W. H., Zhang, S., Wang, H. L. (2024). Retrogradation behaviors of damaged wheat starch with different water contents. *Food Chemistry X,* 28, 22, 101258. https://doi.org/10.1016/j.fochx.2024.101258.

Tang, C. H. (2007). Functional properties and in vitro digestibility of buckwheat protein products: Influence of processing. *Journal of Food Engineering*, 82(4), 568–576.

Tang, C. H., Peng, J., Zhen, D. W., & Chen, Z. (2009). Physicochemical and antioxidant properties of buckwheat (Fagopyrum esculentum Moench) protein hydrolysates. *Food Chemistry*, 115(2), 672–678.

Thakur, P., Kumar, K., Ahmed, N., Chauhan, D., Rizvi, Q. U. E. H., Jan, S., ... & Dhaliwal, H. S. (2021a). Effect of soaking and germination treatments on nutritional, anti-nutritional, and bioactive properties of amaranth (Amaranthus hypochondriacus L.), quinoa (Chenopodium quinoa L.), and buckwheat (Fagopyrum esculentum L.). *Current Research in Food Science*, 4, 917–925.

Thakur, P., Kumar, K., & Dhaliwal, H. S. (2021b). Nutritional facts, bio-active components and processing aspects of pseudocereals: A comprehensive review. *Food Bioscience*, 42. https://doi.org/10.1016/j.fbio.2021.101170

Timilsena, Y. P., Adhikari, R., Barrow, C. J. & Adhikari, B. (2016a). Physicochemical and functional properties of protein isolate produced from Australian chia seeds. *Food Chemistry*, 212, 648–656.

Timilsena, Y. P., Wang, B., Adhikari, R. & Adhikari, B. (2016b). Preparation and characterization of chia seed protein isolate–chia seed gum complex coacervates. *Food Hydrocolloids*, 52, 554–563.

Uncu, O., Jolayemi, O. S., & Unluturk, S. (2017). Evaluation of rheological, textural and thermal properties of quinoa (Chenopodium quinoa Willd) based breakfast puree. *International Journal*, *4*(2), 23.

Usman, M., Patil, P. J., Mehmood, A., Rehman, A., Shah, H., Haider, J., & Li, X. (2022). Comparative evaluation of pseudocereal peptides: A review of their nutritional contribution. *Trends in Food Science & Technology*, 122, 287–313.

Valencia-Chamorro, S. (2003). Quinoa. In B. Caballero (Ed.), *Encyclopedia of Food Science and Nutrition* (Vol. 8, pp. 4895–4902).

Vasco-Méndez, N. L., & Paredes-López, O. C. T. A. V. I. O. (1994). Antigenic homology between amaranth glutelins and other storage proteins. *Journal of Food Biochemistry*, 18(4), 227–238.

Vatansever, S., Whitney, K., Ohm, J. B., Simsek, S., & Hall, C. (2021). Physicochemical and multi-scale structural alterations of pea starch induced by supercritical carbon dioxide + ethanol extraction. *Food Chemistry*, 344. https://doi.org/10.1016/j.foodchem.2020.128699

Vázquez-Ovando, A., Betancur-Ancona, D. & Chel-Guerrero, L. (2013). Physicochemical and functional properties of a protein-rich fraction produced by dry fractionation of chia seeds (*Salvia hispanica* L.). *CyTA-Journal of Food*, 11, 75–80.

Vidaurre-Ruiz, J., Bender, D., & Schönlechner, R. (2023). Exploiting pseudocereals as novel high protein grains. *Journal of Cereal Science*, 114, 103795. https://doi.org/10.1016/j.jcs.2023.103795

Vignesh, A., Amal, T. C., Sarvalingam, A., & Vasanth, K. (2024). A review on the influence of nutraceuticals and functional foods on health. *Food Chemistry Advances*, 5, 100749. https://doi.org/10.1016/j.focha.2024.100749

Villalobos, L. M., Inga, M., & Betalleluz-Pallardel, I. (2024). Properties and emulsifying performance of octenyl succinic anhydride-modified starch from an Andean Pseudocereal: Cañihua (Chenopodium pallidicaulle Aellen). *Carbohydrate Polymer Technologies and Applications*, 100605. https://doi.org/10.1016/j.carpta.2024.100605

Villarreal, M. E., Ribotta, P. D., & Iturriaga, L. B. (2013). Comparing methods for extracting amaranthus starch and the properties of the isolated starches. *LWT - Food Science and Technology*, 51(2), 441–447. https://doi.org/10.1016/j.lwt.2012.11.009

Wang, S., Li, C., Zhang, X. et al. (2016). Retrogradation enthalpy does not always reflect the retrogradation behavior of gelatinized starch. *Scientific Reports,* 6, 20965. https://doi.org/10.1038/srep20965

Wang, C., Li, B., & Corke, H. (2015). Extraction and functional properties of amaranth protein isolates. *Food Chemistry*, 173, 113–120.

Wang, Q., Wang, Y., Huang, M., Hayat, K., Kurtz, N. C., Wu, X., ... & Zheng, F. (2021). Ultrasound-assisted alkaline proteinase extraction enhances the yield of pecan protein and modifies its functional properties. *Ultrasonics Sonochemistry*, 80, 105789. https://doi.org/10.1016/j.ultsonch.2021.105789

Wronkowska, M., & Haros, M. (2014). Wet-milling of buckwheat with hull and dehulled–the properties of the obtained starch fraction. *Journal of Cereal Science*, 60(3), 477–483.

Yadav, P., Bharathi, U., Suruthi, K., & Bosco, S. J. D. (2024). Effect of ultrasonic modification on physiochemical, structural, functional properties and in vitro starch digestibility of Amaranthus paniculatus (Rajgeera) starch. *Biomass Conversion and Biorefinery*, 14(16), 19017–19024. https://doi.org/10.1007/s13399-023-04134-8

Yang, C., Liu, W., Zhu, X., Zhang, X., Wei, Y., Huang, J., ... & Yang, F. (2024). Ultrasound-assisted enzymatic digestion for efficient extraction of proteins from quinoa. *LWT*, 194, 115784. https://doi.org/10.1016/j.lwt.2024.115784

Zhu, F., & Li, H. (2019). Effect of high hydrostatic pressure on physicochemical properties of quinoa flour. *LWT*, 114. https://doi.org/10.1016/j.lwt.2019.108367

8 Role of Pseudocereal Starch and Protein in Industrial Sectors

Mariam Islam, Farhan Saeed, Bushra Niaz, Muhammad Afzaal, Amara Rasheed, Hamna Zafar, Aliza Naeem, Maha Atiq, and Sobia Chashman

8.1 INTRODUCTION

The food group known as pseudocereals is made up of non-grass plant varieties that are similar to cereals in terms of their characteristics, uses, and functions. Pseudocereals are considered as dicotyledonous and they differ from cereal grains, which are monocotyledonous based on physical and botanical traits. The term "pseudocereals" is still used to refer to them because of their physical similarities to cereals, including their starch, appearance, palatable qualities, and cooking procedure (Waisundara, 2024). Furthermore, pseudocereals have the potential to enhance and broaden the diversity of natural resources from an ecological point of view. The usage of pseudocereals is environmentally benign, practical, affordable, and cheap due to their agronomic qualities, adaptability to adverse conditions, and excellent nutritional profile. (Kaur et al., 2023). In this regard, pseudocereals are widely acknowledged grains, and their nutritious qualities are highly valued. Pseudocereals are being used more often for human nutrition, which encourages food manufacturers to come up with new, convenient food items. This calls for basic knowledge of the raw materials' chemical constituents as well as their processing-related physical and functional characteristics (Bender and Schoenlechner, 2021).

According to Zhang et al. (2024a), the United Nations projects that by 2050, there will be 10 billion people on the planet, which will cause significant changes in food consumption habits due to urbanization and rising wages. It calls for a shift in the types and amounts of foods consumed. By 2050, the world's protein needs are predicted to increase, highlighting the importance of assessing food proteins. Consequently, it is necessary to investigate sustainable substitutes, especially plant-based proteins, in order to fulfill world protein requirements properly (Zhang et al., 2024b).

The world is currently experiencing a major food crisis due to a number of issues, including climate change, drought, pests, and illnesses. As a result, it has become essential to search for substitute varieties of crops that may be more productive while offering better nutritional qualities than traditional cereals. Pseudocereals therefore

 DOI: 10.1201/9781003584858-8

have a very high ranking among the competing plants (Usman et al., 2022). Although pseudocereal seeds don't contain gluten, they can be processed to make flour for food products like pasta and baked goods. The three most often used pseudocereals are buckwheat, amaranth, and quinoa. Additionally, the need to find gluten-free food alternatives for those with non-celiac gluten sensitivity, wheat allergies, or celiac disease has sparked interest in pseudocereals as a potential source of gluten-free foods (Graziano et al., 2022).

The most widely acknowledged qualities of pseudocereals are their anticancer, antidiabetic, antioxidant, and antihypertensive properties, despite the fact that many research has shown that they have potential for positive health benefits using both animal and human models (Caeiro et al., 2022). In comparison to grains such as wheat and maize, pseudocereals typically have less carbohydrates and more protein and fat. Pseudocereals, as opposed to cereal grains, are rich in bioactive substances such as dietary fiber, vitamins, minerals, polyphenols, lignans, phytosterols, flavonoids, antioxidants, unsaturated fatty acids, and premium proteins with superior digestibility, accessibility, and balanced profile of amino acid (Reddy et al., 2024).

They provide a significant source of fiber, fats, carbohydrates, and high-quality protein. A variety of important amino acids are also present in pseudocereals, and several of them like lysine, arginine, and methionine are present in greater amounts as compared to other cereals (Thakur & Kumar, 2019). A substantial source of vitamins and minerals can also be found in pseudocereals like quinoa, buckwheat, and amaranth. These grains are beneficial substitutes for gluten-free food products and are frequently used in gluten-free goods for patients having celiac disease (Agregán et al., 2023). Owing to their diverse composition, as shown in Table 8.1, they have been linked to a number of health advantages, including hepatoprotective qualities, anti-inflammatory, anticancer, and antihypertensive effects, as well as advantages against diseases such as obesity and diabetes (Shahbaz et al., 2023). Therefore, it is believed that substituting pseudocereals for carbohydrate-rich cereals (rice, wheat, and maize) could restore the nutrient imbalance caused by these foods (Nandan et al., 2024).

Since pseudocereal-derived proteins are of higher quality than cereal proteins, they have been regarded as a suitable option for plant-based meat substitutes. Plant proteins are rich in antioxidants and can enhance gut health by boosting the formation of short-chain fatty acids (SCFAs) and by the number of *Bifidobacterium* and *Lactobacillus* (Usman et al., 2022). Because of their enhanced nutritional content, pseudocereals have recently caught the interest of numerous food science and

TABLE 8.1
Composition of Pseudocereals

Component	Amaranth	Quinoa	Buckwheat	References
Protein (%)	15–16	9–16	13–16	De Bock et al. (2021)
Fat (%)	5–6	6–7	1–2	Thakur et al. (2021)
Carbohydrate (%)	60–65	62–63	62–75	Thakur et al. (2021)
Dietary fiber (%)	2–17	7–26	17.8	Nandan et al. (2024)
Ash (%)	2.2–2.8	1.9–3.4	1.9–2.3	De Bock et al. (2021)

technology researchers as well as dietitians. These pseudocereals have an even better nutritional profile than true cereals (Langyan et al., 2024).

Fermentation enhances pseudocereals' probiotic and prebiotic properties while reducing anti-nutritional factors and boosting nutrient absorption. The high-quality protein level (10%–20%) of pseudocereals is highly valued, and it is commonly extracted through both dry and wet methods. For gluten-free, plant-based products, an alkaline method of extraction is employed (Poshadri et al., 2024). They can be added to traditional dishes to improve them or used for developing new food products. These pseudocereals lack the baking qualities of dough since they don't include gluten (De Bock et al., 2021).

8.2 PSEUDOCEREALS STARCHES

Pseudocereals such as amaranth, buckwheat, and quinoa are gaining popularity as gluten-free crops. This trend can be linked to their extensive health advantages, which encompass antioxidant effects, hypoglycemic properties, and the ability to lower serum cholesterol levels. Furthermore, the proteins derived from amaranth, buckwheat, and quinoa (referred to as AP, BP, and QP, respectively) possess physicochemical characteristics that are beneficial for producing various food products (Constantino & Garcia-Rojas, 2022b). The composition and characteristics of pseudocereal starches encompass various aspects, including their chemical makeup, granule morphology and dimensions, degree of crystallinity, structural properties of amylose and amylopectin, swelling capacity and solubility, thermal behavior, pasting characteristics, retrogradation tendencies, freeze-thaw resilience, and digestibility, as shown in Figure 8.1. Additionally, modifications to starch are

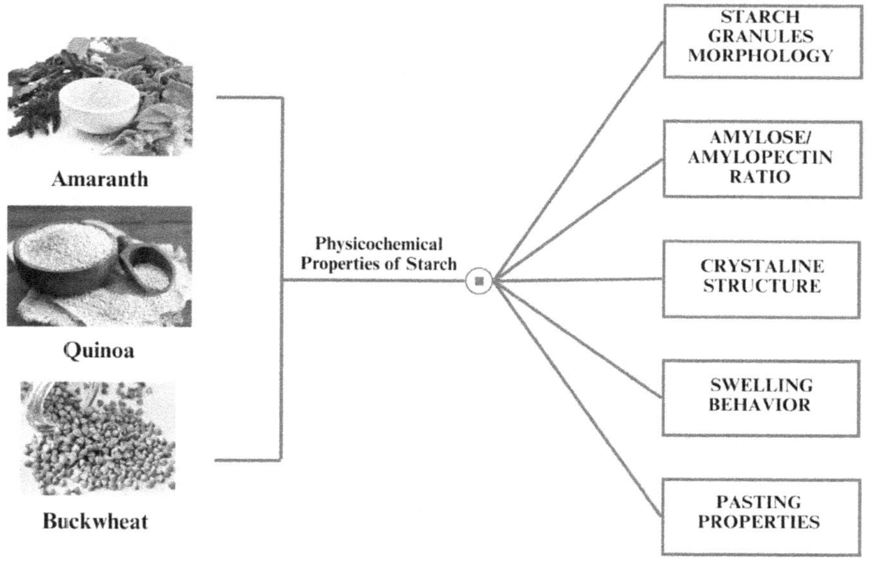

FIGURE 8.1 Physicochemical properties of pseudocereals starches.

employed to enhance their functional properties in food processing contexts, as well as in applications involving starch and flour in food products (Perez-Rea & Antezana-Gomez, 2024).

Starch, being the primary constituent in pseudocereals, is a biopolymer characterized by semicrystalline lamellae. It is composed of amylopectin and amylose. The side chains of amylopectin aggregate into double helices within the crystalline regions, whereas the chains located near branching points predominantly intermingle with amylose in the amorphous state (Zheng et al., 2024). Moreover, the amylose levels in quinoa and amaranth starches are considerably lower than those in cereal starches, generally comprising less than 10% of the overall starch. In comparison, canihua has a slightly elevated amylose content relative to quinoa, which falls between 11% and 20%. On the other hand, buckwheat shows a higher amylose content than cereals, with percentages varying from 21% to 46%. (Schoenlechner & Bender, 2020).

Carbohydrates constitute the primary component of all pseudocereal seeds, accounting for 60%–75% of their composition, and are predominantly made up of starch. Unlike cereal grains, pseudocereals store starch in the perisperm (Bender & Schoenlechner, 2021). In amaranth, the starch granules are among the smallest recorded, typically measuring between 1 and 3 μm in diameter. Conversely, quinoa features polygonal or angular starch granules, with sizes ranging from 0.5 to 3 μm. Additionally, quinoa contains compound starch, where several hundred starch granules aggregate to form clusters measuring between 10 and 30 μm. Similarly, in buckwheat, the starch granules exhibit a polygonal shape and are frequently found in aggregates. Like amaranth, the size of starch granules in buckwheat is relatively small, ranging from 2 to 14 μm, with an average diameter of 6.5 μm (Schoenlechner & Bender, 2020). The water solubility index for pseudocereals varied from 6.22 to 10.22 g/100 g, demonstrating a higher value compared to cereals, yet generally lower than that of legumes. Amaranth flour, noted for its elevated protein content, exhibited the highest water-holding capacity among all pseudocereals, with values ranging from 1.60 to 2.15 g/g (De Bock et al., 2021).

The swelling characteristics vary among different pseudocereals due to the distinct organization of starch granules within the grains, and this behavior is significantly influenced by the amylopectin content. De Bock et al. (2021) conducted a study on the swelling and pasting properties of wholemeal flour from buckwheat, quinoa, and amaranth. Their findings indicated that amaranth possessed the highest swelling power when compared to other pseudocereals. Nevertheless, the pasting properties revealed that the starch from amaranth had the lowest shear stability, a factor attributed to its reduced amylose content. Quinoa and buckwheat demonstrated a lower sensitivity to shearing, which appears to be stabilized by amylose–lipid complexes. This phenomenon was also reported by Bhinder et al. (2020) and Aluwi et al. (2017), who analyzed flour from 23 tartary buckwheat varieties and 28 quinoa varieties, respectively. Bhinder et al. (2020) found that the majority of buckwheat flours had very low setback and breakdown viscosities, indicating a high level of paste stability and reduced retrogradation, making these flours suitable thickeners for sauces and soups.

8.3 MODIFICATION OF PSEUDOCEREALS STARCH

8.3.1 Physical Modifications

Changes in physical properties can be used to break or alter the granular size and/or packing arrangement. As a result, these physical changes have the ability to form starch with properties comparable to chemically modified starch but without any toxins that must be removed subsequently. These changes are easy, inexpensive, and environmentally friendly to a great extent. In conclusion, physical modifications can be classified into two broad categories: thermal and nonthermal techniques. The main thermal modifications are pregelatinization, extrusion, heat-moisturizing, annealing, and microwave treatments. Nonthermal treatments include applications like ultra-high pressure, ultrasonication, and pulsed electric fields. Thermal treatments most widely applied are heat-moisture treatment (HMT) and annealing (Compart et al., 2023).

8.3.1.1 Heat-Moisture Treatment

Such hydrothermal treatments significantly enhance the physicochemical properties of starch with the granular structure maintained. The two treatments differ in the levels of moisture used. Annealing employs, a high amount of water (more than 40%), in addition to a longer processing time and temperatures higher than the glass transition but lower than the gelatinization temperature. HMT, on the other hand, requires high temperatures and only functions in a range of low moisture content between 10% and 35% (Anggreini et al., 2021). HMT alters granules' crystalline and amorphous layers, which brings about a decrease in swelling behavior, solubility, and viscosity but increases the pasting temperature (Chen et al., 2015a). However, its efficiency is greatly affected by process parameters and the source of starch. A decrease in moisture content increases the alteration of the surface of the starch granule, thus making it more susceptible to different modifications. In addition, starch isolated from cereals is less sensitive to alteration than starches from tubers and roots (Bello-Perez et al., 2020).

8.3.1.2 Annealing

Annealing process enhances crystallinity through the facilitation of molecular reorganization and reduction of free energy; hence, allowing for the creation of more ordered structures. Since the heat required for gelatinizing starch is directly proportional to the crystalline region area, this method is better utilized with those starches that exhibit greater amorphous region presence. Starch exhibits greater resistivity against digestibility following the use of annealing technology because of the recrystallization of the structure (Gunaratne, 2023).

8.3.2 Enzymatic Modifications

Enzymatic modification refers to the use of enzymes that alter the starch parameters. It is predominantly utilized in the food sector. In general, modifications of starch through enzymatic reactions are classified into four classes: endoamylases, exoamylases, debranching enzymes, and transferases.

8.3.2.1 A-amylases

A-amylases are a classic example among the endoamylases. These enzymes cause random hydrolysis of starch at any (1,4)-linkage found in the two starch polymers, amylose, and amylopectin (Hua & Yang, 2016). They result in a decrease in chain length. Depending on the type and amount of enzyme used, it may produce pockets on the starch granule surface. Exoamylases include β-amylases, Amyloglucosidases, and α-glucosidases (Del Moral et al., 2018). Exoamylases work by cleaving the surface glucose residues of amylose or amylopectin, thus producing β-limit dextrins (from β-amylase), maltose, and glucose (from glucoamylase and α-glucosidase). The action of the enzymes of the β-amylases results in the creation of a number of cracks on the starch granule surface. When compared to native starch, the starch that has been enzymatically modified demonstrates greater solubility, although it has a diminished swelling capacity and a lower pasting viscosity (Compart et al., 2023).

8.3.2.2 Debranching Enzyme

The debranching enzyme category includes pullulanase and isoamylase. Both of these enzymes are able to hydrolyze the α-1,6 glucosidic bond (Xia & Wu, 2023). One major difference between isoamylase and pullulanase is that pullulanase can hydrolyze the α-1,6 glycosidic bond in starch and pullulan whereas isoamylase can only hydrolyze the α-1,6 bond in glycogen and starch. With the explicit degradation of amylopectin, these enzymes produce linear glucans (Saeid et al., 2024).

8.3.2.3 Transferases

Transferases act by catalyzing the hydrolysis of an α-(1,4) glycosidic bond in a donor molecule, transferring the severed residue to a glycosidic acceptor, thus forming a new glycosidic bond. Enzymes such as amylomaltase and cyclodextrin glycosyltransferase catalyze the formation of a new α-(1,4) glycosidic bond (Martinez & Gómez, 2017). Cyclodextrin glycosyltransferases have little hydrolytic activity and yield cyclic oligosaccharides of 6, 7, or 8 glucose units, together with highly branched, high molecular weight limit dextrins formed by cyclodextrin glycosyltransferase. On the aspect of the nature of the enzymatic reaction, amylomaltase is very similar to cyclodextrin glycosyltransferase. The major difference is the difference in terms of where they catalyze the reaction to yield a linear product for amylomaltase and a cyclic product for cyclodextrin glycosyltransferase (Wang et al., 2020b).

8.3.3 Chemical Modifications

Chemical approaches enable a much wider scope for the functionalization of starch, thus making it more applicable in many areas. These chemical alterations are non-degrading, with the hydroxyl groups found in starches serving as active sites for the introduction of functional groups, such as acetyl-, carboxyl-, and ethyl-groups, which act as electrophilic reactants (Compart et al., 2023). The rate and efficiency of modification, however, are influenced by various factors including the source of starch, the ratio of amylose to amylopectin, the morphology of the granules, the conditions under which the reaction occurs, and the characteristics and quantity of the modifying reagent (Wang et al., 2020a). These modifications can be classified into several categories: cross-linking, oxidation, and acid hydrolysis.

8.3.3.1 Cross-Linking

Cross-linking is the most common technique employed for the modification of starch properties and is used in industries like pharmaceuticals, food processing, wastewater treatment, and bioplastic production. For various applications, some reagents are used, including epichlorohydrin, phosphoryl chloride, adipic acid anhydride mixed with adipic acid, and sodium tripolyphosphate. Cross-linking involves a molecular reaction between the starch reactive hydroxyl groups and multifunctional reagents, forming ether or ester linkages (Amaraweera et al., 2021). These interactions are not limited to between separate chains but also between chains that are in close proximity. Therefore, the modified starch exhibits increased acid, heat, and shear resistance. Besides, improved properties like lower retrogradation, higher viscosity, and better paste clarity are attained (Guo, 2020). Various parameters such as starch nature, the composition and concentration of cross-linking reagents, pH, reaction time, and temperature have been recognized to affect the effectiveness of cross-linking (Shah et al., 2016).

8.3.3.2 Oxidation

Another process through which starch is modified is oxidation, and it is a process that can result in starch depolymerization. This alteration enhances hydrophilicity, destabilizes the linearity and crystallinity of the molecules, and reduces enthalpy and retrogradation. This is achieved through the incorporation of carbonyl and carboxyl groups in native starch that is promoted by an oxidant (Pandiselvam et al., 2019). Several oxidants are available, such as potassium permanganate, sodium hypochlorite, and hydrogen peroxide, among others. The oxidation process is mainly carried out at the hydroxyl groups at C2, C3, and C6, depending on the pH, temperature, and the particular reagent utilized, resulting in the formation of dialdehyde and dicarboxylic acid derivatives (Chen et al., 2015b). In recent studies, the function of ozone as an oxidizing agent in oxidation reactions has gained prominence, as it is able to react with starch without the necessity for a catalyst or special reaction conditions and show negligible effect on the morphology, crystallinity, and internal structure, especially at the amorphous areas of starch (Pandiselvam et al., 2019).

8.3.3.3 Acid Hydrolysis

Hydrolysis of starch by acid takes place at temperatures less than the gelatinization temperature, inducing properties like solubility, gelatinization enthalpy, pasting viscosity, and swelling power. The effect on morphology of the granules differs in different species, tissues, and hydrolysis levels. For instance, no significant morphological changes are obtained by minimal hydrolysis, meaning the starch has not yet been boiled (Wang & Copeland, 2015). In general, different acids can be targeted at the glycosidic bond with nitric, sulfuric, and phosphoric acids most commonly used. The acid mainly acts on the starch granule surface before penetrating the interior of its structure (Pratiwi et al., 2018).

8.4 PSEUDOCEREAL STARCH IN THE FOOD INDUSTRY: APPLICATIONS AND INNOVATIONS

Starch, a dominant carbohydrate source, is a central ingredient in food formulation and processing. Cereals such as corn, wheat, and rice are traditional sources of starch, but pseudocereals such as quinoa (*Chenopodium quinoa*), amaranth (*Amaranthus* spp.), and buckwheat (*Fagopyrum esculentum*) are emerging for their distinctive starch characteristics. In contrast to cereals proper, pseudocereals are dicotyledonous plants but starchy seeds with no gluten, and thus fit for health-conscious and celiac consumers. Pseudocereal starches differ from normal starches in granule morphology, amylose/amylopectin proportion, and gelatinization characteristics (Dan et al., 2025). Teff (*Eragrostis tef*), an ancient Ethiopian grain, has recently become the subject of worldwide attention because of its lack of gluten and balanced nutritional characteristics. Its starch possesses a relatively small to medium granule size (1.2–4.5 μm) and moderate amylose content (19%–22%), with good swelling power and paste clarity. Teff starch has also been investigated for application in porridges, flatbreads, and infant foods, where its moisture-holding capacity and ability to produce smooth gels contribute to both mouthfeel and nutritional quality. Its functional properties render it a potential replacement for wheat starch in gluten-free bakery uses. In addition, teff starch is highly regarded in the traditional fermented foods such as injera, where its paste viscosity and fermentation pattern play a role in determining product texture and shelf life (Aguiar et al., 2021).

Amaranth and quinoa starch granules are considerably smaller (0.5–2 μm) than those of maize or wheat and this influences their digestibility and ability to form films. Pseudocereal starches usually contain less amylose and thus are more viscous and better suited for uses demanding high water-retaining capacity and gelation. The majority of pseudocereal starches show A-type or B-type crystalline arrangement, with consequences for heat stability and retrogradation characteristics (Jiang et al., 2025). Chia (*Salvia hispanica*) starch, being less studied than chia mucilage or oil, is a relatively new functional starch source with special gelling and thickening properties. With fine starch granules (1.0–2.5 μm) and low-to-moderate amylose levels (10%–13%), chia starch produces viscous, stable gels at low concentrations. This characteristic is extremely useful in fat-decreased, egg-free, or plant-based food products like puddings, dressings, and vegan baked goods (Patro & Chowdhury, 2025). Additionally, chia starch shows promise as a clean-label additive in health foods because of its natural thickening capability and compatibility with dietary fiber and protein in whole chia flour. Its antioxidant and water-holding capabilities make it an appropriate choice for functional beverages and sports nutrition applications, particularly in combination with its omega-3 and fiber-rich matrix (Eb et al., 2025).

Pseudocereal starches, suitable for gluten-free products, are summarized in Table 8.2 along with their nutritional properties, functional characteristics, and emerging food industry applications. Quinoa starch, for example, enhances crumb strength and moisture content in gluten-free bread. Amaranth starch increases elasticity and shelf life owing to swelling power and solubility. Owing to high viscosity and gelatinization properties, pseudocereal starches are used as thickeners for soups, gravies, and sauces. Pseudocereal starches possess superior freeze-thaw stability and

TABLE 8.2
Nutritional and Functional Characteristics of Pseudocereal Starch and Emerging Innovations in the Food Industry

Properties	Properties of Pseudocereal Starch	Food Industry Applications	Innovations and Emerging Trends
Nutritional profile	*Gluten-Free*: Essential for celiac diets and gluten-sensitive consumers. *High-Quality Protein*: Rich in essential amino acids (e.g., lysine). *Rich in Bioactive Compounds*: Including flavonoids, phenolic acids, and antioxidants. *Minerals and Fiber*: Good source of dietary fiber, iron, calcium, and magnesium.	*Gluten-Free Products*: Breads, pasta, biscuits, and baked goods. *Nutritional Fortification*: Enhancing the nutritional profile of conventional cereal products. *Functional Foods*: Baby foods, sports nutrition, and health-focused snacks.	*Development of Low-Glycemic Index Foods*: Utilizing starches with a high resistant starch content, particularly in buckwheat. *Bioactive Compound Delivery*: Using starch as a carrier for enhanced bioavailability of antioxidants and other health-promoting compounds.
Physicochemical properties	*Small Granule Size*: Generally 0.5–3.0 μm, significantly smaller than cereal starches. *Waxy Characteristics*: Amaranth and quinoa starches often have very low amylose content (high amylopectin), similar to waxy cereal starches. *Unique Gelatinization and Pasting Behavior*: • Varies by pseudocereal, impacting texture and viscosity. • High water-binding capacity and swelling power. • Good freeze-thaw stability: Due to low retrogradation tendency.	*Thickening and Gelling Agents*: In soups, sauces, and desserts. *Stabilizers and Emulsifiers*: In mayonnaise, dressings, and processed foods (especially amaranth and quinoa). *Fat Replacers*: Small granule size can mimic the texture of fats, providing a creamy mouthfeel.	*Modified Starches*: Utilizing physical (e.g., heat-moisture treatment) and chemical (e.g., octenyl succinic anhydride modification) methods to tailor starch properties for specific applications. *Pickering Emulsions*: Small pseudocereal starch granules (especially quinoa and amaranth) are being explored as natural stabilizing agents for emulsions.

(Continued)

TABLE 8.2 (*Continued*)
Nutritional and Functional Characteristics of Pseudocereal Starch and Emerging Innovations in the Food Industry

Properties	Properties of Pseudocereal Starch	Food Industry Applications	Innovations and Emerging Trends
Functional characteristics	*Quinoa*: Low amylose content. Rapid gelatinization due to small granules. *Amaranth*: Extremely small granules (often <1.5 µm). Waxy starch (high amylopectin). *Buckwheat*: • Higher amylose content than quinoa and amaranth. • High resistant starch content (particularly Tartary buckwheat).	*Quinoa*: Used in infant foods due to easy digestibility. Gluten-free bakery products. *Amaranth*: Used as a thickener and stabilizer. Popcorn-like snacks ("popping"). *Buckwheat*: • Noodles (soba), pancakes, and gluten-free bread. • Used in foods aimed at blood glucose management.	*Targeted Extraction and Isolation*: Developing efficient and sustainable methods for isolating high-purity starch from pseudocereals. *Utilization of By-products*: Extracting valuable components from the outer layers of pseudocereals. *Flavor and Texture Improvement:* Innovations in blending pseudocereal flours and starches with other ingredients to improve the sensory qualities of gluten-free products.
Industrial applications	Pseudocereals respond well to various processing methods.	*Extrusion*: Used for snack foods, breakfast cereals, and pasta. *Fermentation*: Enhancing digestibility and flavor in baked goods and beverages. *Roasting and Puffing*: Improving flavor and texture.	*Advanced Extrusion Techniques*: Optimizing extrusion parameters to improve the nutritional profile and texture of pseudocereal-based snacks. *Bioprocessing (Germination and Fermentation)*: Enhancing nutrient bioavailability and modifying starch properties through controlled biological processes. *3D-Printing*: Utilizing pseudocereal flours and starches in 3D food printing for customized nutrition and novel food structures.

retrogradation resistance compared to regular cereal starches; hence they can be used in refrigerated and frozen food products. For instance, modified quinoa starch has high paste stability and clarity, and it contributes to maintaining texture and structure in frozen foods and ready-to-eat foods (Lupu et al., 2025). Amaranth starch, with its small granule size (1–3 μm), disperses well in water and produces smooth, stable pastes, suitable for creamy sauces and instant mixes. Additionally, buckwheat starch with its comparatively higher amylose content (~22%–24%) is responsible for stronger gel structures, which find application in pie fillings and jelly products, while maintaining its functionality under acidic and high-heat conditions. Buckwheat starch has been reported to be superior to rice starch under acidic and high-temperature conditions. Pseudocereal starches in low-fat dairy and bakery foods replicate the mouthfeel and texture of fat. Their fine granule size provides a rich texture without affecting flavor drastically (Al Zoghbi et al., 2025). Additionally, the property of pseudocereal starches to simulate fat renders them extremely important in creating low-calorie and functional foods. Their specific gelatinization and swelling properties make them deliver a rich and smooth mouthfeel that is very close to the sensory character of fat, especially in low-fat yogurts, custards, salad dressings, and baked foods. For instance, amaranth and chia starches can enhance stability and viscosity in thin mayonnaise or cheese spreads without the use of synthetic thickeners or emulsifiers. Moreover, since they are naturally gluten-free and contain high levels of resistant starch (RS), they make positive contributions to glycemic control, providing functional and nutritional benefits to the product (Yurtseven & Özbay, 2025).

Chemical (acetylation, oxidation), enzymatic, and physical treatments (ultrasound, HMT) have been utilized to enhance the functionality of pseudocereal starches. Quinoa starch with modified properties has enhanced freeze-thaw stability and gel strength and can be used in frozen and refrigerated food. Nanostructured pseudocereal starch provides enhanced solubility, controlled bioactive release, and increased film-forming ability. The latest developments in dual- and multi-step modification methods – like the combination of enzymatic hydrolysis with HMT or acid thinning – have produced starches with highly specific functionalities (Raghunathan et al., 2025). For example, dual-modified amaranth starch exhibits greater resistance to retrogradation and better paste clarity, which are appealing in transparent sauces, fruit fillings, and jelly products. Enzymatic conversions may selectively hydrolyze amylopectin's long chains, lowering molecular weight and permitting greater control of viscosity and texture in particular uses. Such customized properties not only improve the shelf life and appearance of processed foods but also provide improved freeze–thaw stability, rendering the modified starch appropriate for frozen doughs, sauces, and ready meals (Sarkar et al., 2024).

In the field of nanotechnology, pseudocereal starch nanoparticles or nanocrystals are picking up pace due to their multifunctionality in food packaging, encapsulation, and delivery systems. The nanostructures, typically achieved through acid hydrolysis, ultrasonication, or high-pressure homogenization, are characterized by higher surface area, crystallinity, and dispersion (Nicole et al., 2014). For instance, quinoa starch nanocrystals have been effectively utilized as carriers for encapsulating delicate bioactives such as vitamins, antioxidants, and probiotics with targeted release and enhanced shelf stability for functional foods. Furthermore, films from amaranth or buckwheat nano-starch mixed with biopolymers (such as chitosan or gelatin) show remarkable oxygen and moisture barrier properties, making them good candidates for edible coatings and green food packaging (Jurkaninová et al., 2024).

Pseudocereal starch is rich in RS, which is not digested and functions as dietary fiber. RS-enriched foods from amaranth and quinoa support glycemic regulation and health of the gut, recommending these for use in diabetic products. Bioplastics derived from starch in pseudocereals provide a sustainable replacement for fossil fuel-derived plastics. Quinoa starch films exhibit excellent oxygen barrier and tensile strength, and in combination with natural antimicrobials (such as essential oils), can promote shelf-life extension of packaged foods. Pseudocereal starches, especially from quinoa and amaranth, are becoming more known for their high levels of RS types II and III, which are not digested in the small intestine and are fermented in the colon (Pirzadah & Malik., 2020). This fermentation yields SCFAs like butyrate, which are important for gut health through their promotion of colonocyte function, alleviation of inflammation, and enhancement of intestinal barrier integrity. RS-rich pseudocereal food intake has been linked with lowered postprandial glycemic peaks, so these foods are well suited to diabetic diets and weight management. Furthermore, their low glycemic index (GI) and dietary fiber level ensure extended satiety, which is useful for appetite regulation and obesity avoidance. Furthermore, pseudocereal starches provide a functional benefit in synbiotic and prebiotic products, where they are used as fermentable substrates for friendly gut microbiota such as *Lactobacillus* and *Bifidobacterium* (Micheal & Jagadhesan, 2025).

Quinoa and chia starches, used in fermented dairy substitutes or probiotic drinks, enhance not only texture and stability but also microbial viability upon storage. With the double function of being both a prebiotic fiber and a structural element, pseudocereal starches are extremely suitable for applications in functional foods aimed at digestive health, particularly for individuals with lactose intolerance or gluten sensitivity (Vajdovich et al., 2025). From a packaging and environmental aspect, pseudocereal starch is proving to be a major material in the production of biodegradable films and coatings. Quinoa and amaranth starch-based bioplastics, because of their excellent film-forming capability, flexibility, and transparency, are being researched as green alternatives for traditional plastic films. They are compostable and biodegradable and can include natural additives like nanoparticles, plant extracts, and essential oils, which provide antimicrobial and antioxidant activity. These kinds of films are applied to wrap fresh fruits, bakery items, and minimally processed foods, increasing shelf life while supporting world sustainability programs and plastic pollution reduction (Zhang et al., 2024b).

8.5 PSEUDOCEREAL PROTEINS

Pseudocereals are untapped sources of seeds or fruits that are high in vital amino acids, vitamins, and minerals. They provide food security and health advantages. The most common pseudocereals are amaranth, quinoa, buckwheat, chia, album, and wattle seeds (Malik & Singh, 2022). Because of their excellent nutritional value, antioxidant qualities, and health advantages, pseudo cereals like quinoa, amaranth, and buckwheat are gaining popularity as gluten-free crops. Although these proteins are helpful in the synthesis of food, their makeup makes them poorly soluble in water (Constantino & Garcia-Rojas, 2022a). Pseudocereal proteins are a sustainable source of protein because they can be altered by heat treatment, enzymatic hydrolysis, high pressure, or ultrasound processing to improve their characteristics and resilience to challenging environmental circumstances (Vidaurre-Ruiz et al., 2023). The protein content of pseudocereals varies from 10% to 20%. The dietary profile of proteins in pseudocereals is also quite high in comparison to cereals and other legumes. Amaranthus species are native to Central and South America, with the notable exception of *Amaranthus spinosus* L., which thrives in tropical and subtropical regions of India (Alonso-Miravalles & O'Mahony, 2018).

8.5.1 Extraction of Proteins

Grains are used in food formulations with protein isolates, like texturized vegetable protein and fitness supplements. Protein extraction from pseudocereals is increasingly relevant, using physical, enzymatic, and chemical methods (Malik & Singh, 2022). Either increasing the protein's solubility or dissolving the protein-containing matrix will increase the extraction yield. Changes in composition and structure, as well as isomerization, bonding, and degradation, can have implications for physiological and techno-functional processes as shown in Figure 8.2 (Deleu et al., 2019).

8.5.1.1 Physical Methods

Physical techniques are safer and more convenient, but they sacrifice desired qualities and require more time and resources, but they yield higher financial returns. They can produce protein-embellished fragments with natural characteristics by using wet extraction instead of dry fractionation (Rao & Poonia, 2023). Two-step

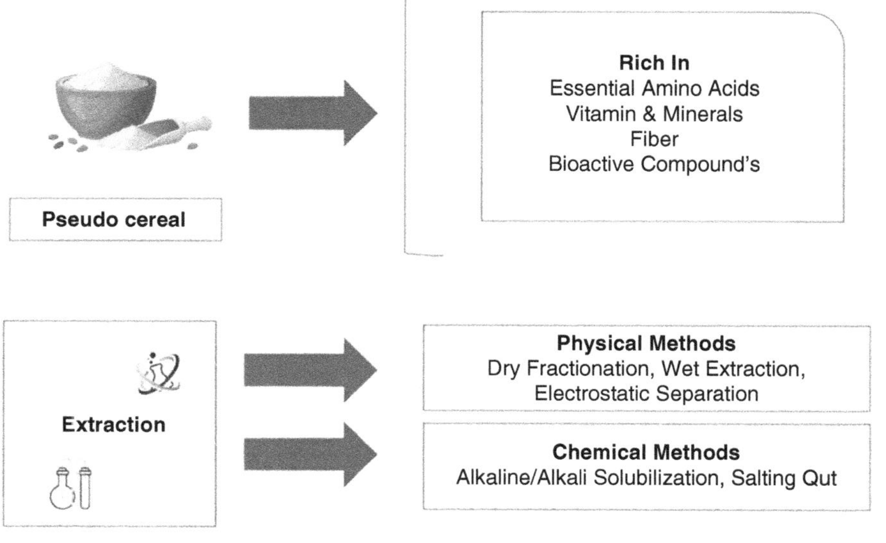

FIGURE 8.2 Extraction of pseudocereal proteins.

dry fractionation provides a water-free substitute for conventional dry fractionation techniques, which entail grinding and aggregating starch molecules and isolating contents using turbo-segregation or electrostatic segregation (Laguna et al., 2018). Despite being a basic food, cereals and pseudocereals (CPC) are frequently disregarded as a source of protein. Protein extraction from CPC is examined in this paper with an emphasis on bioavailability and nutritional considerations (Bharti et al., 2024). Mondor & Hernández-Álvarez (2023) described both wet and dry methods for extracting plant proteins, including aqueous two-phase system extraction, reverse micelle, electrostatic separation, and subcritical water. However, in order to extract protein from protein bodies within plant cells, these methods must first cause cell disruption. Because proteins are extremely sensitive, conventional mechanical, thermal, and chemical treatment techniques do not produce effective results. Because to protein degradation, traditional methods (aqua, salt, solvent, detergent, and alkali-based) have lower extraction yields. In order to increase protein output and decrease degradation while maintaining cost-effectiveness and environmental safety, researchers are investigating cutting-edge plant protein extraction methods such as enzymes, microwaves, high pressure, pulse fields, homogenization, and sonication (Thakur et al., 2024). Mullaiselvan, et al. (2024), examines the many characteristics and uses of proteins made from pseudocereals. Physical characteristics, extraction, isolation, and purification are all covered. For scholars, professionals, and hobbyists curious about the potential of pseudocereal proteins in a variety of domains, such as food science and biotechnology.

8.5.1.2 Chemical Methods

Proteins from most pseudo-cereals are often split using chemicals to isolate specific components; quinoa is one of the few that naturally provides a high-quality protein fulfilling complete dietary needs (Rollán et al., 2019). Proteins may be isolated from the seed by steeping delipidized quinoa flour in deionized water with an alkaline media. After mixing and centrifuging the blend to produce overlaying liquid, the blend was acid which is influenced and centrifuged again to produce solid residue (Tian et al., 2020). Among the latest most well-known approaches for isolated plant proteins including those from pseudocereals like buckwheat, amaranth, and quinoa is chemical removal. These ways mostly use brine or alkaline solvents to disperse proteins, degrade of cellular structures, and make it easier to separate them from other substances like starch, fiber, and lipids. Chemical extraction remains a popular method because of its high protein yield and ease for scaling operations, even though there are capacity environmental and dietary issues (such as denaturation and residual solvents) (Reddy et al., 2024). The physicochemical characteristics and nutritional value of quinoa proteins are examined in this paper, with an emphasis on isoelectric precipitation and alkaline extraction. The final QPI (Quinoa Protein Isolate) has a high nutritional content, contains important amino acids, and maintains its functional properties after extraction (Arguello-Hernández et al., 2024). Protein with a pH of 90%–93% is produced by high pH values, whereas proteinaceous substances with a pH of 82%–88% are produced by low pH values. Quinoa protein components are impacted by processing temperature. Chemical extraction with membrane-assisted salting out is economical (Wu et al., 2018). Yadav and Yadav (2024), examined the use of alkaline solubilization for protein isolation. Alkaline derivation enhanced protein output but partially denatured functional groups.

8.5.2 Functional Properties

Proteins from pseudocereals, such as quinoa, buckwheat, and amaranth are not very water soluble, their specific physical traits make them perfect for food preparation (Constantino & Garcia-Rojas, 2022b). Pseudocereals are important suppliers of dietary and energy since they are rich in protein, fiber, and unsaturated fatty acids. Because they contain vitamins, minerals, and bioactive substances, they can be used as cereal substitute. A green food source, pseudocereals such as quinoa, buckwheat, and amaranth may be grown in unsuitable soil and are appropriate for gluten exclusion diets (Sindhu & Khatkar, 2019). Although they have been farmed for thousands of years, quinoa, amaranth, and buckwheat are unusual protein sources that have been looked over in recent years. Pseudocereals are gluten exclusion dicotyledonous grains that are similar to cereals like rice and corn in terms of composition and dietary component (Alonso-Miravalles & O'Mahony, 2018). Rich in essential proteins, pseudocereals are important for correcting dietary shortage and offering health benefits, but their production is confined worldwide (Nandan et al., 2024). Moreover, pseudocereals are excellent in stimulated Protein Digestibility-Corrected Amino Acid Score (PDCAAS): buckwheat (0.78), amaranth (0.70), and quinoa (0.85), whereas those for wheat, rice, and maize are 0.42, 0.56, and 0.47 respectively. The latest research insights trends in the production and consumption of plant-based functional beverages, especially based on pseudocereals (Vajdovich et al., 2025). Moreover, pseudocereal proteins have better functional qualities and are plentiful in essential amino acids, but their limited commercial application limits their potential in food products. Their physicochemical, functional, and nutritional qualities can be improved (Mir et al., 2024). Pseudocereal seeds have gained appeal in modern diets due to their gluten-excluded nature and impressive dietary and nutraceuticals qualities. Recent research have highlighted the potential health benefits related with pseudocereals (Nandan et al., 2024).

8.5.3 Nutritional Value

Pseudocereals are applied to produce functional food products and are rich in protein, carbs, vitamins, minerals, and fiber as shown in Figure 8.3. They also have health-enhancing properties against diabetes, high blood pressure, cancer, and cardiovascular conditions (Priego-Poyato et al., 2021). Moreover, pseudocereals are widely used in gluten-excluded mixtures. In comparison to cereals, pseudocereals have much elevated concentrations of parameters like protein efficiency ratio or net protein use, bioavailability or digestibility (Langyan et al., 2024). Some countries that grow buckwheat, a gluten-excluded pseudocereal, include China, Russia, Japan, France, Brazil, Poland, Kazakhstan, and the United States. Its dietary attributes include protein, fiber, vitamins, and bioactive compounds and are applied in celiac disease gluten-excluded dishes (Vani & Soni, 2024). Carbohydrates, including starch, make up 68%–73% of the dietary status, with buckwheat having 54%–55%. Its high

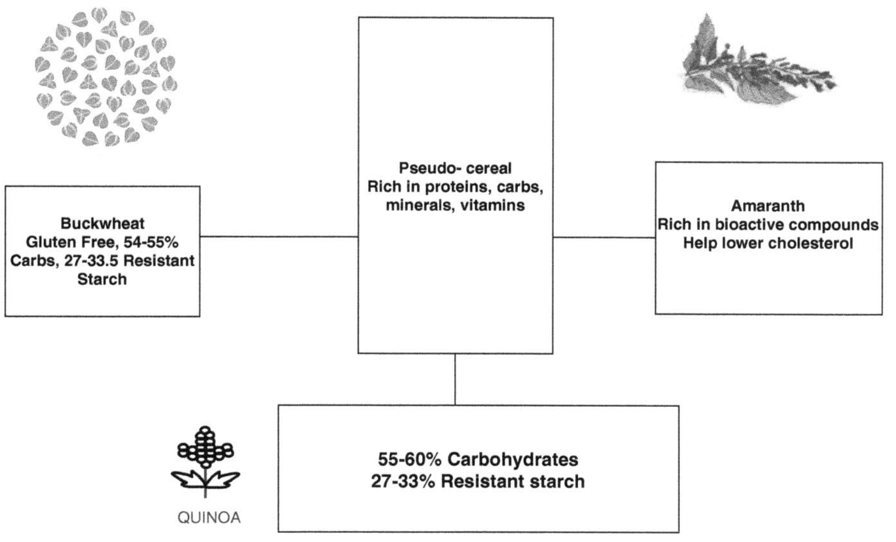

FIGURE 8.3 Nutritional value of pseudocereals.

amylose content causes a medium rise in GI. Buckwheat contains 27%–33.5% tolerate starch, providing health benefits. It also contains 17.8% dietary fiber, including lignin, cellulose, and hemicelluloses (Rupa et al., 2024). Because of their unique viscosity and freezing stability, these whole-grain protein and whole-vegetable 55%–60% carbohydrates are used in a variety of culinary products (Coţovanu et al., 2020). Amaranth is a plant with over 400 types that is rich in proteins, carbs, fats, fiber, vitamins, minerals, and bioactive compounds (Chungkham & Singh, 2021).

8.5.4 Applications of Pseudocereal Proteins in the Food Industry: Plant-Based Solutions

Pseudocereals such as quinoa, amaranth, buckwheat, chia, and album are gaining interest in food industry due to their high-quality protein content (10%–20%) and amino acid profiles, which make them superior to traditional cereals and legumes. Pseudocereals proteins primarily made of albumins and globulins, offering superior nutritional value suitable for plant-based food formulations (Poshadri et al., 2024; Vidaurre-Ruiz et al., 2023). They are also considered as good option for creating gluten-free baked goods, meat substitutes, breakfast cereals, and functional beverages due to their gluten-free nature and useful qualities, such as high water-binding capacity, swelling, and freeze-thaw stability (Poshadri et al., 2024, Schoenlechner & Bender, 2020).

Plant-based proteins have the potential to substitute animal proteins in a number of food products; they have drawn a lot of interest due to their exceptional nutritional qualities. Amaranth protein is becoming a more and more popular component in the food business. Amaranth protein has a limited concentration of anti-nutritional components and high quantities of important amino acids that can help fulfill daily requirements (Das et al., 2021)

8.5.4.1 Used of Pseudocereal Proteins in Bakery Industry

The intention of adding pseudocereals to food formulations is to enhance the functional and nutritional properties of the finished product. Many techno-functional qualities provided by pseudocereals have the ability to greatly enhance product quality and consumers' experience. In order to produce appealing baked goods, considerable amounts of wheat flour can be substituted with 25%–50% pseudocereal flour. However, the viscoelastic network of gluten becomes weaker when gluten-forming proteins are not present, resulting in bread with a firm and irregular crumb structure with reduced volume (Bender & Schoenlechner, 2021). Panettone, the famous bread consumed in Peru and Italy, has recently been reported to contain quinoa flour. Quinoa flour can be used 26% more as an alternative to wheat flour. For the purpose of attaining an acceptable crumb texture, it was discovered that the proteins and fibers in quinoa support the use of a decreased fat level (7.6%) in the formulation (Huánuco-Azabache et al., 2022).

8.5.4.2 Used of Pseudocereal Proteins in Formulation of Gluten Free Food

Buckwheat is not an essential source of protein for the food industry, the protein-rich fraction is gluten-free and has properties that are comparable to some commonly found food proteins, indicating it may find use in food applications. For example, Buckwheat protein concentrates produced through alkaline extraction have reduced solubility in water than bovine casein at neutral and alkaline pH, and are more soluble in liquid than soy proteins and bovine casein at acidic pH (Tomotake et al., 2002). In contrast to widely accessible gluten-free breads manufactured from simple starches or rice flour, the use of pseudocereal flours in gluten-free baked goods can range from 5 to 50% and significantly increase the compositions of protein, mineral, and fiber content (Repo-Carrasco-Valencia & Vidaurre-Ruiz, 2022).

8.5.4.3 Pseudocereal Protein Use in Beverage Industry

Fermented beverages may also be made with pseudocereal flours. According to reports, processing and the qualities of the finished product may be influenced by numerous quinoa varieties. For instance, due to increased protein content, lower saponin concentration, and reduced viscosity loss during fermentation, the red quinoa variety known as Pasankalla showed improved fermentation characteristics. Heating amaranth flour by extrusion improves the viscosity and organoleptic properties of beverage (Bender & Schoenlechner, 2021). The nutritional value of Quinoa is well recognized as it has well-balanced amino acid composition and 12%–17% protein (Wang et al., 2020a). The nutritional and functional advantages of meals and drinks made with pseudocereals protein can be improved by utilizing the protein content of quinoa (Abbas, 2025).

8.5.5 Nonfood Industrial Applications of Pseudocereal Proteins

The potential of pseudocereal proteins is also exploited for non-food uses, such as packaging or coatings, and for pharmaceutical purposes. Amaranth protein has been applied for the development of edible films; a recent study indicated that it was possible to microencapsulate betanin by complex coacervation, using ultrasound-treated amaranth protein isolate and sodium carboxymethylcellulose (Constantino & Garcia-Rojas, 2022a). Also, quinoa proteins were investigated for edible films in combination with other components. Recent research has reported that quinoa protein isolate and gum Arabic can form an appropriate material for microencapsulating essential oils of spices. The optimum pH was 3.6 and the appropriate ratio of quinoa protein isolate to gum Arabic was 4:1, this material showed high encapsulation efficiency (82%–88%) for essential oils of Sichuan, ginger, and star anise. They kept these essential oils stable during the oral and gastric digestion stages, and only released them during the intestinal stage, which improved their absorption (Chen et al., 2023).

8.5.5.1 Bioplastic and Biopolymer Production from Pseudocereal Proteins

Functional and film-forming characteristics of pseudocereal proteins (examples are proteins extracted by quinoa, amaranth, and chia) are superior and thus they could become viable sources of bioplastics and biopolymers. The major components in these proteins are globulins that make the protein produce a strong, flexible film to be used as a biodegradable packaging material among other industrial uses (Pawde et al., 2025). Lately, there has been a burst of breakthroughs showing that bioplastics can be produced out of plant proteins in a nontoxic way through eco-friendly processes that do not use solvents and crosslinkers that come with adverse effects on the environment. As an example, a new method employs plant proteins that are impermeable to water to produce bioplastics through self-assembly under aqueous conditions to fabricate materials with desirable mechanical properties and high rates of biodegradation, compliant with properties of green chemistry (Sarkar et al., 2025). Distillers dried grains containing buckwheat protein have been used in development of composite film-edible food packaging (Wang et al., 2017). Canihua flour has demonstrated the capability of the formation of sufficient biofilm to be utilized in the aspect of food packaging. They featured high blocking capacity, compared to the polyethylene films, and are less soluble and less permeable to water than those in some types of polysaccharide films (Salas-Valero et al., 2015).

8.5.5.2 Pseudocereals Protein Use in Cosmetics Industry

Pseudocereals contain a good amount of squalene; a natural substance commonly used in cosmetic products as a moisturizer and as an antioxidant. For example, the squalene of quinoa and amaranth is popular in cosmeceutical products, thus having an impact on skin moisture and antioxidant activity. They contain saponins, which are known for their surfactant properties and are widely used in cosmetic formulations as foaming and cleansing agents. Moreover, saponins possess hemolytic and antilipemic effects, which could be useful for skin health applications (Dhondiram et al., 2023).

8.5.5.3 Pseudocereals in Agro-Pharmacology

Pseudocereals are a source of nutraceuticals and agro-pharmaceuticals such as quinoa, amaranth, chia, and buckwheat that have been investigated in regard to their biological activities, which can be useful in disease prevention and health promotion.

Pseudocereals contain various compounds with biologically active effects including saponins, peptides, flavonoids, and phenolic acids, among others, which have been characterized to show pharmacological properties such as anticancer, antioxidant, and anti-inflammatory (Morales et al., 2021). Pseudocereals contain saponins which can reduce blood serum cholesterol and can be used for managing heart disease (Dhondiram et al., 2023).

8.6 CHALLENGES AND FUTURE PERSPECTIVES IN PSEUDOCEREAL PROCESSING AND APPLICATION

Pseudocereals are harder to digest than true cereals because they are nutritious plants, contain different proteins and starch organization. They are particularly accommodative of the hot and dry climate conditions due to the fact they can adapt and are cultured in several locations, thus they are very important in regard to food security. Their usage has continued to draw more research which has availed new information and profound knowledge. A variety of goods contain pseudocereals and, most importantly, they contain quinoa, which as a crop is traded in the international market. Nevertheless, there is more work to be done and a systematic approach to trade, production, and political domination to ensure that they are underutilization. The production and cultivation of quinoa are no longer the traditional crops but the goods traded at the international arena and necessity of more intensifications and the holistic approach in all spheres of production, marketing, and political dominance is needed now (Bender & Schoenlechner, 2021).

Pseudocereals constitute a type of grain that is rich in phenolic phytochemicals and has numerous treatment and health benefits. These grains have been established to contain anticancerous, anti-inflammatory, antidiabetic, and hypocholesterolemic activities. They also come in handy in ensuring food plant biodiversity and in diversifying diets. Due to consumer demand of nutrient-rich gluten-free products, there is a growing trend in the use of pseudocereals in both wheat-free and gluten-free formulation. Pseudocereals have been used to develop many gluten-free products like bread, biscuits, spaghetti, and crackers, among others. But the ones available are not numerous. Generation of knowledge about the composition and the use of pseudocereals in gluten-free products and staple foods of a dough-based diet widely used in the developing world is also important to understand the chemical composition of the storage proteins in these plants and increase the overall awareness about the existence of quinoa and amaranths. Consequently, a greater number of individuals will be in a position of enjoying the nutritional benefits of pseudocereals.

8.7 CONCLUSION

Pseudocereal-derived protein and starch provide substantial industrial potential because of their unique functional and nutritional qualities. They are suitable for a variety of industries, including food, pharmaceuticals, packaging, cosmetics, and nutraceuticals, due to their exceptional protein content, gluten-free quality, and versatile starch qualities. The high-water retention, thermal resistance, and biodegradability of pseudocereal proteins' starches, along with their exceptional emulsifying, gelling, and film-forming capabilities, make it easier to use them in the creation of environmentally friendly and health-conscious goods. Pseudocereals become valuable raw materials as companies move more and more toward greener, plant-based, and eco-friendly components. Their importance will be further enhanced by ongoing research and technical advancement, establishing them as crucial components of global food security and sustainable industrial innovation.

REFERENCES

Abbas, A. (2025). Production of protein isolate from quinoa and studying of its some physiochemical properties. *International Journal of Health & Medical Research, 04*(1), 1–12.

Agregán, R., Guzel, N., Guzel, M., Bangar, S. P., Zengin, G., Kumar, M., & Lorenzo, J. M. (2023). The effects of processing technologies on nutritional and anti-nutritional properties of pseudocereals and minor cereal. *Food and Bioprocess Technology, 16*(5), 961–986.

Aguiar, E. V., Santos, F. G., Krupa-Kozak, U., & Capriles, V. D. (2021). Nutritional facts regarding commercially available gluten-free bread worldwide: Recent advances and future challenges. *Critical Reviews in Food Science and Nutrition, 63*(5), 693–705.

Al Zoghbi, M., El Deghel, M., Serhan, M., & Mnayer, D. (2025). Development of healthy gluten-free double-layer pita bread: Nutritional, technological, and sensorial properties. *Journal of Food Processing and Preservation, 2025*(1), 3811480.

Alonso-Miravalles, L., & O'Mahony, J. A. (2018). Composition, protein profile and rheological properties of pseudocereal-based protein-rich ingredients. *Foods, 7*(5), 73.

Alonso-Miravalles, L., & O'Mahony, J. A. (2022). Composition, protein profile and rheological properties of pseudocereal-based protein-rich ingredients. *Foods, 7*(5), 73.

Aluwi, N. A., Murphy, K. M., & Ganjyal, G. M. (2017). Physicochemical characterization of different varieties of quinoa. *Cereal Chemistry, 94*(5), 847–856.

Amaraweera, S. M., Gunathilake, C., Gunawardene, O. H., Fernando, N. M., Wanninayaka, D. B., Dassanayake, R. S., … & Manipura, A. (2021). Development of starch-based materials using current modification techniques and their applications: A review. *Molecules, 26*(22), 6880.

Anggreini, R. A., Choiriyah, N. A., & Athennia, A. (2021). Modification of sorghum bicolor (L) moench starch: Review of HMT (heat moisture treatment), autoclaving cooling, and annealing methods. *International Journal of Advance Tropical Food, 3*(2), 57–66.

Arguello-Hernández, P., Samaniego, I., Leguizamo, A., Bernalte-García, M. J., & Ayuso-Yuste, M. C. (2024). Nutritional and functional properties of quinoa (Chenopodium quinoa Willd.) Chimborazo ecotype: Insights into chemical composition. *Agriculture, 14*(3), 396.

Bello-Perez, L. A., Flores-Silva, P. C., Agama-Acevedo, E., & Tovar, J. (2020). Starch digestibility: Past, present, and future. *Journal of the Science of Food and Agriculture, 100*(14), 5009–5016.

Bender, D. & Schoenlechner, R. 2021. Recent developments and knowledge in pseudocereals including technological aspects. *Acta Alimentaria, 50*, 583–609.

Bender, D., & Schoenlechner, R. (2021). Recent developments and knowledge in pseudocereals including technological aspects. *Acta Alimentaria, 50*(4), 583–609.

Bender, D., & Schoenlechner, R. (2021). Recent developments and knowledge in pseudocereals including technological aspects. *Acta Alimentaria, 50*(4), 583–609.

Bharti, P., Mohapatra, M., Sah, R. P., Pradhan, A., Priyadarsani, S., & Rayaguru, K. (2024). Effect of green extraction techniques on the quality and functional attributes of protein isolates from cereals and pseudo cereals: A review. *International Journal of Food Science and Technology, 59*(11), 8045–8057.

Bhinder, S., Kaur, A., Singh, B., Yadav, M. P., & Singh, N. (2020). Proximate composition, amino acid profile, pasting and process characteristics of flour from different Tartary buckwheat varieties. *Food Research International*, *130*, 108946.

Caeiro, C., Pragosa, C., Cruz, M. C., Pereira, C. D., & Pereira, S. G. (2022). The role of pseudocereals in celiac disease: Reducing nutritional deficiencies to improve well-being and health. *Journal of Nutrition and Metabolism*, *2022*(1), 8502169.

Chen, K., Zhang, M., Mujumdar, A. S., & Wang, M. (2023). Encapsulation of different spice essential oils in quinoa protein isolate-gum Arabic coacervates for improved stability. *Carbohydrate Polymers*, *300*, 120250.

Chen, Q., Yu, H., Wang, L., ul Abdin, Z., Chen, Y., Wang, J., … & Chen, X. (2015a). Recent progress in chemical modification of starch and its applications. *RSC Advances*, *5*(83), 67459–67474.

Chen, X., He, X., Fu, X., & Huang, Q. (2015b). In vitro digestion and physicochemical properties of wheat starch/flour modified by heat-moisture treatment. *Journal of Cereal Science*, *63*, 109–115.

Chungkham, N., & Singh, N. (2021). Utilization of amaranth grain flour at different products and its acceptability. *The Pharma Innovation Journal*, *10*(4), 436–443.

Compart, J., Singh, A., Fettke, J., & Apriyanto, A. (2023). Customizing starch properties: A review of starch modifications and their applications. *Polymers*, *15*(16), 3491.

Constantino, A. B. T., & Garcia- Rojas, E. E. (2022a). Microencapsulation of betanin by complex coacervation of carboxymethylcellulose and amaranth protein isolate for application in edible gelatin films. *Food Hydrocolloids*, *133*, 107956.

Constantino, A. B. T., & Garcia-Rojas, E. E. (2022). Proteins from pseudocereal seeds: solubility, extraction, and modifications of the physicochemical and techno-functional properties. *Journal of the Science of Food and Agriculture*, *102*(7), 2630–2639.

Constantino, A. B. T., & Garcia-Rojas, E. E. (2022). Proteins from pseudocereal seeds: solubility, extraction, and modifications of the physicochemical and techno-functional properties. *Journal of the Science of Food and Agriculture*, *102*(7), 2630–2639.

Constantino, A. B. T., & Garcia-Rojas, E. E. (2022b). Proteins from pseudocereal seeds: Solubility, extraction, and modifications of the physicochemical and techno-functional properties. *Journal of the Science of Food and Agriculture*, *102*(7), 2630–2639.

Coţovanu, I., Batariuc, A., & Mironeasa, S. (2020). Characterization of quinoa seeds milling fractions and their effect on the rheological properties of wheat flour dough. *Applied Sciences*, *10*(20), 7225.

Dan, H., Li, H., Li, C., Fang, Z., Hu, B., Chen, H., & Liu, Y. (2025). Application of sourdough in gluten-free bakery products. *Critical Reviews in Food Science and Nutrition*, *65*(15), 3048–3068.

Das, D., Mir, N. A., Chandla, N. K., & Singh, S. (2021). Combined effect of pH treatment and the extraction pH on the physicochemical, functional and rheological characteristics of amaranth (Amaranthus hypochondriacus) seed protein isolates. *Food Chemistry*, *353*, 129466.

De Bock, P., Daelemans, L., Selis, L., Raes, K., Vermeir, P., Eeckhout, M., & Van Bockstaele, F. (2021). Comparison of the chemical and technological characteristics of wholemeal flours obtained from amaranth (Amaranthus sp.), quinoa (chenopodium quinoa) and buckwheat (fagopyrum sp.) seeds. *Foods*, *10*(3), 651.

De Bock, P., Daelemans, L., Selis, L., Raes, K., Vermeir, P., Eeckhout, M., & Van Bockstaele, F. (2021). Comparison of the chemical and technological characteristics of wholemeal flours obtained from amaranth (Amaranthus sp.), quinoa (Chenopodium quinoa) and buckwheat (Fagopyrum sp.) seeds. *Foods*, *10*(3), 651.

Del Moral, S., Barradas-Dermitz, D. M., & Aguilar-Uscanga, M. G. (2018). Production and biochemical characterization of α-glucosidase from Aspergillus niger ITV-01 isolated from sugar cane bagasse. *3 Biotech*, *8*(1), 7.

Deleu, L. J., Lambrecht, M. A., Van de Vondel, J., & Delcour, J. A. (2019). The impact of alkaline conditions on storage proteins of cereals and pseudo-cereals. *Current Opinion in Food Science, 25*, 98–103.

Dhondiram, D., Sawant, A., Mamoon, F., Aneja, I., & Duggal, M. (2023). Nutritional characterization of pseudocereals: A review. *The Pharma Innovation Journal, 12*, 109–118.

Eb, C., Vy, I., & Mi, C. (2025). Nutritional and functional characterization of chia expeller and gluten-free flours as ingredients for premixes. *Plant Foods for Human Nutrition, 80*(1), 1–8.

Graziano, S., Agrimonti, C., Marmiroli, N., & Gullì, M. (2022). Utilisation and limitations of pseudocereals (quinoa, amaranth, and buckwheat) in food production: A review. *Trends in Food Science & Technology, 125*, 154–165.

Gunaratne, A. (2023). Heat-moisture treatment of starch. In *Physical modifications of starch* (pp. 49–72). Singapore: Springer Nature Singapore.

Guo, P. (2020). Botanical sources of starch. In S. Wang (Ed.), *Starch Structure, Functionality and Application in Foods* (pp. 9–27). Singapore: Springer.

Hua, X., & Yang, R. (2016). Enzymes in starch processing. In *Enzymes in Food and Beverage Processing* (pp. 139–170). CRC Press.

Huánuco-Azabache, E., Melgarejo-Cabello, S., Vidaurre-Ruiz, J., & Repo-Carrasco-Valencia, R. (2022). Simultaneous optimization for the elaboration of a low-fat panettone with wheat and quinoa flour. *Journal of Food Processing and Preservation, 46*, e17199.

Jiang, H., Ma, R., Pan, X., Liu, C., Zhan, J., Yang, T., & Tian, Y. (2025). Molecular structure and pasting, rheological, and digestive properties of a novel small granular starch obtained from Chlorella sorokiniana cultivated under a nutrition-limited condition. *Food Hydrocolloids, 170*, 111669.

Jurkaninová, L., Dvořáček, V., Gregusová, V., & Havrlentová, M. (2024). Cereal β-D-glucans in food processing applications and nanotechnology research. *Foods, 13*(3), 500.

Kaur, H., Shams, R., Dash, K. K. & Dar, A. H. (2023). A comprehensive review of pseudocereals: Nutritional profile, phytochemicals constituents and potential health promoting benefits. *Applied Food Research, 3*, 100351.

Kaur, H., Shams, R., Dash, K. K., & Dar, A. H. (2023). A comprehensive review of pseudocereals: Nutritional profile, phytochemicals constituents and potential health promoting benefits. *Applied Food Research, 3*(2), 100351.

Laguna, O., Barakat, A., Alhamada, H., Durand, E., Baréa, B., Fine, F., … & Lecomte, J. (2018). Production of proteins and phenolic compounds enriched fractions from rapeseed and sunflower meals by dry fractionation processes. *Industrial Crops and Products, 118*, 160–172.

Langyan, S., Khan, F. N., & Kumar, A. (2024). Advancement in nutritional value, processing methods, and potential applications of Pseudocereals in dietary food: A review. *Food and Bioprocess Technology, 17*(3), 571–590.

Langyan, S., Khan, F. N., & Kumar, A. (2024). Advancement in nutritional value, processing methods, and potential applications of Pseudocereals in dietary food: A review. *Food and Bioprocess Technology, 17*(3), 571–590.

Lupu, A., Banu, I., Grigore-Gurgu, L., Vasilean, I., & Aprodu, I. (2025). Gluten-free sourdough based on quinoa and sorghum: Characterization and applications in breadmaking. *Applied Sciences, 15*(10), 5468.

Malik, A. M. Singh, A. (2021) Pseudocereals Proteins-A comprehensive review on its isolation, composition and quality evaluation techniques. *Food Chemistry Advances*, 1, 100001.

Malik, A. M., & Singh, A. (2022). Pseudocereals proteins-A comprehensive review on its isolation, composition and quality evaluation techniques. *Food Chemistry Advances, 1*, 100001.

Martinez, M. M., & Gómez, M. (2017). Starch-active debranching and α-glucanotransferase enzymes. Edited by Ramesh C. Ray, Cristina M. Rosell. In *Microbial Enzyme Technology in Food Applications* (pp. 46–68). CRC Press, Boca Raton, USA.

Michael, A. M., & Jagadhesan, S. (2025). Nanotechnology: An emerging field for enhancing micronutrient enrichment in millets via biofortification strategies-present knowledge and prospects for the future. *Journal of Food Science and Technology*, *62*(2), 213–220.

Mir, N. A., Yousuf, B., & Gani, A. (2024). Recent advances in the modification of underutilized pseudocereal proteins for improving their functionality: A practical, comprehensive and up-to-date review. *Food Reviews International*, *40*(9), 2844–2872.

Mondor, M., & Hernández-Álvarez, A. J. (2023). Emerging solvent extraction technologies for plant protein extraction: aqueous two-phase extraction; deep eutectic solvent; subcritical water extraction. Edited by Hernández-Álvarez, A.J., Mondor, M., Nosworthy, M.G. In *Green protein processing technologies from plants: novel extraction and purification methods for product development* (pp. 111–130). Cham: Springer International Publishing.

Morales, D., Miguel, M., & Garcés- Rimón, M. (2021). Pseudocereals: A novel source of biologically active peptides. *Critical Reviews in Food Science and Nutrition*, *61*, 1537–1544.

Mullaiselvan, I., Nelluri, P., Nisha, B., & Saha, D. (2024). Properties and applications of proteins from pseudocereals. Edited by Sanju Bala Dhull, Aarti Bains, Prince Chawla, Sawinder Kaur. In *Pseudocereals* (pp. 132–153). CRC Press, Boca Raton, USA.

Nandan, A., Koirala, P., Tripathi, A. D., Vikranta, U., Shah, K., Gupta, A. J., ... & Nirmal, N. (2024). Nutritional and functional perspectives of pseudocereals. *Food Chemistry*, *448*, 139072.

Nandan, A., Koirala, P., Tripathi, A. D., Vikranta, U., Shah, K., Gupta, A. J., ... & Nirmal, N. (2024). Nutritional and functional perspectives of pseudocereals. *Food Chemistry*, 139072.

Nandan, A., Koirala, P., Tripathi, A. D., Vikranta, U., Shah, K., Gupta, A. J., ... & Nirmal, N. (2024). Nutritional and functional perspectives of pseudocereals. *Food Chemistry*, 139072.

Nicole, L., Laberty-Robert, C., Rozes, L., & Sanchez, C. (2014). Hybrid materials science: A promised land for the integrative design of multifunctional materials. *Nanoscale*, *6*(12), 6267–6292.

Pandiselvam, R., Manikantan, M. R., Divya, V., Ashokkumar, C., Kaavya, R., Kothakota, A., & Ramesh, S. V. (2019). Ozone: An advanced oxidation technology for starch modification. *Ozone: Science & Engineering*, *41*(6), 491–507.

Patro, A., & Chowdhury, A. (2025). Pseudocereals: Nutritional Powerhouse for a Healthier Future. 6(3), 2582–9467.

Pawde, S. V., Kaewprachu, P., Kingwascharapong, P., Sai- Ut, S., Karbowiak, T., Jung, Y. H., & Rawdkuen, S. (2025). A comprehensive review on plant protein-based food packaging: Beyond petroleum-based polymers. *Current Research in Food Science*, 10, 101104.

Perez-Rea, D., & Antezana-Gomez, R. (2024). The functionality of pseudocereal starches. In *Starch in Food* (pp. 377–403). Woodhead Publishing.

Pirzadah, T. B., & Malik, B. (2020). Pseudocereals as super foods of 21st century: Recent technological interventions. *Journal of Agriculture and Food Research*, *2*, 100052.

Poshadri, A., Deshpande, H. & Kshirsagar, R. 2024. Pseudocereals: Development of functional foods, their properties, challenges, and opportunities in food processing industry. *International Food Research Journal*, *31*.

Poshadri, A., Deshpande, H. W., & Kshirsagar, R. B. (2024). Pseudocereals: Development of functional foods, their properties, challenges, and opportunities in food processing industry. *International Food Research Journal*, *31*(2), 266–275.

Pratiwi, M., Faridah, D. N., & Lioe, H. N. (2018). Structural changes to starch after acid hydrolysis, debranching, autoclaving-cooling cycles, and heat moisture treatment (HMT): A review. *Starch-Stärke*, *70*(1–2), 1700028.

Priego-Poyato, S., Rodrigo-Garcia, M., Escudero-Feliu, J., Garcia-Costela, M., Lima Cabello, E., Carazo-Gallego, A., … & Jiménez-López, J. C. (2021). Current advances research in nutraceutical compounds of legumes, pseudocereals and cereals. doi:10.5772/intechopen.97782

Raghunathan, R., Farahnaky, A., Majzoobi, M., Chandrapala, J., Brennan, C., Eri, R., & Pandiselvam, R. (2025). Slowly digestible starch: A biochemical engineering perspective on functional food development and metabolic health. *Journal of Food Measurement and Characterization*, 19, 5197–5221. 1–25.

Rao, V., & Poonia, A. (2023). Protein characteristics, amino acid profile, health benefits and methods of extraction and isolation of proteins from some pseudocereals—a review. *Food Production, Processing and Nutrition*, *5*(1), 37.

Reddy, M. V. K., Dubey, P. K., Mishra, A. A., & Sabeel, V. A. (2024). Potential processing techniques for safe utilisation of pseudo cereals in the food system. *Journal of Food Composition and Analysis*, *135*, 106609.

Reddy, M. V. K., Dubey, P. K., Mishra, A. A., & Sabeel, V. A. (2024). Potential processing techniques for safe utilisation of pseudo cereals in the food system. *Journal of Food Composition and Analysis*, 106609.

Repo-Carrasco-Valencia, R., & Vidaurre-Ruiz, J. M. (2022). Quinoa, kañiwa, amaranth, and lupin as ingredients in gluten-free baking. In Native Crops in Latin America (pp. 35-59). CRC press.Repo-Carrasco-Valencia, R., & Vidaurre-Ruiz, J. M. (2022). Quinoa, kañiwa, amaranth, and lupin as ingredients in gluten-free baking. In *Native Crops in Latin America*. CRC Press.

Rollán, G. C., Gerez, C. L., & LeBlanc, J. G. (2019). Lactic fermentation as a strategy to improve the nutritional and functional values of pseudocereals. *Frontiers in Nutrition*, *6*, 98.

Rupa, A. Z., Akhter, K. T., & Rahim, A. T. M. (2024). Nutritive, polyphenolic, and antioxidative properties of locally produced cereals and pseudograins. *Journal of Science and Technology Research*, *6*(1), 89–98.

Saeid, A., Akter, F., Ali, M. A., & Rahman, M. H. (2024). Enzymatic modification of starch: Amylases and pullulanase. Edited by Nirmal Mazumder, Md. Hafizur Rahman. In *Advanced Research in Starch* (pp. 95–115). Singapore: Springer Nature Singapore.

Salas-Valero, L. M., Tapia-Blácido, D. R., & Menegalli, F. C. (2015). Biofilms based on canihua flour (Chenopodium pallidicaule): design and characterization. *Química nova*, *38*(1), 14–21.

Sarkar, A. K., Yang, Z., Astruc, T., & Amdursky, N. (2025). Aqueous-based assembly of plant-derived proteins yields a crosslinker-free biodegradable bioplastic consistent with green chemistry principles. *ChemSusChem*, *18*, e202401567.

Sarkar, A., Saha, D., Pradhan, A., & Tripathy, P. P. (2024). Innovative processing technologies for pseudocereals. In Pseudocereals (pp. 191-215). CRC Press.

Schoenlechner, R. & Bender, D. 2020. Pseudocereals for global food production. *Cereal Foods World*, *65*.

Schoenlechner, R., & Bender, D. (2020). Pseudocereals for global food production. *Cereal Foods World*, *65*(2), 1–8.

Shah, N., Mewada, R. K., & Mehta, T. (2016). Crosslinking of starch and its effect on viscosity behaviour. *Reviews in Chemical Engineering*, *32*(2), 265–270.

Shahbaz, M., Raza, N., Islam, M., Imran, M., Ahmad, I., Meyyazhagan, A., … & Wan, C. (2023). The nutraceutical properties and health benefits of pseudocereals: A comprehensive treatise. *Critical Reviews in Food Science and Nutrition*, *63*(29), 10217–10229.

Sindhu, R., & Khatkar, B. S. (2019). Pseudocereals: Nutritional composition, functional properties, and food applications. In *Food Bioactives* (pp. 129–147). Apple Academic Press.

Thakur, P., & Kumar, K. (2019). Nutritional importance and processing aspects of pseudocereals. *Journal of Agricultural Engineering and Food Technology*, *6*(2), 155–160.

Thakur, P., Kumar, K., Ahmed, N., Chauhan, D., Rizvi, Q. U. E. H., Jan, S., ... & Dhaliwal, H. S. (2021). Effect of soaking and germination treatments on nutritional, anti-nutritional, and bioactive properties of amaranth (Amaranthus hypochondriacus L.), quinoa (Chenopodium quinoa L.), and buckwheat (Fagopyrum esculentum L.). *Current Research in Food Science*, *4*, 917–925.

Thakur, S., Pandey, A. K., Verma, K., Shrivastava, A., & Singh, N. (2024). Plant-based protein as an alternative to animal proteins: A review of sources, extraction methods and applications. *International Journal of Food Science and Technology*, *59*(1), 488–497.

Tian, R., Feng, J., Huang, G., Tian, B., Zhang, Y., Jiang, L., & Sui, X. (2020). Ultrasound driven conformational and physicochemical changes of soy protein hydrolysates. *Ultrasonics Sonochemistry*, *68*, 105202.

Tomotake, H., Shimaoka, I., Kayashita, J., Nakajoh, M., & Kato, N. (2002). Physicochemical and functional properties of buckwheat protein product. *Journal of Agricultural and Food Chemistry*, *50*, 2125–2129.

Usman, M., Patil, P. J., Mehmood, A., Rehman, A., Shah, H., Haider, J., ... & Li, X. (2022). Comparative evaluation of pseudocereal peptides: A review of their nutritional contribution. *Trends in Food Science & Technology*, *122*, 287–313.

Usman, M., Patil, P. J., Mehmood, A., Rehman, A., Shah, H., Haider, J., ... & Li, X. (2022). Comparative evaluation of pseudocereal peptides: A review of their nutritional contribution. *Trends in Food Science & Technology*, *122*, 287–313.

Vajdovich, D. K., Csajbókné Csobod, É., & Benedek, C. (2025). Pseudocereal-based functional beverages: Main properties and nutritional evaluation with an emphasis on amino acid content: A review. *Foods*, *14*(12), 2080.

Vajdovich, D. K., Csajbókné Csobod, É., & Benedek, C. (2025). Pseudocereal-based functional beverages: Main properties and nutritional evaluation with an emphasis on amino acid content: A review. *Foods*, *14*(12), 2080.

Vani, S. N., & Soni, D. (2024). Nutritional profile and health benefits of buckwheat: A miracle pseudo cereal. *Research Journal of Agricultural Sciences*, *15*(5), 1204–1209.

Vidaurre-Ruiz, J., Bender, D. & Schönlechner, R. 2023. Exploiting pseudocereals as novel high protein grains. *Journal of Cereal Science*, *114*, 103795.

Vidaurre-Ruiz, J., Bender, D., & Schönlechner, R. (2023). Exploiting pseudocereals as novel high protein grains. *Journal of Cereal Science*, *114*, 103795.

Waisundara, V. Y. (2024). Introductory chapter: Pseudocereals as subexploited food. Edited by Viduranga Y. Waisundara. In *Pseudocereals-Recent Advances and New Perspectives*. IntechOpen, London, UK.

Wang, S., & Copeland, L. (2015). Effect of acid hydrolysis on starch structure and functionality: A review. *Critical reviews in food science and nutrition*, *55*(8), 1081–1097.

Wang, S., Wang, J., Liu, Y., & Liu, X. (2020a). Starch modification and application. Edited by Shujun Wang. In *Starch Structure, Functionality and Application in Foods* (pp. 131–149). Singapore: Springer Singapore.

Wang, X., Ullah, N., Sun, X., Guo, Y., Chen, L., Li, Z., & Feng, X. (2017). Development and characterization of bacterial cellulose reinforced biocomposite films based on protein from buckwheat distiller's dried grains. *International Journal of Biological Macromolecules*, *96*, 353–360.

Wang, X., Zhao, R., & Yuan, W. (2020b). Composition and secondary structure of proteins isolated from six different quinoa varieties from China. *Journal of Cereal Science*, *95*, 103036.

Wu, X., Sun, Y., Shi, Y., Tang, Y., Tian, J., & Zhao, G. (2018, July). Study on kinetics of protein forward extraction from quinoa by AOT/isooctane reverse micelles system. In *2018 3rd International Conference on Modelling, Simulation and Applied Mathematics (MSAM 2018)* (pp. 334–339). Atlantis Press.

Xia, W., & Wu, J. (2023). Classification and enzyme properties of starch debranching enzymes. In *Industrial Starch Debranching Enzymes* (pp. 21–39). Singapore: Springer Nature Singapore.

Yadav, A., & Yadav, K. (2024). From humble beginnings to nutritional powerhouse: The rise of amaranth as a climate-resilient superfood. *Tropical Plants*, *3*(1), 1–8.

Yurtseven, G., & Özbay, G. (2025). Gluten-free product development: Cold fermented bread with buckwheat grain. *Ordu* Üniversitesi *Sosyal Bilimler Enstitüsü Sosyal Bilimler Araştırmaları Dergisi*, *15*(2), 961–988.

Zhang, W., Boateng, I. D., & Xu, J. (2024a). How does ultrasound-assisted ionic liquid treatment affect protein? A comprehensive review of their potential mechanisms, safety evaluation, and physicochemical and functional properties. *Comprehensive Reviews in Food Science and Food Safety*, *23*(1), e13261.

Zhang, W., Boateng, I. D., Xu, J., & Zhang, Y. (2024). Proteins from legumes, cereals, and pseudo-cereals: composition, modification, bioactivities, and applications. *Foods*, *13*(13), 1974.

Zhang, W., Boateng, I. D., Xu, J., & Zhang, Y. (2024b). Proteins from legumes, cereals, and pseudo-cereals: Composition, modification, bioactivities, and applications. *Foods*, *13*(13), 1974.

Zheng, X., Wang, Q., Li, L., Liu, C., & Ma, X. (2024). Recent advances in germinated cereal and pseudo-cereal starch: Properties and challenges in its modulation on quality of starchy foods. *Food Chemistry*, *458*, 140221.

9 Safety Profiles of Pseudocereals

Anti-nutritional Factors and Their Reduction Strategies

Dayanand Peter,
Thottiam Vasudevan. Ranganathan,
Aanchal Sharma, and Gulzar Ahmad Nayik

9.1 INTRODUCTION

Food grains are essential to lead an active and healthy lifestyle. Each food has a unique composition that includes a wide range of macro and micronutrients in various amounts. Certain dietary grains are rich in nutrients. While some are rich sources of proteins (legumes, pulses, etc.), some others are sources of lipids (oilseeds), some of minerals (pearl millet, etc.), and others of carbohydrates (rice, wheat, maize, etc.), and some have ideal nutritional combinations with high digestibility (quinoa, most minor millet, etc.) (Graziano et al., 2022).

With a few exceptions, the production of maize, wheat, and rice tripled globally due to the adoption of high-yielding cultivars, fertilizers, and insecticides, as well as better irrigation techniques. Such a focus on these cereal crops unintentionally led to a large rise in their consumption, accounting for about half of all calories consumed worldwide. Large amounts of rice, wheat, and maize were added to the meals, which signaled a change to a diet high in detrimental carbohydrates and raised the chances of getting heart disease and obesity (Bekkering and Tian, 2019). Important nutritional elements such as dietary fiber, premium proteins, minerals including iron, zinc, and calcium, and bioactive compounds like polyphenols were absent from this diet which is high in carbohydrates.

It is reported that over 1.3 billion individuals worldwide suffered from micronutrient deficiencies, a matter of great concern (Awuchi et al., 2020). Possible approaches to solve this problem could be (a) to fortify the diets with added micronutrients and/or (b) to popularize the cultivation of crops that are rich in micronutrients. The first option, i.e., fortification is an option widely followed over a long period of time. However, this approach has the drawback as these formulations contain synthetic vitamins and minerals, which may have issues with low absorption or overconsumption. An alternative could be to add more food choices. One way to do this is by using pseudocereals in cooking (Nandan et al., 2024). The percentage cultivable land is drastically reducing, the world over, due to the ever-rising population. It is speculated that by the year 2050, there will be less land

 DOI: 10.1201/9781003584858-9

to grow main foods because of an extra 1.5 billion people. So, farming pseudocereals in less suitable areas for staple crops should be promoted. Using different types of crops can help improve food, feed, and nutrition security (Mir et al., 2022).

The phrase "sub-exploited foods" refers to the foods that have been consumed by the people at large, with documented history. However, in the 1900s, due to reasons including advent or popularization of "new" foods, modifications in agricultural practices or lack of economic incentives, cultivation and consumption of such crops were discontinued abruptly. They are not true cereals but share similar features and uses. They are dicotyledonous plants, which makes them different from monocotyledonous cereals (Waisundara, 2024; Nandan et al., 2024; Upasana and Yadav, 2022). The word "pseudocereals" is used because these grains and seeds resemble cereals in aspects like texture, starch levels, taste, and cooking methods.

Pseudocereals have superior protein quality, better fiber content and higher content of micro-nutrients like iron and calcium, as compared to the conventional cereals like rice and wheat (Kaur et al., 2023; Martinez-Villaluenga et al., 2020). A comparative assessment of the nutrient profile of pseudocereals with the widely consumed cereals is given in Table 9.1. Similarly a comparative assessment of the amino acids and protein composition of the pseudo-cereals is given in Table 9.2. The carbohydrate, fat and mineral profile is presented in Table 9.3. However, they also contain a higher amount of anti-nutritional factors which are elements that can harm a nutrient's absorption and lessen its nutrition value. Due to their high nutrient content and potential health benefits, pseudocereals are well-known. But their availability and consumption can be limited by the presence of anti-nutritional factors such as phytates, tannin, saponins, and oxalates. Knowledge of these characteristics is essential for the implementation of proper processing and preparation techniques to enhance the health benefits of pseudocereals.

This chapter focuses on the health effects; therefore, the main objective of this chapter is to summarize the safety profiles of the pseudocereals with an emphasis on the anti-nutritional factors to human health. Moreover, it intends to address the strategies for the reduction of the anti-nutritional factors to maximize the overall potential of pseudocereals in the food system.

TABLE 9.1
Nutritional Composition of Pseudocereals and Cereals

	Amaranth	Quinoa	Rice	Corn	Wheat
Carbohydrates (%)	62	60	74.9	70	69.7
Fats (%)	7.5	5	2.8	7	3.5
Fiber (%)	12	4.1	1.4	1.7	2.7
Proteins (%)	16	14	7.7	9.8	11.6
Lysine (%)	0.89	0.89	–	0.22	0.36
Tryptophan (%)	0.15	0.15	0.1	0.07	0.14
Methionine (%)	0.32	0.32	0.27	0.27	0.17
Energy (kcal/100 g)	333	341	353	338	309

TABLE 9.2
Protein Content and Essential Amino Acids of Various Pseudocereals and Cereals

Cereal Type	Protein Content (g/100g)	Essential Amino Acids
	Pseudocereals	
Quinoa	8.0	Lysine: 5.1% Leucine: 5.3% Methionine: 2.1%
Amaranth	14.6	Lysine: 5%–10% Leucine: 4.8% Methionine: 2.6%
Buckwheat	13.3	Lysine: 2.1% Leucine: 4.4%
	Common Cereals	
Rice (White)	2.36	Lysine: 2.7% Leucine: 4.3% Methionine: 3.1%
Wheat	12.5	Lysine: 2.1% Leucine: 2.67% Methionine: 3.9%
Barley	3.6	Lysine: 2.7% Leucine: 6.3%
Oats	26.35	Lysine: 0.8% Leucine: 5.9%
Rye	11.0	Lysine: 2.2% Leucine: 3.4%

TABLE 9.3
Carbohydrate, Fat, and Mineral Content of Various Cereals and Pseudocereals

Cereal Type	Carbohydrate (%)	Fat (%)	Fe (mg/100 g)	Mg (mg/100 g)	Ca (mg/100 g)	Zn (mg/100 g)
			Common Cereal			
Wheat	69.7	3.5	3.3 ± 0.1	96.4 ± 3.7	34.8 ± 0.0	1.2 ± 0.1
Corn	70	7	2.7 ± 0.2	127 ± 4.5	7.2 ± 0.3	2.2 ± 0.1
Rice	74.9	2.8	0.8 ± 0.1	35.0 ± 1.8	10.0 ± 0.5	1.2 ± 0.1
			Pseudocereal			
Quinoa	60	5	5.5 ± 0.5	206.8 ± 6.4	32.9 ± 3.3	1.8 ± 0.0
Amaranth	62	7.5	9.2 ± 0.2	279.2 ± 1.1	180.1 ± 6.1	1.6 ± 0.0

9.2 TYPES OF PSEUDOCEREALS

Pseudocereals consist of an extensive range of nutrients such as vitamins, lipids, minerals, proteins, carbohydrates, and nutraceuticals, and they play a crucial role in meeting the needs of the population with various health consequences.

9.2.1 Quinoa

Quinoa belongs to the Chenopodiaceae family and the *Chenopodium* genus. Quinoa is a crop which started in the Andes and may thrive under different climatic conditions, soil, and environment (Navruz-Varli and Sanlier, 2016). It has high protein, minerals, fiber, minerals, and fats. It contains phytochemicals that enhance well-being which include the phytoecdysteroids, phytosterols, and saponins (Navruz-Varli and Sanlier, 2016). Compared to conventional cereals, pseudocereals contain premium proteins such as lysine and methionine (Filho et al., 2017). Moreover, earlier studies showed that the fatty acid composition of quinoa is virtually similar to that of soy oil (Filho et al., 2017). Quinoa contains high levels of antioxidants and fiber, low saturated fats, and salt, and is cholesterol-free. It is a low risk for disease such as kidney stones and gallbladder stones. It also aids digestion. Quinoa is a good source of vitamins like B, C, and E (Tang and Tsao, 2017). Some phytochemicals in quinoa are produced during metabolism (Flórez-Martínez et al., 2024). Cooked quinoa is digestible; it is very nutrient dense and offers a range of health benefits. Almost all people everywhere eat leaves because they are very nutrient dense and easy to digest, and they have numerous health benefits (Thakur et al., 2021b; Bender and Schönlechner, 2021).

9.2.2 Amaranth

Amaranth is a dicotyledonous quickly growing plant that can grow well in different soils and climates. Amaranth is a form of healthy pseudocereal because it contains a good amount of protein with potential for generation of bioactive peptides (Aderibigbe et al., 2020; Tovar-Pérez et al., 2019). It is thought to have been created long ago in the United States of America. Amaranth is free of gluten, fats, and sodium and helps reduce weight; amaranth also maintains healthy cholesterol levels. Amaranth contains some of the plant proteins which include globulin, albumin, prolamins, and glutelin. Quinoa and amaranth both contain folic acid in high amounts. Amaranth flour is used in the preparation of pasta, bread, and other food products (Thakur et al., 2021b; Schmidt et al., 2021). Amaranth has a high quantity of iron and calcium (Singh et al., 2020a). Athletes, those with type 2 diabetes, those with celiac disease, and undernourished students can be supplied with products containing amaranth as a nutrient supplement (Mukuwapasi et al., 2024; Singh et al., 2024a,b). Preliminary investigations reveal a statistically non-correlated improvement in hemoglobin concentration after nutritional interventions with amaranth-based products (Yilma et al., 2025).

9.2.3 Buckwheat

Buckwheat belongs to the Polygonaceae family and *Fagopyrum* genus, and it is a gluten-free crop. Common buckwheat's which are cultivated for human consumption purposes are *Fagopyrum tartaricum* and *Fagopyrum esculentum.* Buckwheat has huge amount of dietary fibers, polyphenols, organic acids, flavonoids, proteins, lipids, vitamins (like B6, B1, B2, and B3), minerals (like copper, zinc, manganese, selenium, potassium, calcium, and magnesium), and inositol (Yilmaz et al., 2020). Buckwheat is rich in albumin and globulin (Thakur et al., 2021b). Buckwheat lowers blood lipid levels and protects against chronic illnesses including type 2 diabetes due to its high fiber content (Yılmaz et al, 2025; Gogoi and Ahmed, 2025). The process of fermentation is reported to decrease the level of anti-nutritional factors including phytic acid and protease inhibitors, while increasing the level of amino acids, vitamins, minerals and bioactives (Gua et al., 2025).

9.2.4 Chia Seeds

Chia seeds (*Salvia hispanica* L.) are native to Mexico and are members of the Lamiaceae family. Mucilage, seed oil, and flour are all made from chia seeds. This superfood contains nutrients such as gluten-free protein, omega-3 fatty acids, and antioxidants that protect seeds from microbiological and chemical degradation (Huang et al., 2025, Paarakh et al., 2025; Kulczyński et al., 2019). Chia seeds are abundant in lipids, carbs, minerals, vitamins, dietary fibers, proteins, and phytochemicals. Germination or sprouting is reported to improve the nutritional benefits of chia seeds (Huang et al., 2025). Consumption of products containing chia seeds is reported to lower serum cholesterol, increase satiety index, protect from cardiovascular diseases and diabetes.

9.3 ANTI-NUTRITIONAL FACTORS IN PSEUDOCEREALS

Quinoa, amaranth, and buckwheat, though rich in essential nutrients contain various anti-nutritional factors that lower the utilization and bioavailability. In quinoa, key compounds that can interfere with nutrient absorption and metabolism are nitrates, oxalates, tannins, trypsin inhibitors, saponin and phytic acid (Melini and Melini, 2021). Anti-nutritional factors in amaranth include notable amounts of phenolic compounds, chymotrypsin inhibitors, trypsin inhibitors, amylase, and phytate that impact protein and carbohydrate digestibility (Jimoh et al., 2020; Jan et al., 2023). Buckwheat contains, amongst other nutrients, trypsin activity inhibitors, alpha-amylase activity inhibitors, polyphenols, phytic acid, and tannins, all of which add to its anti-nutritional status (Sofi et al., 2023). Awareness of these anti-nutritional factors and how to circumvent them can enable the nutritional value and digestibility of these pseudocereals to be optimized. It should be noted that simple unit operations like germination, fermentation or soaking of the grains in water has a telling effect on the reduction of anti-nutritional factors. This greatly enhances their application to nutrition-rich foods including beverages (Guo et al., 2025; Li et al., 2025; Toimbayeva et al., 2025).

9.3.1 Saponin

Saponin acts as a natural detergent; it is water-soluble and has a bitter taste. It can be effectively eliminated through wet methods such as cold water rinsing or through dry method including browning and abrasion (Sharma et al., 2023).

The quinoa grain is shielded from insects and birds by naturally existing coating which is saponin. This often exists in the outer layers of the grain (episperm). These compounds are triterpenoid glycosides that are present in plants, especially those in the Leguminosae family, and have undefined chemical formula, however fundamental chemical formula is $C_nH_{2n-8}O_{10}$ ($n \geq 5$). Quinoa contains four primary sapogenin structures; they are 30-O-methyl- espergulagenate, phytolaccagenic acid, oleanolic acid, and hederagenin. Glucose, arabinose, and galactose are the main available carbohydrates (Filho et al., 2017). Saponins are not just present in seeds; they are also present in other plant components and protect the plant from the harsh environmental conditions. Throughout the plant's growth, the amount of total saponins in the leaves may rise, although it is lower than that of the seeds, which can possess dry matter contents ranging from 0.2 to 0.4 g/kg in sweet genotypes and 4.7 to 11.3 g/kg in bitter genotypes. The main sapogenin present in the leaves are hederagenin and oleanolic acid is present in seeds.

Low amounts of saponin are present in amaranth grains. A minimal concentration of saponin, averaging 0.09%, is found in various species of amaranth. Saponins can serve beneficial roles, including potential antioxidant properties and anticancer effects (Tang and Tsao, 2017).

9.3.2 Phytic Acid

Phytic acid is mostly present in the peels of the majority of legumes and cereals, where it makes up 1%–3% of the dry mass and serves as the primary source of phosphate necessary for these seeds (Thakur et al., 2021a,b; Agregán et al., 2023, Chen and Xy, 2023). In quinoa grains it is present within the endosperm, contrasting with its concentration in the wheat and rye outer layer. The bran of quinoa accounts for about 38%–41% of the total phytate present in the seeds, exhibiting a higher phytate concentration compared to both the flour and the peel.

Phytate of amaranth is present as a naturally occurring storage form of phosphorus, primarily in the form of phytic acid or its salts (phytin) in the seeds. It is evenly distributed throughout the grain relative to its appearance in concentration in the outer layers, like that of tannins or saponins. Phytate plays important functions in the plant like a phosphorus and cation reserve and as a supporter of seed viability and growth (Castro-Alba et al., 2019).

Buckwheat is one of the seed, grain and legumes that occur naturally with phytic acid, myo-inositol hexakisphosphate. It is the primary way that plants store phosphorus and also plays a crucial role in seed germination and plant development. The content of phytic acid in buckwheat ranges from 35 to 38 g/kg and the highest content is found in bran (Klepacka et al., 2020). Although phytic acid is nutritious and possesses antioxidant activity, excessive content of phytic acid can lead to insoluble complexes that hinder the absorption of minerals such as Ca, Zn, and Fe.

9.3.3 Tannins

Tannins are a class of polyphenols that occur in plant species of most kinds in great amounts. Their ability to form complexes with proteins and other macromolecules like starch that lower food nutritional value is one of the most significant biological activities they perform. Involvement by hydrogen bonding and hydrophobic interactions and their occurrence according to environment pH define tannins in their interaction with proteins (Zhu, 2019). Tannins are more concentrated in the seed coat or pericarp of amaranth, and their levels can vary across cultivars. Tannins have antioxidant, antimicrobial, and anti-inflammatory properties. They may help in lowering the likelihood of developing specific chronic diseases.

The bran of buckwheat has been discovered to contain the majority of the tannins. Although they are the plant's natural defensive mechanisms, they can also hinder human protein and carbohydrate digestion, which reduces the availability of nutrients (Priego-Poyato et al., 2021).

9.3.4 Trypsin Inhibitor

Proteins found in nature which produce extremely stable compound with proteolytic enzymes are called protease inhibitors. Seeds of quinoa have a protease inhibitor content of less than 50 ppm, which does not provide a significant risk because of the very low levels of trypsin inhibitors compared to cereals that are often ingested (Nandan et al., 2024). The quinoa seeds had greater levels of trypsin inhibitor and phytates in its embryo. Comparing the entire grain with embryo, it showed 89% of the trypsin inhibitory and 60% of the total phytate action.

9.3.5 Nitrates

All plants contain nitrates, which are vital sources of nitrogen for the growth of plants. When absorption surpasses their metabolic requirements, certain plants store these nutrients in their roots and shoots. Petioles and stems are the areas where accumulation is greatest but leaves, particularly mesophyll, have high levels of nitrogen. Furthermore, phloem has low nitrate levels which supplies amino acids to reproductive organs, fruits, and seeds.

The vegetables which contain nitrogen may have come by means of mineralization or nitrification on the substrate, or it may have come from utilized fertilizers (Shahbaz et al., 2023; Ňorbová et al., 2022). It is reported that NO_3 in whole meal quinoa were twice as low as those found in vegetables like lettuce, spinach, radishes, and beets in which the values are above 100 mg/100 g (Filho et al., 2017). This proves that quinoa has no negative nutritional value. The World Health Organization suggests an ADI of about 3.7 mg/kg of body weight for nitrate ions and 0.06 mg/kg of body weight for nitrite. Therefore, an adult weighing 60 kg should not consume more than 222 mg of nitrate/day and 3.6 mg of nitrite/day.

9.3.6 Oxalates

Vegetables like chard, beets, rhubarb, spinach, tomatoes, cocoa, and almonds contain oxalate. The Chenopodiaceae, Polygonaceae, and Amaranthaceae families comprise the majority of vegetables with high oxalate concentration (Gupta and Mishra, 2024).

Among all these plants, oxalate builds up in all plant tissues, including stems, roots, leaves, and hypocotyl seeds. Roots and seeds had soluble oxalate levels between 59 and 131 mg/100 g, whereas leaves and stems had levels between 258 and 1,029 mg/100 g. The overall concentration of oxalate in leaves and stems was found in the range between 874 and 1,959 mg/100 g whereas in seeds and roots, it varies between 143 and 232 mg/100 g (Pathan and Siddiqui, 2022). The quinoa wholemeal variety BRS Piabiru contains a lower amount of oxalic acid (380 mg/100 g) compared to other food within the similar category like beets (328 mg/100 g) and spinach (822 mg/100 g) (Gelinas and Seguin, 2007; López-Moreno et al., 2022).

The authors believe that 50–200 mg of oxalates per day should be included in a typical diet. The vegetables which belong to the Chenopodiaceae and Polygonaceae family including beets, spinach, and rhubarb have high amount of oxalates.

There are usually a lot of oxalates in amaranth, especially the types used for grain. The average total oxalate content found in several amaranth genotypes is approximately 229 mg/100 g, with levels varying from 178 to 278 mg/100 g (Gelinas and Seguin, 2007). About 80% of the oxalate in amaranth is insoluble, making up the majority of the plant's oxalate content.

Buckwheat contains naturally occurring chemical oxalate. It is mostly found as soluble oxalic acid, which can aid in the kidneys' production of calcium oxalate crystals. A number of variables, such as the type of buckwheat, soil type, and growth-related environmental factors, might affect the amount of oxalate in buckwheat (Li et al., 2022). In comparison to other grains and legumes, buckwheat typically has moderate amounts of oxalate.

9.3.7 Protease Inhibitors

Protease inhibitors are substances that prevent proteolytic enzyme activity, which is the breaking down of proteins. Both trypsin and chymotrypsin inhibitors are among the particular protease inhibitors found in amaranth (Kamal et al., 2021). By attaching to digestive enzymes, these inhibitors can lower the protein digestibility and perhaps reduce the nutrient absorption. Protease inhibitors are known to possess therapeutic properties (Nardo et al., 2020; Kamal et al., 2021). In some dietary circumstances, protease inhibitors' anti-nutritional qualities may outweigh their potential advantages, such as lowering the risk of some cancers by preventing tumor growth (Laparra and Haros, 2019; You et al., 2022).

9.4 IMPACT OF ANTI-NUTRITIONAL FACTORS ON HUMAN HEALTH

9.4.1 Saponin

There is no evidence that saponin complexes with vitamins like A, D3, and E, although some saponins create complex with Fe and Zn that lower their absorption. This interaction can negatively impact mineral nutrition. Additionally, the distinctively bitter taste of saponins can influence the sensory qualities of quinoa, affecting its acceptance among consumers.

High concentration of saponin can create a negative effect on human health, especially when it comes to nutrient absorption. They can cause digestive problems and decrease nutrient absorption by inhibiting digestive enzymes including glucosidase and amylase. It decreases absorption by forming compounds with mineral elements like Zn and Fe (Suárez-Estrella et al., 2021; Sharma et al., 2023; Mroczek, 2015).

9.4.2 Phytic Acid

Phytic acid has a strong negative charge because it is believed to have anti-nutritional properties. Phytic acid can chelate bivalent minerals like Ca, Fe, Mg, Zn, and Cu, and it can also chelate enzymes, proteins, and starches reducing their bioavailability. Although quinoa has a greater phytic acid content than cereals, neither the absorption of iron nor the incorporation of calcium in the bones were found to be negatively impacted (Gupta and Mishra, 2024).

While these compounds play protective roles in plants, they interfere with absorption of nutrients in humans. Phytate cannot be digested by humans, and it cannot act as a dietary source of inositol or phosphate. Phytic acid creates a complex structure with proteins and it reduces their bioavailability. Additionally, both phytates and oxalates have been reported to inhibit starch digestion. Dietary phytic acid has a variety of health effects. As a strong chelator, phytic acid can decrease the bioavailability of essential minerals, which over time may result in deficiencies, particularly in populations that rely mostly on foods high in phytate without getting enough minerals. Phytic acid can also offer health protection. In relation to the findings, it possesses antioxidant effects, which may benefit the prevention of inflammatory diseases and oxidative stress. Also, by reducing the amount of prooxidants, its ability to inhibit individual digestive enzymes can decrease the incidence of individual cancers. Therefore, the net health impact of phytic acid can be significantly different based on the dietary scenario and individual nutritional requirements (Thakur et al., 2021b; Mir et al., 2018).

9.4.3 Tannin

Intestinal mucosa is damaged due to the presence of tannin in food and it also disrupts the Fe, glucose, and vitamin B12 absorption, tannins also create unpleasant coloration in foods due to enzymatic browning reactions and astringency causes poor palatability (Das et al., 2020).

Iron and zinc uptake can be inhibited by the capacity of tannins to chelate minerals and proteins. Deficiency can be caused by this activity, particularly in low dietary diversity communities. The good news is that tannins possess antioxidant activities that can inhibit oxidative stress and related diseases. While more research needs to be done to demonstrate an immediate preventive factor, these characteristics have been linked with a reduced incidence of chronic disease such as CVD and certain cancers (Priego-Poyato et al., 2021). Tannins, according to some studies, are said to counteract the mutagenic action of some carcinogens while others mention that excessive tannin intake may be contributory to certain cancers, notably because they impair digestive enzymes.

The tannin content of buckwheat can have an array of health effects. Tannins, however, are renowned for their health-enhancing properties that include their anti-inflammatory, anticancerous, and antioxidant activities that can reduce the risk of developing chronic diseases. They can also regulate blood sugar and cholesterol levels (Sofi et al., 2023). However, tannins negatively impact the absorption of nutrients, particularly nonheme iron, which is critical in risk groups for iron deficiency, including vegetarians, pregnant women, and children. According to research, by forming complexes with substances that inhibit absorption, tannins reduce iron bioavailability significantly (Sofi et al., 2023). If buckwheat serves as the central component of someone whose diet lacks sufficient iron contained foodstuffs or counteractions to this process, this will lead to potential deficiencies.

9.4.4 Trypsin Inhibitor

Trypsin, which breaks down proteins, is less active when an inhibitor of trypsin is present in the digestive tract. This results in an increase in enzyme production by the pancreas, which causes the organ to enlarge and expand less (Agregán et al., 2023).

9.4.5 Nitrates

Nitrates can be converted to nitrite and once absorbed, cause cyanosis because of metmyoglobin formation. They also interact with secondary and tertiary amines to form carcinogenic N-nitrous compounds. Nitrates disrupt metabolism of vitamin A and functions of thyroid gland (Reddy et al., 2024).

9.4.6 Oxalates

It is a poisonous substance which creates serious risk to one's health. Humans are unable to metabolize it, and it is eliminated in the urine. The absorption of minerals and trace elements is affected by high dietary intake of oxalates. Calcium oxalate forms an insoluble complex with divalent cation in the gastrointestinal tract, and it plays an important role in hyperoxaluria, which is responsible for development of calcium oxalate stones in the kidneys (Stepanova, 2023).

Oxalic acid is associated with numerous detrimental effects in both humans and animals, including decreased mineral bioavailability, gastrointestinal distress, tetany or muscle contraction with other neurological symptoms, impaired blood coagulation, and potential damage to the excretory organs as a result of deposition of high levels of calcium oxalate crystals (Zayed et al., 2025).

Amaranth has a high oxalate content, which may affect its nutritional bioavailability, especially in terms of the absorption of calcium and magnesium (Terzieva et al., 2026). High dietary oxalate intake increases the urinary oxalate excretion, which may increase the risk of stone formation (Bargagli et al., 2020). Restricting foods that are rich in oxalate is recommended for individuals with a history of kidney stones. Oxalates can bind to necessary minerals such as calcium and magnesium preventing them from being absorbed through the gastrointestinal tract. This can also happen particularly in populations who are consuming high amounts of foods that have high oxalates but are not receiving sufficient calcium. Amaranth is dense with healthy nutrients such as protein, vitamins, and minerals but oxalates are certain to have negative consequences.

Dietary oxalate can potentially have a significant impact on human health. Its association with kidney stone formation is one of the central problems, particularly in those most at risk of kidney stone which is calcium oxalate stones. Excessive dietary intake of soluble oxalate can potentially increase urine oxalate concentration that enhances the risk of stone development.

In addition, oxalates bind to Ca in the gut reducing its absorption and leading to deficiencies if ingested excessively (Holmes and Assimos, 2004). Extrusion of amaranth grain is reported to increase calcium bioavailability (Ferreria and Areas, 2010). Oxalates have numerous health advantages despite these potential adverse effects. They can lower the occurrence of chronic diseases through the induction of antioxidant activity and potentially providing protection against inflammation and oxidative stress (Avila-Nava et al., 2021).

9.4.7 Protease Inhibitors

Ingestion of amaranth's protease inhibitors has several adverse health consequences. Conversely, the compounds can exert anticarcinogenic activity given that certain research shows that certain inhibitors can inhibit cancer cell growth (Angel Huerta-Ocampo and Paulina Barba de la Rosa, 2011; Usman et al., 2022). For instance, a trypsin-chymotrypsin inhibitor from amaranth has been found to possess potential in inhibiting anchorage-independent growth of cancer cells and thus has potential use in cancer therapy.

Conversely, malabsorption syndromes may be caused by excessive intake of protease inhibitors, particularly in societies that have a high intake of raw or undercooked amaranth.

The inhibitors may bind to minerals and reduce the bioavailability of essential amino acids which reduces nutritional absorption (Manyelo et al., 2020). This is a very serious problem for children and individuals with special dietary requirements.

9.5 REDUCTION STRATEGIES FOR ANTI-NUTRITIONAL FACTORS

9.5.1 Saponin: Effective Processing and Reduction Techniques

To improve acceptance and sensory quality among consumers the amount of saponin is typically decreased or eliminated from the grain's exterior due to its distinctively bitter flavor. Removal is achieved by either by choosing sweet cultivars or using wet method which is strong washing in cold alkaline water or dry method like extrusion, heat treatment, roasting or mechanical abrasion, or a combination of two (Agregán et al., 2023). The most commonly used method used at the home level is the wet method, which can also be applied commercially. In addition to contaminating the water with saponin waste, it causes both ecological and economic difficulty because of the increased consumption of water and the requirement to dry the grains after washing. Saponins are removed on a commercial scale through a dry technique known as abrasive peeling. The disadvantage of this process is loss of nutrients along with the bran. To reduce nutritional loss and obtain production of quinoa derivatives with considerably reduced saponin content, the wet process combined with polishing is recommended.

Fermentation: Using microorganisms that break down anti-nutritional factors during fermentation, this process can potentially speed up the breakdown of saponins (Woiciechowski et al., 2023; Arsov et al., 2024).

Germination: Enzymatic activity caused when seeds germinate can successfully reduce the quantity of saponin while enhancing nutritional bioavailability simultaneously (Arsov et al., 2024; Samtiya et al., 2020).

Debranning: Saponins are largely restricted within the seed coat and stripping off outer coats of grains has the potential to reduce the quantities of these substances (Samtiya et al., 2020).

Soaking and Boiling: Water-soluble saponins in amaranth grains can be removed by soaking them prior to boiling. Additionally, since saponins are heat denatured, boiling can also reduce their amount (Samtiya et al., 2020).

9.5.2 Phytic Acid: Effective Processing and Reduction Techniques

The washing and brushing procedure utilized to remove saponins resulted in a significant reduction of nearly 30% phytic acid content in quinoa grains. Phytic acid in raw quinoa seed is 10.4 mg/g and in processed seed is 7.8 mg/g while in whole flour grains the contents are rye (7.7 mg/g), wheat (8.7 mg/g), lentils (8.4 mg/g), and faba beans (8.0 mg/g).

Steeping, germination, and fermentation are a few methods that can very easily reduce phytic acid. Processes like fermentation and cooking that enhance the activation of phytase demonstrate improved degradation efficiencies (Samtiya et al., 2020; Nandan et al., 2024; Thakur et al., 2021b).

The phytic acid content of buckwheat groat considerably decreased due to soaking in water for 6–8 hours or overnight. By activating the enzymes that break down phytates, this process improves the bioavailability of minerals. One of the proper cooking methods that can decrease phytic acid by a considerable percentage is boiling.

Buckwheat is boiled for 15–20 minutes to enhance mineral absorption by reducing its phytic acid by around 60%–80%.

Germination of buckwheat seed enhances the overall nutritional value and enhances the phytase enzyme activity that hydrolyzes the phytic acid. Protein and carbohydrate digestibility is enhanced, and phytic acid is reduced effectively by germinating for one to three days. Phytic acid in buckwheat may also be reduced by adhering to the fermentation procedures. Beneficial microbes are used for the process that produce phytase and hydrolyze the phytic acid, making the minerals accessible.

9.5.2.1 Traditional Methods of Phytate Reduction

Soaking: Soaking of seeds in water for a specified time can lead to a decrease in the content of phytate in amaranth seeds. Studies have shown that such a process significantly decreases phytate content as well as enhances minerals absorption (Samtiya et al., 2020).

Sprouting: Amaranth seed nutrient content changes during germination. Sprouting has been reported to improve nutritional value and phytate reduction by 37%–81% in various cereals and legumes. Germination induces phytase enzyme synthesis that hydrolyzes phytate to assimilable myoinositol phosphates.

Fermentation: Fermentation is an effective conventional procedure that is very effective in decreasing the phytate content. Phytate is hydrolyzed by microorganisms in particular by lactic acid bacteria such as *Lactobacillus plantarum* during fermentation. Dephosphorylation of as much as 88% has been achieved provided amaranth is fermented long enough (Castro Alba et al., 2019). The fermentation yield of lactic acid further facilitates a fall in pH levels hence also enhancing phytase activity.

9.5.2.2 Modern Processing Techniques

Hydrothermal Processing: In the application of hydrothermal treatments grains are subjected to steam or boiling water. This treatment has been proven to stimulate the phytate content to reduce. Depending on the length of time and temperature of the treatment boiling in particular might result in a notable decrease in phytate levels (Jan et al., 2023; Motta et al., 2022).

Extrusion and Enzymatic Treatment: In extrusion, the grain is cooked at high pressure and temperature which in turn lower phytate levels (Ali et al., 2024). To further break down phytate in processed amaranth products phytase enzymes can be used in enzymatic treatments.

9.5.3 Tannin: Effective Processing and Reduction Techniques

Quinoa seeds have 0.53% tannin content, which is significantly less than that of rice which is 1.3% and this concentration is even more decreased after washing and water rinsing. The tannin content in peeling was found to be higher (0.92%) when compared to that in bran or flour. Nonetheless, 46%–50% of the total tannin of the seeds are still present in the bran of the peeled grain (Agregán et al., 2023;

Langyan et al., 2024). The appropriate amount of washing and cooking of quinoa reduce the negative impact of tannin and this improves the digestibility (Thakur et al., 2021b).

The tannin content of buckwheat groats can be minimized by soaking them for 6–8 hours. This procedure increases the nutrient bioavailability of the groats while at the same time reducing anti-nutritional factors. Boiling them is one of the best ways of minimizing tannins. The tannin content of buckwheat is greatly reduced by boiling it for 15–20 minutes, which enhances its digestibility and nutrient utilization.

Buckwheat seed germination can also reduce tannin content. This action increases the nutritional content as well as activates enzyme activities that hydrolyze anti-nutrients thereby increasing digestibility. Soaked buckwheat fermentation can significantly reduce its tannin content. Since helpful bacteria assist in the hydrolysis process of tannins and other anti-nutrients it increases the nutrition content of the end product.

Soaking grains in water can lower their tannin content by as much as 32.30% (Thakur et al., 2021b; Jan et al., 2023). Such treatments increase the bioavailability of minerals and at the same time decrease tannins. Germination has been found to increase the nutritional value of amaranth considerably. According to studies, germination can increase the nutritious compounds such as proteins and phenolic acids and decrease the tannin content by around 32.30% (Thakur et al., 2021b; Jan et al., 2023). The fermentation process improves the nutritional quality of amaranth through the reduction of phytic acid and tannins hence improving nutrient bioavailability and digestibility (Jan et al., 2023). Traditional cooking processes such as boiling and roasting also assist in reducing tannins quite effectively while improving the overall nutritional content of amaranth (Thakur et al., 2021b).

9.5.4 Trypsin Inhibitors: Effective Processing and Reduction Techniques

Since these anti-nutritional constituents are thermolabile and usually destroyed under normal household or industrial food processing conditions they have little effect on human nutrition (Agregán et al., 2023; Priego-Poyato et al., 2021). By exposing different legumes including beans and peas to 18–22 hours of soaking and water washing and subsequent heat treatments such as boiling, roasting, autoclaving, and microwave, complete elimination of trypsin component can be achieved. This observation proves that techniques commonly employed in household cooking can be employed for inactivation of this anti-nutrient in quinoa.

9.5.5 Oxalates: Effective Processing and Reduction Techniques

The very effective approach to reducing the content of oxalates is by cooking amaranth leaves and then washing away the cooking water. Experimental research has shown that this treatment lowers total oxalate content by 90%, primarily due to the fact that water-soluble oxalates exist in raw leaves. Calcium uptake from foods other than milk is less adversely affected when amaranth leaves are cooked and the liquid discarded later (Zayed et al., 2025). It should be cooked in such a way because in addition to reducing oxalates, calcium is not lost and remains absorbed well because it is insoluble.

Germination is also a method of reducing the content of oxalates in amaranth. Certain anti-nutritional factors such as oxalates are destroyed through soaking seeds in water for several days prior to germination. Nutrients become more available by germination and this has the potential to increase digestibility considerably (Valadez Vega et al., 2022). Through protein digestibility improvement, this process enhances the nutritional quality of the seeds, thereby making it more beneficial.

Amaranth seeds processed through the standard popping process have been successful in reducing oxalates and other anti-nutritional compounds. In comparison with other methods of cooking, popping which involves rapid heat the seeds to an extreme temperature can potentially lower the absorption of oxalates. While further research is required to determine just how effective the process is at reducing oxalate specifically, it indicates an effective process to ready amaranth for consumption by humans (Valadez Vega et al., 2022).

One of the methods to decrease the oxalate content of foods is by soaking the raw leaver or seeds in water. This operation leaches out the soluble oxalates, by the process of hydrolysis. Once this water is drained off, there appears to be a significant reduction in the oxalate content of food materials. This is a beneficial process since it is able to effectively lower the oxalate level through the integration of soaking and boiling (Valadez Vega et al., 2022).

With these interventions such as eating foods that are high in calcium will be able to minimize the damage done by dietary oxalates. Calcium in the intestines can bind to soluble oxalates and cause them to be expelled from the body instead of being absorbed into the bloodstream. Those who have sensitivity toward dietary oxalates that is those who are prone to forming kidney stones may find this part of dietary management very helpful.

Oxalate content in the buckwheat grain can be reduced by soaking them overnight or a few hours. The process reduces the soluble oxalate concentration in the groats by causing them to leach into soaking water (Huynh et al, 2022). Boiling is one of the more reliable methods to reduce the level of oxalate. Heat has the ability to lower the content of oxalates significantly as well as tenderize the grain (Zayed et al., 2025). Oxalate content can significantly decrease after 15–20 minutes of boiling.

Buckwheat seeds are able to lower their oxalate content and enhance nutritional quality by germinating. Phytase works during germination and that is helpful in hydrolyzing oxalates and increasing the nutritive content of nutrients. Buckwheat can further be improved using fermentation, which reduces its content of oxalates. Probiotic bacteria that decompose oxalates are used during fermentation and that improves the nutritional value of the foodstuff (Al-Kabe and Niamah, 2024; Anumudu et al., 2024).

9.5.6 Protease Inhibitors: Effective Processing and Reduction Techniques

Boiling and roasting are examples of heat processing technique that dramatically lowers protease inhibitor levels. Cooking techniques can improve protein digestibility and nutrient absorption by reducing trypsin inhibitor activity (Faizal et al., 2023). Boiling amaranth seeds and leaves is an effective cooking method that can lower the amount of protease inhibitors by upto70% (Faizal et al., 2023).

Amaranth seeds are water soaked to initiate sprouting and this can enhance protein digestibility and reduce anti-nutritional factors (Faizal et al., 2023). It has been proven that germination enhances the bioavailability of nutrients while reducing the content of protease inhibitors. Germination has been shown through research to lead to a significant reduction in tannins and phytic acid that can complement the reduction in protease inhibitors (Faizal et al., 2023; Gunathunga et al., 2024).

Protease inhibitors can also be efficiently broken down by using microbial fermentation which significantly improves amaranth's overall nutritional profile (Faizal et al., 2023). Products made from fermented amaranth can maintain their healthful qualities making them a useful dietary choice. To make a puffed product, amaranth seeds are heated to high temperatures in a traditional preparation procedure. Although popping has been shown to lower some anti-nutrients, it might not be as successful as other techniques like fermentation or cooking at reducing the amounts of protease inhibitors (Valadez Vega et al., 2022).

9.6 CONCLUSION

The significance of understanding the anti-nutritional characteristics of pseudocereals and the ways of minimizing them is highlighted by their safety profiles, which include a range of nutrient rich grains such as amaranth, quinoa, and buckwheat. The presence of anti-nutritional factors like tannins, oxalates, phytic acid, and saponins may adversely affect nutrient bioavailability and absorption even though these pseudocereals are rich in essential nutrients and provides several health benefits. This highlights the significance of thoroughly evaluating their safety profiles in terms of nutritional interactions and other health risks.

To enhance the safety and nutritional quality of pseudocereals reduction methods are necessary. It's been proven techniques such as germination, soaking, boiling, and fermentation can effectively reduce anti-nutrient content enhancing nutrient absorption and digestibility. Customers can greatly reduce the negative effects of anti-nutritional components by putting these strategies into practice, thereby guaranteeing that the positive qualities of pseudocereals are fully fulfilled.

In summary, although pseudocereals are a great complement to a balanced diet, it's critical to identify and address any anti-nutritional aspects of them using the right preparation techniques. In addition to increasing the health advantages of these grains, the quest for dietary safety and enhanced nutritional availability will increase public awareness of correct use of pseudocereals. Encouraging the successful integration of pseudocereals in modern diets which will ultimately enhance nutritional wellbeing and food security will involve continued research and education on the safety profiles and anti-nutrient reduction options.

REFERENCES

Aderibigbe, O. R., Ezekiel, O. O., Owolade, S. O., Korese, J. K., Sturm, B., & Hensel, O. (2020). Exploring the potentials of underutilized grain amaranth (*Amaranthus* spp.) along the value chain for food and nutrition security: A review. *Critical Reviews in Food Science and Nutrition*, *62*(3), 656–669. https://doi.org/10.1080/10408398.2020.1825323

Agregán, R., Guzel, N., Guzel, M. et al. (2023). The effects of processing technologies on nutritional and anti-nutritional properties of pseudocereals and minor cereal. *Food and Bioprocess Technology*, *16*, 961–986. https://doi.org/10.1007/s11947-022-02936-8

Ali, I. M., Forsido, S. F., Kuyu, C. G., Ahmed, E. H., Andersa, K. N., Chane, K. T., & Regasa, T. K. (2024). Effects of extrusion process conditions on nutritional, anti-nutritional, physical, functional, and sensory properties of extruded snack: A review. *Food Science & Nutrition*, *12*(11), 8755–8761.

Al-Kabe, S. H., & Niamah, A. K. (2024). Current trends and technological advancements in the use of oxalate-degrading bacteria as starters in fermented foods—A review. *Life*, *14*(10), 1338.

Angel Huerta-Ocampo, J., & Paulina Barba de la Rosa, A. (2011). Amaranth: A pseudo-cereal with nutraceutical properties. *Current Nutrition & Food Science*, *7*(1), 1–9.

Anumudu, C. K., Miri, T., & Onyeaka, H. (2024). Multifunctional applications of lactic acid bacteria: Enhancing safety, quality, and nutritional value in foods and fermented beverages. *Foods*, *13*(23), 3714.

Arsov, A., Tsigoriyna, L., Batovska, D., Armenova, N., Mu, W., Zhang, W., … & Petrova, P. (2024). Bacterial degradation of antinutrients in foods: The genomic insight. *Foods*, *13*(15), 2408.

Avila-Nava, A., Isabel Medina-Vera, I., Rodríguez-Hernández, P., Guevara-Cruz, M., Canton, P. K. H. G., Armando R. Tovar A. R., & Nimbe, T. (2021). Oxalate content and antioxidant activity of different ethnic foods. *Journal of Renal Nutrition*, *31*(1), 73–79. https://doi.org/10.1053/j.jrn.2020.04.006

Awuchi, C. G., Igwe, V. S., Amagwula, I. O., & Echeta, C. K. (2020). Health benefits of micronutrients (vitamins and minerals) and their associated deficiency diseases: A systematic review. *International Journal of Food Sciences*, *3*(1), 1–32.

Bargagli, M., Tio, M. C., Waikar, S. S., & Ferraro, P. M. (2020). Dietary oxalate intake and kidney outcomes. *Nutrients*, *12*(9), 2673.

Bekkering, C. S., & Tian, L. (2019). Thinking outside of the cereal box: Breeding underutilized (pseudo) cereals for improved human nutrition. *Frontiers in Genetics*, *10*. https://doi.org/10.3389/fgene.2019.01289

Bender, D., & Schönlechner, R. (2021). Recent developments and knowledge in pseudocereals including technological aspects. *Acta Alimentaria*, *50*(4), 583–609. https://doi.org/10.1556/066.2021.00136

Castro-Alba, V., Lazarte, C. E., Perez-Rea, D., Carlsson, N. G., Almgren, A., Bergenståhl, B., & Granfeldt, Y. (2019). Fermentation of pseudocereals quinoa, canihua, and amaranth to improve mineral accessibility through degradation of phytate. *Journal of the Science of Food and Agriculture*, *99*(11), 5239–5248.

Chen, W., & Xu, D. (2023). Phytic acid and its interactions in food components, health benefits, and applications: A comprehensive review. *Trends in Food Science & Technology*, *141*, 104201. ISSN 0924-2244. https://doi.org/10.1016/j.tifs.2023.104201

Chung, K. T., Wong, T. Y., Wei, C. I., Huang, Y. W., & Lin, Y. (1998). Tannins and human health: A review. *Critical Reviews in Food Science and Nutrition*, *38*(6), 421–464.

Das, A. K., Islam, M. N., Faruk, M. O., Ashaduzzaman, M., & Dungani, R. (2020). Review on tannins: Extraction processes, applications and possibilities. *South African Journal of Botany*, *135*, 58–70.

El Hazzam, K., Hafsa, J., Sobeh, M., Mhada, M., Taourirte, M., EL Kacimi, K., & Yasri, A. (2020). An insight into saponins from quinoa (*Chenopodium quinoa* Willd): A review. *Molecules*, *25*(5), 1059. https://doi.org/10.3390/molecules25051059

Faizal, F. I., Ahmad, N. H., Yaacob, J. S., Halim-Lim, S. A., & Rahim, M. A. (2023). Food processing to reduce antinutrients in plant-based foods. *International Food Research Journal*, *30*(1), 25–45.

Ferreira, T. A., & Areas, J. A. G. (2010). Calcium bioavailability of raw and extruded amaranth grains. *Food Science and Technology*, *30*(2), 532–538.

Filho, A. M., Pirozi, M. R., Borges, J. T., Pinheiro Sant'Ana, H. M., Chaves, J. B., & Coimbra, J. S. (2017). Quinoa: Nutritional, functional, and antinutritional aspects. *Critical Reviews in Food Science and Nutrition*, *57*(8), 1618–1630. https://doi.org/10.1080/10408398.2014.1001811

Flórez-Martínez, D. H., Rodríguez-Cortina, J., Chavez-Oliveros, L. F., Aguilera-Arango, G. A., & Morales-Castañeda, A. (2024). Current trends and prospects in quinoa research: An approach for strategic knowledge areas. *Food Science & Nutrition*, *12*(3), 1479–1501.

Gelinas, B., & Seguin, P. (2007). Oxalate in grain amaranth. *Journal of Agricultural and Food Chemistry*, *55*(12), 4789–4794.

Gogoi, M., & Ahmed, P. (2025). Nutriceral buckwheat as a potential source of nutrients. In *Disease and Health: Research Developments*, 1. Ed., Magiorkinis E. BP International Publications. ISBN 978-93-49473-63-8 (Print) ISBN 978-93-49473-40-9 (eBook). https://doi.org/10.9734/bpi/dhrd/v5

Graziano, S., Agrimonti, C., Marmiroli, N., & Gullì, M. (2022). Utilisation and limitations of pseudocereals (quinoa, amaranth, and buckwheat) in food production: A review. *Trends in Food Science & Technology*, *125*, 154–165. ISSN 0924-2244. https://doi.org/10.1016/j.tifs.2022.04.007

Gunathunga, C., Senanayake, S., Jayasinghe, M. A., Brennan, C. S., Truong, T., Marapana, U., & Chandrapala, J. (2024). Germination effects on nutritional quality: A comprehensive review of selected cereal and pulse changes. *Journal of Food Composition and Analysis*, *128*, 106024.

Guo, H., Li, H., Xiao, Y., Wu, D. T., Gan, R. Y., Kang, Z., … & Gao, H. (2025). Revisiting fermented buckwheat: A comprehensive examination of strains, bioactivities, and applications. *Critical Reviews in Food Science and Nutrition*, 1–22.

Gupta, V., & Mishra, A. A. (2024). Different minerals and their bioavailability in pseudocereals. In *Pseudocereals* (pp. 154–170). CRC Press.

Holmes, R. P., & Assimos, D. G. (2004). The impact of dietary oxalate on kidney stone formation. *Urological Research*, *32*, 311–316.

Hosseini, S. M., Soltanizadeh, N., Mirmoghtadaee, P., Banavand, P., Mirmoghtadaie, L., & Shojaee-Aliabadi, S. (2018). Gluten-free products in celiac disease: Nutritional and technological challenges and solutions. *Journal of Research in Medical Sciences*, *23*(1), 109.

Huang, M., Xu, H., Zhou, Q., Xiao, J., Su, Y., & Wang, M. (2025). The nutritional profile of chia seeds and sprouts: Tailoring germination practices for enhancing health benefits-a comprehensive review. *Critical Reviews in Food Science and Nutrition*, *65*(12), 2365–2387.

Huynh, N. K., Nguyen, D. H., & Nguyen, H. V. (2022). Effects of processing on oxalate contents in plant foods: A review. *Journal of Food Composition and Analysis*, *112*, 104685.

Jan, N., Hussain, S. Z., Naseer, B., & Bhat, T. A. (2023). Amaranth and quinoa as potential nutraceuticals: A review of anti-nutritional factors, health benefits and their applications in food, medicinal and cosmetic sectors. *Food Chemisty X*, *18*, 100687. https://doi.org/10.1016/j.fochx.2023.100687

Jimoh, M. O., Afolayan, A. J., & Lewu, F. B. (2020). Nutrients and antinutrient constituents of Amaranthus caudatus L. Cultivated on different soils. *Saudi Journal of Biological Sciences*, *27*(12), 3570–3580. https://doi.org/10.1016/j.sjbs.2020.07.029

Kamal, H., Mudgil, P., Bhaskar, B., Fisayo, A. F., Gan, C. Y., & Maqsood, S. (2021). Amaranth proteins as potential source of bioactive peptides with enhanced inhibition of enzymatic markers linked with hypertension and diabetes. *Journal of Cereal Science*, *101*, 103308.

Kaur, H., Shams, R., Dash, K. K., & Dar, A. H. (2023). A comprehensive review of pseudo-cereals: Nutritional profile, phytochemicals constituents and potential health- promoting benefits. *Applied Food Research*, *3*(2), 100351.

Klepacka, J., Najda, A., Klimek, K. (2020). Effect of buckwheat groats processing on the content and bioaccessibility of selected minerals. *Foods*, *9*(6), 832. https://doi.org/10.3390/foods9060832. PMID: 32630374; PMCID: PMC7353638.

Kowalski, S., Mikulec, A., Mickowska, B., & Buksa, K. (2022). Nutritional properties and amino acid profile of buckwheat bread. *Journal of Food Science and Technology*, *59*(8), 3020–3030.

Kulczyński, B., Kobus-Cisowska, J., Taczanowski, M., Kmiecik, D., & Gramza-Michałowska, A. (2019). The chemical composition and nutritional value of chia seeds-current state of knowledge. *Nutrients*, *11*(6), 1242. https://doi.org/10.3390/nu11061242. PMID: 31159190; PMCID: PMC6627181.

Langyan, S., Khan, F.N. & Kumar, A (2024). Advancement in nutritional value, processing methods, and potential applications of pseudocereals in dietary food: A review. *Food and Bioprocess Technology*, *17*, 571–590. https://doi.org/10.1007/s11947-023-03109-x

Laparra, J. M., & Haros, C. M. (2019). Plant seed protease inhibitors differentially affect innate immunity in a tumor microenvironment to control hepatocarcinoma. *Food & Function*, *10*(7), 4210–4219.

Li, H., Zhu, F., & Li, G. (2025). Beverages developed from pseudocereals (quinoa, buckwheat, and amaranth): Nutritional and functional properties. *Comprehensive Reviews in Food Science and Food Safety*, *24*(1), e70081.

Li, P., Liu, C., Luo, Y., Shi, H., Li, Q., PinChu, C., … & Fan, W. (2022). Oxalate in plants: Metabolism, function, regulation, and application. *Journal of Agricultural and Food Chemistry*, *70*(51), 16037–16049.

López-Moreno, M., Garcés-Rimón, M., & Miguel, M. (2022). Antinutrients: Lectins, goitrogens, phytates and oxalates, friends or foe? *Journal of Functional Foods*, *89*, 104938.

Manyelo, T. G., Sebola, N. A., van Rensburg, E. J., & Mabelebele, M. (2020). The probable use of genus Amaranthus as feed material for monogastric animals. *Animals*, *10*(9), 1504.

Martínez-Villaluenga, C., Peñas, E., & Hernández-Ledesma, B. (2020). Pseudocereal grains: Nutritional value, health benefits, and current applications for the development of gluten-free foods. *Food and Chemical Toxicology*, *137*, 111178.

Melini, V.; Melini, F (2021). Functional components and anti-nutritional factors in gluten-free grains: A focus on quinoa seeds. *Foods*, *10*, 351. https://doi.org/10.3390/ foods10020351

Mir, N. A., Riar, C. S., & Singh, S. (2018). Nutritional constituents of pseudo cereals and their potential use in food systems: A review. *Trends in Food Science & Technology*, *75*, 170–180.

Mir, R. R., Rustgi, S., Zhang, Y. M., & Xu, C. (2022). Multi-faceted approaches for breeding nutrient-dense, disease-resistant, and climate-resilient crop varieties for food and nutritional security. *Heredity*, *128*(6), 387–390.

Morales, D., Miguel, M., & Garcés-Rimón, M. (2021). Pseudocereals: A novel source of biologically active peptides. *Critical Reviews in Food Science and Nutrition*, *61*(9), 1537–1544.

Motta, C., Castanheira, I., Matos, A. S., Nascimento, A. C., Assunção, R., Martins, C., & Alvito, P. (2022, October). Effect of cooking on the content and bioaccessibility of minerals in pseudocereals. *Biology and Life Sciences Forum*, *17*(1), 17.

Mroczek, A. (2015). Phytochemistry and bioactivity of triterpene saponins from Amaranthaceae family. *Phytochemistry Reviews*, *14*, 577–605.

Mukuwapasi, B., Mavengahama, S., & Gerrano, A. S. (2024). Grain amaranth: A versatile untapped climate-smart crop for enhancing food and nutritional security. *Discover Agriculture*, *2*(1), 44.

Nandan, A., Koirala, P., Tripathi, A. D., Vikranta, U., Shah, K., Gupta, A. J., Agarwal, A., & Nirmal, N. (2024). Nutritional and functional perspectives of pseudocereals. *Food Chemistry*, *448*, 139072. ISSN 0308-8146. https://doi.org/10.1016/j.foodchem.2024.139072

Nardo, A. E., Suárez, S., Quiroga, A. V., & Añón, M. C. (2020). Amaranth as a source of antihypertensive peptides. *Frontiers in Plant Science*, *11*, 578631.

Navruz-Varli, S., & Sanlier, N. (2016). Nutritional and health benefits of quinoa (Chenopodium quinoa Willd.) *Journal of Cereal Science*, *69*, 371–376. ISSN 0733-5210. https://doi.org/10.1016/j.jcs.2016.05.004

Ňorbová, M., Vollmannová, A., Harangozo, Ľ., Franková, H., Čeryová, N., Jančo, I., & Fandrová, A. (2022). Risk elements, antioxidant activity and polyphenols in pseudocereal grains. *Agrobiodiversity for Improving Nutrition, Health and Life Quality*, *6*(1), 20–30.

Paarakh, P. M., Mishra, S., Kavya, R. S., Pasha, M., & Kumar, A. M. (2025). Chia seeds: A comprehensive review. *Research Journal of Pharmacy and Technology*, *18*(2), 912–918.

Pathan, S., & Siddiqui, R. A. (2022). Nutritional Composition and Bioactive Components in Quinoa (*Chenopodium quinoa* Willd.) Greens: A Review. *Nutrients*, *14*(3), 558. https://doi.org/10.3390/nu14030558

Priego-Poyato, S., Rodrigo-Garcia, M., Escudero-Feliu, J., Garcia-Costela, M., Lima-Cabello, E., Carazo-Gallego, A., … C. Jimenez-Lopez, J. (2021). Current advances research in nutraceutical compounds of legumes, pseudocereals and cereals. In *Grain and Seed Proteins Functionality*. IntechOpen. https://doi.org/10.5772/intechopen.97782

Reddy, M. V. K., Dubey, P. K., Mishra, A. A., & Sabeel, V. A. (2024). Potential processing techniques for safe utilisation of pseudo cereals in the food system. *Journal of Food Composition and Analysis*, *135*, 106609.

Samtiya, M., Aluko, R. E., & Dhewa, T. (2020). Plant food anti-nutritional factors and their reduction strategies: An overview. *Food Production, Processing and Nutrition*, *2*, 1–14.

Schmidt, D., Verruma-Bernardi, M. R., Forti, V. A., & Borges, M. T. M. R. (2021). Quinoa and amaranth as functional foods: A review. *Food Reviews International*, *39*(4), 2277–2296. https://doi.org/10.1080/87559129.2021.1950175

Shahbaz, M., Raza, N., Islam, M., Imran, M., Ahmad, I., Meyyazhagan, A., … & Wan, C. (2023). The nutraceutical properties and health benefits of pseudocereals: A comprehensive treatise. *Critical Reviews in Food Science and Nutrition*, *63*(29), 10217–10229.

Sharma, K., Kaur, R., Kumar, S., Saini, R. K., Sharma, S., Pawde, S. V., & Kumar, V. (2023). Saponins: A concise review on food related aspects, applications and health implications. *Food Chemistry Advances*, *2*, 100191. ISSN 2772-753X. https://doi.org/10.1016/j.focha.2023.100191

Singh, N., Samarth, R. M., Vashishth, A., & Pareek, A. (2024a). Amaranthus as a potential dietary supplement in sports nutrition. *CyTA-Journal of Food*, *22*(1), 2375253.

Singh, N., Ujinwal, M., & Paikray, D. (2024b). Pseudocereals as treasures of nutritional and nutraceutical compounds. In *Pseudocereals-Recent Advances and New Perspectives*. IntechOpen, 113369.

Stepanova, N. (2023). Oxalate homeostasis in non-stone-forming chronic kidney disease: A review of key findings and perspectives. *Biomedicines*, *11*(6), 1654. https://doi.org/10.3390/biomedicines11061654

Sofi, S. A., Ahmed, N., Farooq, A., Rafiq, S., Zargar, S. M., Kamran, F., … & Mousavi Khaneghah, A. (2023). Nutritional and bioactive characteristics of buckwheat and its potential for developing gluten-free products: An updated overview. *Food Science & Nutrition*, *11*(5), 2256–2276. https://doi.org/10.1002/fsn3.3166. PMID: 37181307; PMCID: PMC10171551.

Suárez-Estrella, D., Borgonovo, G., Buratti, S., Ferranti, P., Accardo, F., Pagani, M. A., & Marti, A. (2021). Sprouting of quinoa (Chenopodium quinoa Willd.): Effect on saponin content and relation to the taste and astringency assessed by electronic tongue. *LWT*, *144*, 111234.

Tang, Y., & Tsao, R. (2017). Phytochemicals in quinoa and amaranth grains and their antioxidant, anti-inflammatory, and potential health beneficial effects: A review. *Molecular Nutrition & Food Research*. *61*(7), 1600767. https://doi.org/10.1002/mnfr.201600767

Terzieva, S., Baycheva, S., Tzanova, M., Ivanova, T., Dimitrova, D., & Grozeva, N. H. (2026). Multifunctional Edible Amaranths: A Review of Nutritional Benefits, Anti-Nutritional Factors, and Potential in Sustainable Food Systems. *Foods*, *15*(1), 130. https://doi.org/10.3390/foods15010130

Thakur, P., Kumar, K., & Dhaliwal, H. S. (2021a). Nutritional facts, bio-active components and processing aspects of pseudocereals: A comprehensive review. *Food Bioscience*, *42*, 101170.

Thakur, P., Kumar, K., Ahmed, N., Chauhan, D., Rizvi, Q. U. E. H., Jan, S., … & Dhaliwal, H. S. (2021b). Effect of soaking and germination treatments on nutritional, anti-nutritional, and bioactive properties of amaranth (Amaranthus hypochondriacus L.), quinoa (Chenopodium quinoa L.), and buckwheat (Fagopyrum esculentum L.). *Current Research in Food Science*, *4*, 917–925.

Toimbayeva, D., Saduakhasova, S., Kamanova, S., Kiykbay, A., Tazhina, S., Temirova, I., … & Ospankulova, G. (2025). Prospects for the use of amaranth grain in the production of functional and specialized food products. *Foods*, *14*(9), 1603.

Tovar-Pérez, E. G., Lugo-Radillo, A., & Aguilera-Aguirre, S. (2019). Amaranth grain as a potential source of biologically active peptides: A review of their identification, production, bioactivity, and characterization. *Food Reviews International*, *35*(3), 221–245.

Upasana, & Yadav, L. (2022). *Pseudocereals: A Novel Path towards Healthy Eating*. IntechOpen. https://doi.org/10.5772/intechopen.103708

Usman, M., Patil, P. J., Mehmood, A., Rehman, A., Shah, H., Haider, J., … & Li, X. (2022). Comparative evaluation of pseudocereal peptides: A review of their nutritional contribution. *Trends in Food Science & Technology*, *122*, 287–313.

Valadez-Vega, C., Lugo-Magaña, O., Figueroa-Hernández, C., Bautista, M., Betanzos-Cabrera, G., Bernardino-Nicanor, A., … & González-Cruz, L. (2022). Effects of germination and popping on the anti-nutritional compounds and the digestibility of Amaranthus hypochondriacus seeds. *Foods*, *11*(14), 2075.

Waisundara, V. Y. (2024). Chapter 1: Pseudocereals as subexploited food. In *Pseudocereals – Recent Advances and New Perspectives* (pp. 1–6). Intechopen Publ. ISBN 978-1-83768-795-4.

Woiciechowski, A. L., Pagnoncelli, M. G. B., Scapini, T., Prado, F. G., Mezzalira, F. K., Savian, C. M., & Soccol, C. R. (2023). Microbial enzymes for reduction of antinutritional factors. In Rai, A. K., Sirohi, R., de Souza Vandenberghe, L. P. & Parameswaran, B. (Eds.), *Microbial Enzymes in Production of Functional Foods and Nutraceuticals* (pp. 195–217). CRC Press.

Yilmaz, H. Ö., Ayhan, N. Y., & Meriç, Ç. S. (2020). Buckwheat: A useful food and its effects on human health. *Current Nutrition & Food Science*, *16*(1), 29–34.

Yilma, M. T., Eifa, A., Belayneh, M., & Orsango, A. Z. (2025). Effect of amaranth-containing dietary intervention in improving hemoglobin concentration: A systematic review and meta-analysis. *Public Health Reviews*, *45*, 1607597.

Yılmaz, B., Sırbu, A., Altıntaş Başar, H. B., Goksen, G., Chabı, I. B., Kumagaı, H., & Ozogul, F. (2025). Potential roles of cereal bioactive compounds in the prevention and treatment of type 2 diabetes: A review of the current knowledge. *Critical Reviews in Food Science and Nutrition*, *65*(7), 1326–1343.

You, H., Wu, T., Wang, W., Li, Y., Liu, X., & Ding, L. (2022). Preparation and identification of dipeptidyl peptidase IV inhibitory peptides from quinoa protein. *Food Research International*, *156*, 111176.

Zayed, A., Adly, G. M., & Farag, M. A. (2025). Management strategies for the anti-nutrient oxalic acid in foods: A comprehensive overview of its dietary sources, roles, metabolism, and processing. *Food and Bioprocess Technology*, *18*, 4280–4300. https://doi.org/10.1007/s11947-024-03726-0 1–21

Zhu, F. (2019). Proanthocyanidins in cereals and pseudocereals. *Critical Reviews in Food Science and Nutrition*, *59*(10), 1521–1533.

Index

Note: **Bold** page numbers refer to tables and *italic* page numbers refer to figures.

R

S

T

U

V

W

For Product Safety Concerns and Information please contact our EU
representative GPSR@taylorandfrancis.com
Taylor & Francis Verlag GmbH, Kaufingerstraße 24, 80331 München, Germany

www.ingramcontent.com/pod-product-compliance
Ingram Content Group UK Ltd.
Pitfield, Milton Keynes, MK11 3LW, UK
UKHW052053230926
475702UK00026B/1210
9781032954233